Vibration Problems in Machines

Vibration Problems
in Machines

Diagnosis and Resolution

Second Edition

Arthur W. Lees

CRC Press
Taylor & Francis Group
Boca Raton London New York

CRC Press is an imprint of the
Taylor & Francis Group, an **informa** business

MATLAB® is a trademark of The MathWorks, Inc. and is used with permission. The MathWorks does not warrant the accuracy of the text or exercises in this book. This book's use or discussion of MATLAB® software or related products does not constitute endorsement or sponsorship by The MathWorks of a particular pedagogical approach or particular use of the MATLAB® software.

Second edition published 2020
by CRC Press
6000 Broken Sound Parkway NW, Suite 300, Boca Raton, FL 33487-2742

and by CRC Press
2 Park Square, Milton Park, Abingdon, Oxon, OX14 4RN

© 2020 Taylor & Francis Group, LLC

First edition published by CRC Press 2016

Second edition published by CRC Press 2020

CRC Press is an imprint of Taylor & Francis Group, LLC

International Standard Book Number-13: 978-0-367-36774-9 (Hardback)
International Standard Book Number-13: 978-0-367-36775-6 (Paperback)
International Standard Book Number-13: 978-0-429-35137-2 (eBook)

Visit the companion website/eResources: [https://www.routledge.com/9780367367749]

Contents

Preface xi
Preface to the Second Edition xiii
MATLAB® xv
About the Author xvii

1. **Introduction** 1
   1.1   Monitoring and Diagnosis. 1
       1.1.1   Monitored Parameters 3
           1.1.1.1   Vibration. 3
           1.1.1.2   Pressure, Flow, Temperature 3
           1.1.1.3   Voltage and Current 3
           1.1.1.4   Acoustic Emission 4
       1.1.2   Fault Localization. 4
       1.1.3   Root Cause 5
       1.1.4   Remaining Life 5
   1.2   Instrumentation 6
       1.2.1   Piezo-electric Accelerometers 6
           1.2.1.1   Attachment of Accelerometers. 7
       1.2.2   Velocity Sensors 7
       1.2.3   Proximity Sensors. 8
       1.2.4   Linear Voltage Differential Transformer (LVDT). . . . 8
       1.2.5   Strain Gauges 9
       1.2.6   Acoustic Emission Sensors 9
       1.2.7   Storage of Data 10
   1.3   Mathematical Models. 10
   1.4   Machine Classification. 12
       1.4.1   Bearing Types 13
       1.4.2   The Rotor 14
   1.5   Considerations for a Monitoring Scheme 15
   1.6   Outline of the Text 17
   1.7   Software. 20
   References. 21

2. **Data Presentation** 23
   2.1   Introduction 23
   2.2   Presentation Formats 23
       2.2.1   Time and Frequency 24

2.2.2   Waterfall Plots 26

2.2.3   Scatter Plots (or Carpet Plots). .29

2.2.4   Order Tracking in Transient Operation. 36

        2.2.4.1   Short-Term Fourier Transform 37

        2.2.4.2   The Vold–Kalman Method. 37

        2.2.4.3   System Response 52

        2.2.4.4   Comparison of Methods. 56

2.2.5   Shaft Orbits 57

2.2.6   Polar Plots 58

2.2.7   Spectragrams. 60

2.3   Comparison with Calculations 61

2.4   Detection and Diagnosis Process 62

2.5   Concluding Remarks 63

Problems. 63

References. 64

3.   **Modeling and Analysis** 67

3.1   Introduction 67

3.2   Need for Models. 67

3.3   Modeling Approaches 68

3.3.1   Beam Models 68

3.3.2   Finite Element Method 71

        3.3.2.1   Element Formulation 71

        3.3.2.2   Matrix Assembly 75

3.3.3   Modeling Choices. 79

3.4   Analysis Methods. 80

3.4.1   Imbalance Response. 80

3.4.2   Campbell Diagram. 81

3.4.3   Analysis of Damped Systems .85

3.4.4   Root Locus and Stability 89

3.4.5   Overall Response 90

3.5   Further Modeling Considerations 93

3.5.1   Mode Shapes. 93

3.5.2   Perturbation Techniques 98

3.6   Summary. 102

Problems. 103

References. 103

4.   **Faults in Machines (1)** 105

4.1   Introduction 105

4.2   Definitions: Rigid and Flexible Rotors. 105

4.3   Mass Imbalance 113
     4.3.1   General Observations 113
     4.3.2   Rotor Balancing 117
         4.3.2.1   Influence Coefficient Approaches –
                   Single Plane 118
         4.3.2.2   Two-Plane Balancing 123
         4.3.2.3   Modal Balancing 132
     4.3.3   A Comparison of Approaches 134
     4.3.4   Nonlinear Effects 135
     4.3.5   Recent Developments 137
4.4   Rotor Bends 137
4.5   Concluding Remarks 142
Problems 144
References 145

5.   **Faults in Machines (2)** 147
5.1   Introduction 147
5.2   Misalignment 147
     5.2.1   Key Phenomena 148
     5.2.2   Flexible Couplings 150
     5.2.3   Solid Couplings 154
         5.2.3.1   The Catenary 155
     5.2.4   Misalignment Excitation – A New Model 158
         5.2.4.1   Moments due to Imbalance 161
         5.2.4.2   The Effect of Bending Moments 163
         5.2.4.3   Influence of the Bolts 170
     5.2.5   Toward an Overall Model 175
5.3   Cracked Rotors 175
     5.3.1   Change in Natural Frequencies 177
     5.3.2   Forced Response 183
5.4   Torsional Excitation 187
     5.4.1   Dynamics of Gearboxes 189
         5.4.1.1   Tooth Stiffness Effects 192
     5.4.2   Mixing of Flexural, Axial, and Torsional 193
5.5   Nonlinearity 193
5.6   Instability 195
     5.6.1   Oil-Whirl and Oil-Whip 195
     5.6.2   Rotor Asymmetry 196
     5.6.3   Rotor Damping 197
5.7   Interactions and Diagnostics 199
     5.7.1   Synchronous Excitation 199

        5.7.2   Twice per Revolution Excitation 199
        5.7.3   Asynchronous Vibration 200
    5.8   Closing Remarks. 201
Problems. 201
References. 202

6. **Rotor–Stator Interaction**. 205
    6.1   Introduction 205
    6.2   Interaction through Bearings 205
        6.2.1   Oil Journal Bearings. 206
        6.2.2   Rolling Element Bearings .209
        6.2.3   Other Types of Bearing 211
            6.2.3.1   Active Magnetic Bearings. 211
            6.2.3.2   Externally Pressurized Bearings 212
            6.2.3.3   Foil Bearings. 212
    6.3   Interaction via Working Fluid 212
        6.3.1   Pump Bushes and Seals. 212
            6.3.1.1   Influence of System Pressure
                       Distribution 216
            6.3.1.2   An Example of an Idealized System 218
        6.3.2   Other Forms of Excitation. 220
        6.3.3   Steam Whirl 222
    6.4   Direct Stator Contact 222
        6.4.1   Extended Contact. 222
            6.4.1.1   Physical Effects .222
            6.4.1.2   Newkirk Effect. .223
        6.4.2   Collision and Recoil 226
            6.4.2.1   Physical Effects .226
            6.4.2.2   Simulation 227
        6.4.3   Acoustic Emission 232
    6.5   The Morton Effect. 235
    6.6   Harmonics of Contact 236
    6.7   Concluding Remarks 237
Problems. 238
References. 238

7. **Machine Identification**. 241
    7.1   Introduction 241
    7.2   Current State of Modeling. 241
    7.3   Primary Components. 243

7.4 Sources of Error/Uncertainty 245

7.5 Model Improvement 247

    7.5.1 System Identification 248

    7.5.2 Error Criteria 250

    7.5.3 Regularization 251

    7.5.4 Singular Value Decomposition 253

7.6 Application to Foundations 254

    7.6.1 Formulating the Problem 254

    7.6.2 Least Squares with Physical Parameters 255

        7.6.2.1 Formulation with Shaft Location Data ... 258

        7.6.2.2 Applying Constraints 259

    7.6.3 Modal Approach with Kalman Filters 262

    7.6.4 Essentials of Kalman Filtering 265

7.7 Imbalance Identification 266

7.8 Extension to Misalignment 268

7.9 Future Options 268

    7.9.1 Implementation 269

    7.9.2 Benefits 270

7.10 Concluding Remarks 270

Problems 270

References 271

8. Some Further Analysis Methods 275

8.1 Introduction 275

8.2 Standard Approaches 275

8.3 Artificial Neural Networks 280

8.4 Merging ANNs with Physics-Based Models 284

8.5 Singular Value Decomposition 284

8.6 Other Useful Techniques 287

    8.6.1 The Hilbert Transform 288

    8.6.2 Time–Frequency and Wigner–Ville 291

    8.6.3 Wavelet Analysis 293

    8.6.4 Cepstrum 295

    8.6.5 Cyclostationary Methods 296

    8.6.6 Higher-Order Spectra 296

    8.6.7 Empirical Mode Decomposition 299

    8.6.8 Kernel Density Estimation 300

8.7 Concluding Remarks 303

Problems 303

References 304

9. Case Studies 307
   9.1   Introduction 307
   9.2   A Crack in a Large Alternator Rotor 307
   9.3   Workshop Modal Testing of a Cracked Rotor 313
   9.4   Gearbox Problems on a Boiler Feed Pump 317
   9.5   Vibration of Large Centrifugal Fans 326
   9.6   Low-Pressure Turbine Instabilities 329
   9.7   Concluding Remarks 330
  Problems 330
  References 331

10. Overview and Outlook 333
   10.1   Progress in Instrumentation 333
   10.2   Progress in Data Analysis and Handing 333
   10.3   Progress in Modeling 334
   10.4   Expert Systems 336
   10.5   Future Prospects 337
       10.5.1   Machine Diagnostics 337
       10.5.2   Self-Correction ("Smart Machines") 338
              10.5.2.1   Magnetic Bearings 339
              10.5.2.2   Electrorheological Bearings 339
              10.5.2.3   Externally Pressurized Bearings 339
              10.5.2.4   Other Approaches 340
       10.5.3   Shaft Modification 340
   10.6   Summary 345
  References 345

Solutions to Problems 347

Index 353

Preface

The understanding of vibration signals from turbo-machinery is an important feature of many industries. It forms the basis of condition monitoring and is of crucial importance in the rapid diagnosis and rectification of faults. The field calls for a wide range of skills ranging from instrumentation expertise to mathematical manipulation and signal processing. The focus is on gaining a physical understanding of the processes giving rise to the vibration signal. This may be viewed as a process of seeking to infer internal conditions from (sometimes limited) external measured data.

This often requires some form of mathematical model and the approach taken is to use some basic finite element analysis to study system behavior. These methods were first applied to rotating machinery several decades ago but with progress in computing and the advent of packages such as MATLAB, the ease with which the concepts may be applied, has dramatically improved. It is now a relatively straightforward matter to examine proposed design changes – in effect, to perform numerical experiments.

The material in this book stems from my experiences both in industry and, in more recent years, from teaching and research in academia. Inevitably, my own experiences have to some degree influenced my choice of examples. In my various posts, I have had fruitful interactions with many people who have significantly enhanced my perception of problems in machinery. There are too many people involved to thank individually, but I must mention three. Professor Mike Friswell and I have worked together closely for 20 years and I must acknowledge innumerable helpful and illuminating discussions. I must also thank Professor John Penny for his help in proofreading and a number of helpful suggestions on the structure of the text. Both of these people were my coauthors, together with Professor Seamus Garvey, of the book I refer to in many places which gives more basic theory of rotor dynamics. My deepest thanks must go to my wife Rita. Not only has she given me grammatical advice, but has also helped maintain my sanity throughout the final months of preparing this text.

Arthur W Lees
May, 2015

Preface to the Second Edition

In producing this second edition, some minor additions and amendments to the text have been made aimed at improving clarity. In Chapter 1, a section has been added to briefly describe the principal types of instrumentation commonly used on rotating machines. Such an introduction is appropriate as the major part of the book concerns the interpretation of data from these instruments. The discussion of the processing of transient data has been brought together in Chapter 2 with an enhanced discussion.

Later, in Chapter 4, the relative merits of different balancing approaches are assessed more fully and further consideration has been given to rotor bends. In Chapter 5, the latest advances in assessing misalignment problems are reported and a section has been added on the main sources of rotor instability. References to some recent papers have been added.

The set of lecture slides (available on the website) have been completely revised. These are MS Powerpoint files organized by chapter. They can, of course, be edited to suit specific requirements.

It is hoped all the material in this edition will help advanced students, researchers, and industrial engineers.

Arthur W Lees
April 2020

MATLAB®

Modeling can be carried out using the package of MATLAB scripts freely available at www.rotordynamics.info. A further toolbox has been constructed for the following studies:

a) Rigid rotor analysis

b) Flexible rotor analysis

c) Single-plane balancing

d) Two-plane balancing

e) Modal balancing

f) Gear torsional analysis

g) Transient predictions

h) Catenary calculations

These modules may be found at www.crcpress.com/product/isbn/9781498726764.

In addition, MATLAB scripts are available on the website for the solution of the majority of the set problems. These may be readily modified to analyze similar situations.

MATLAB® is a registered trademark of The MathWorks, Inc. For product information, please contact:

The MathWorks, Inc.
3 Apple Hill Drive
Natick, MA 01760-2098, USA
Tel: 508-647-7000
Fax: 508-647-7001
E-mail: info@mathworks.com Web: www.mathworks.com

About the Author

Professor **Arthur W. Lees**
B.Sc., Ph.D., D.Sc., C.Eng., C.Phys., F.I.Mech.E., F.Inst.P., L.R.P.S.

Professor Arthur W. Lees graduated in Physics and remained in Manchester University, Manchester, United Kingdom, for three years research. After completing his Ph.D., he joined the Central Electricity Generating Board, Leatherhead, United Kingdom, initially developing finite element codes, then later resolving plant problems.

After a sequence of positions, he was appointed head of the Turbine Group for Nuclear Electric Plc., Gloucester, United Kingdom. He moved to Swansea University in 1995 and has been active in both research and teaching.

He is a regular reviewer of many technical journals and was, until his retirement, on the editorial boards of the *Journal of Sound and Vibration* and *Communications on Numerical Methods in Engineering*. His research interests include structural dynamics, rotor dynamics, inverse problems, and heat transfer. Professor Lees is a fellow of the Institution of Mechanical Engineers and a fellow of the Institute of Physics, a Chartered Engineer, and a Chartered Physicist. He was a member of Council of the Institute of Physics, 2001–2005. He is now Professor Emeritus at Swansea University, but remains an active researcher.

1

Introduction

The general field of condition monitoring has received substantial attention over the last few decades, and it is worth reflecting on the state of the topic. Although it has always been practiced at some level, the manner of assessing the condition is constantly under review in the light of recent developments in understanding.

In assessing the condition of a piece of equipment, the operator gathers data such as vibration, operating temperature, noise, performance, and electrical parameters where appropriate. At one time, the comparison with normal condition was achieved largely on the basis of staff experience, but the general trend has been toward a more precise quantified approach. This has been required by revised patterns of working and increasing plant complexity, but it is, in essence, the same operation. A fundamental question arises as to how one can "codify" the knowledge of an experienced engineer and focus the knowledge on a specific area of plant. This presents a challenge that has shown significant progress in recent years, although the issue cannot be regarded as completely resolved. Important progress has been made in computational modeling (both finite element (FE) analysis and computational fluid dynamics), artificial neural networks (ANNs), statistical approaches, expert systems, and identification methods. All of these have a role to play in assessing the condition of a piece of equipment, and their role will be outlined in subsequent chapters. First of all, however, the general field of condition monitoring, as applied to rotating machines, is reviewed.

1.1 Monitoring and Diagnosis

Generally these two terms are linked under the general heading of condition monitoring, but, in fact, they are two quite distinct functions. In both areas, the first requirement is to gather and record all salient details of the operation of the piece of equipment; however, as will be discussed, the choice as to what details are salient is far from a trivial task. However, that discussion is deferred for the present.

To illustrate this point with a specific example, let us consider a centrifugal pump driven by an electric motor. In such a case, the monitored parameters may well include bearing vibration levels, temperature, water pressure, water flow rate, motor current, and voltage. Note that although this is a fairly long list, it is by no means exhaustive. In some circumstances, one may wish to record the rotor vibration (as opposed to that of the bearings) and bearing oil temperature. In fact, even a relatively simple piece of machinery may have a significant number of parameters, which may be useful for monitoring purposes, and a judicious choice is required to limit the measured set to cost-effective proportions: however, making this choice requires some appreciable physical insight.

Having decided on the set of monitored parameters, the method of recording is the next choice to be made, and this ranges from regular spot checks to some form of continuous monitoring, now almost invariably computer based. While the latter represents a more expensive option, it does offer more flexibility in terms of the ways in which data can be manipulated to offer insight into the underlying features of machine operation. Here again decisions demanding physical insight into machine operation and the relative failure scenarios are required.

We now consider some of the ways in which plant data may be analyzed, and how this may be used to form judgments about plant operation. Clearly, any general trend in the plant data, or indeed a sudden change suggests that the equipment has altered in some way and, subjected to some checks, may require the removal of the plant from service. Note that, realistically, all pieces of equipment are subjected to some random perturbations, and so statistical techniques are needed to form a valid decision as when to remove the plant from service.

To proceed further, we examine these three basic questions that are posed in Condition Monitoring systems:

a) Has something gone wrong?
b) What has gone wrong?
c) How long can the plant run safely?

To address the first of these questions, it is often sufficient to adopt a purely statistical analysis of monitored data and seek trends and changes. At the most basic level, no knowledge of the internal operation of the machinery is required. For example, a monitoring system may simply plot the overall vibration levels at the bearings (or elsewhere) and not examine the (short-term) time variation. Some of the ways in which data can be examined are discussed in Chapter 2. As discussed in later chapters, a great deal depends on the frequency with which measurements are monitored, and provided several measurements are recorded during each rotor revolution, the orbit of the shaft may be traced and this gives some further insight into the machine's behavior. In many cases, rudimentary measurements give some indications that something has happened to the machine and straightforward comparison with records will suffice to answer the first of the three questions. To answer the second, generally it requires considerably more insight, which may be provided by extensive experience, a detailed theoretical analysis or, very often, a combination of these. The third question, as to how long the plant will/can run safely is more difficult, and to a large extent, is still at the research stage. Nevertheless, an in-depth understanding of the machine's operation is an essential prerequisite. Later chapters discuss the interpretation of plant data, but first we give a brief survey of quantities that may be useful in assessing a machine's behavior.

1.1.1 Monitored Parameters

1.1.1.1 Vibration

In a rotating plant, vibration is the most commonly monitored parameter. There are two reasons for this: first, it is readily measured with convenient instrumentation and, perhaps more importantly, vibrations give a comprehensive reflection of the state of a rotating machine. The disadvantage is that much of the information for diagnosis can only be gained after extensive data processing, but standard techniques have now been developed, which go some way to relieving this burden. For basic monitoring, vibration levels can be, and are, used to a great effect.

1.1.1.2 Pressure, Flow, Temperature

In the case of pumps, pressure and flow represent the main performance parameters, and hence it is important to monitor them regularly, although clearly these will not be expected to vary as rapidly as the vibration. There may, of course, be fluctuations in pressure, but the flow rate will not track these, owing to substantial effects of inertia. Chapter 6 discusses some of the ways in which the pressure field within a typical centrifugal pump has a direct influence on the vibration characteristics. This treatment is by no means comprehensive and the interested reader is referred to the work of Childs (1993). The important point to emphasize here is that various pieces of information are interrelated; taken together, they present a complex picture. Temperature may be included in this section as another slowly varying parameter. It is the role of the diagnostic engineer to form an overall view from this complex pattern.

1.1.1.3 Voltage and Current

Electrical measurements also form an important part of this picture. A large number of machines are driven by electric motors, and the electrical measurements can be used to yield information on the overall machine efficiency, a key indicator of general deterioration. Being used at this level, periodic checks would suffice, but a more detailed analysis of fluctuations gives additional information on both the motor and the auxiliary machine that is being driven.

1.1.1.4 Acoustic Emission

Acoustic emission (AE) is, in one sense, simply vibration at very high frequency, but this description fails to give due recognition to its important distinction from conventional vibration data. As the name suggests in looking at AE, the engineer is monitoring the acoustic energy emitted by the material as changes (e.g., crack formation and propagation) take place. In effect, one is "listening" to the high-frequency waves generated by the breaking of intermolecular bonds within a component, rather than the direct consequences of externally applied forces. Until very recently, AE has been used as a rather course monitor to count so-called events as a guide to the presence or otherwise of cracking activity. For many years, it has been used to monitor the integrity of structures but only much more recently (Price *et al.*, 2005, Sikorska and Mba, 2008) it has been applied to rotating machinery. Improving hardware and software has facilitated the examination of the frequency resolution of AE signals and this too has enhanced our understanding. Measurements are often made in the range of several hundred kHz as this is reasonably easy to obtain and process with modern equipment and, this in turn, yields information of interest. A common misunderstanding is that these frequencies are characteristics of the breaking bonds within the material but this is not the case: such bonds give frequencies that are several orders of magnitude higher. The more accurate picture is that a breaking bond generates a very short pulse, which contains a wide range of frequencies. As the waves travel, those which resonate within some part of the body become predominant and hence the spectrum of AE may be used to tell the operator the nature of the body around the fault. It is the essence of AE that it provides a good approach to the identification of highly localized phenomena.

1.1.2 Fault Localization

The effective use of condition monitoring and condition-based maintenance encompasses a whole hierarchy of interrelated disciplines. At level 1, purely statistical approaches can be applied to measured data to detect if there has been some underlying change to the condition of the machine, and some insight into such approaches is given in Chapter 2. This type of analysis can be carried out without any knowledge of the machine's construction or operation and this is both a strength and a weakness: it means that the operator can determine if a fault/change has occurred without any assumptions as to the physical processes, but nothing further can be gleaned about the location or nature of the fault.

It may appear at first sight that the requirement to localize a fault is largely academic but this is far from the case. On large machines, such as turbo-alternators, many days of production can be saved by focusing maintenance work on the appropriate part of the machine. However, to make any valid progress on localization requires insight or prior knowledge into the machine's design and dynamic properties. Foremost among the requirements is a knowledge of the machine's natural frequencies (or more precisely, critical speeds) and mode shapes. This knowledge may be developed by operational experience, plant tests, and/or validated mathematical models, most commonly finite element (FE) models. More details on the requirements of these models are given in Section 1.3. The essential point at this stage is to emphasize that although condition monitoring is conducted without any reference to models for some simple machines, the potential is greatly extended by the use of models because their key role is to relate the external measurements to the internal operating conditions.

Having established a model, the determination of natural frequencies and mode shapes is a straightforward matter and is discussed in many text books (see, for example, Inman, 2008 or Friswell *et al.*, 2010). The mode shapes will often give some clues as to the important locations for particular types of fault. Conclusive location identification may require further analysis.

1.1.3 Root Cause

The issue of root cause is a theme running throughout this text and embraces the issues of fault localization and frequency composition of the vibration signals. Chapters 4, 5, and 6 discuss a range of common machine faults and the types of vibration signal they give rise to. The process of fault identification is basically one of recognizing the appropriate patterns of response and matching them to fault types. More importantly, the process may be seen as one that is gaining insight into the physical processes within the machine.

One might imagine that it would be easy to automate this process, but while there has been progress, there is still need for a human expert on more complex machines. Recent research in this area has focused on two areas, the development of extended models as discussed in Chapter 7 and the development of neural network and expert systems as discussed in Chapters 8 and 10, respectively. For the most complex machines however, no fully automated system is likely to be available in the foreseeable future.

1.1.4 Remaining Life

While the issues of fault localization and root cause raise some problems, the identification of remaining life remains the "Holy Grail of all Condition Monitoring." It is an extremely complex topic and involves input from a number of disciplines. While in some instances reasonable predictions can be made, in most case these rely substantially on practical experience of the plant involved. Predictions can be made on, for example, crack growth rates, but these rely heavily on materials data which is, in some instances, subject to substantial error.

1.2 Instrumentation

This book is primarily concerned with the understanding and interpretation of vibration signals from rotating machinery, but it is appropriate to give a brief summary of the main types of instrumentation used in the monitoring and investigation of machine operation.

1.2.1 Piezo-electric Accelerometers

Undoubtedly, the device used most commonly to measure machine vibration is the piezo-electric accelerometer. The basis of this device is a piece of piezoelectric material having the property that the dipoles within the crystal align when subjected to a strain.

This being so, when the device undergoes acceleration, the resulting strains will align the dipoles and this will lead to a build-up of charge on the surface of the crystal. The sensitivities of accelerometers vary in the range of 1–10,000 pC/g. It is important to recognize the fact that the quantity being measured is indeed acceleration rather than displacement.

Choosing the appropriate accelerometer requires some care: at first sight, one may be tempted to imagine that the highest sensitivity will give the best results, but there are some difficulties associated with this. Higher sensitivities result from large piezo-electric crystals and accompanying structures. This all means that the overall device will have a higher mass which in turn may modify the dynamics of the structure to which it is mounted. Consequently, accelerometer mass is a significant factor in choosing the appropriate device.

There are, however, a number of other factors to be considered. In some accelerometers, the internal structure is such that the piezo-electric crystal is subjected to a compressive strain, whilst in others the strain is shear. The latter tends to have better temperature stability whilst the compressive type tends to have higher resonant frequency. This is another important consideration. Because each accelerometer is a structure, it has natural frequencies, only the lowest of which is of interest in practice. Clearly signals will be distorted if the frequency of motion approaches the accelerometer's natural frequency and for this reason the upper practical frequency limit is typically about a quarter of the natural frequency. All manufacturers quote a usable frequency range for each device.

Having a signal in the form of charge representing the vibration level, this must be converted to a voltage and amplified to give a useful electrical signal. This is achieved by the signal conditioning unit which may either incorporated within the structure of the accelerometer or in a separate, remote device. There has recently been a marked tendency to incorporate the electronics within the unit and this is very convenient for many applications. However, where high temperatures are involved, such as on steam or gas turbines, this can lead to problems and a separate conditioning unit is desirable.

1.2.1.1 Attachment of Accelerometers

Having chosen the appropriate sensors, it is important to connect it to the machine or structure being observed. There are several ways in which the accelerometer can be attached; the main methods being handheld, magnet, epoxy bond, and a fixed stud. It is, however, essential to choose an applicable method to avoid unreliable data.

Handheld probes are, of course, quick and convenient to use. They can produce good results if used with appropriate care but the circumstances of their usage are restricted. In general, reliable results can be obtained within the frequency range of 10–100 Hz. The low-frequency limit is set by the ability to hold the probe still for a period, whilst the upper limit concerns the manner in which the probe and the sample remain in contact. Of course, high temperatures may also preclude handheld measurement.

Magnetic contact is also very convenient and can be used with confidence to slightly higher frequencies. These are used in the investigation of many problems in the rotating plant .

When high frequencies are of interest, it is appropriate to install a stud using epoxy resin, with the accelerometer then screwed onto the stud. For cases in which stud mounting is not viable, the accelerometer may be held in place directly with appropriate adhesive. Care must be taken, however, to achieve bonding whilst avoiding ground loop problems by adequate electrical insulation.

The mounting of accelerometers has an important influence on the results obtained, particularly at higher frequencies. More detailed discussion of the issue can be found on the website of the instrument manufacturers.

1.2.2 Velocity Sensors

Although superficially monitoring a closely related quantity, velocity transducers are fundamentally distinct from accelerometers. The velocity transducer is an electromagnetic device in which an electrical coil moves through a permanent magnetic field. Although the usage is declining, it does have some advantages over the piezoelectric accelerometer. It remains sensitive to lower-frequency motion and is rather less sensitive to temperature and interference signals.

Of course, signals only need to be integrated once to yield the displacement and this may be a consideration in some occasions.

The disadvantages are that they tend to be larger and heavier than accelerometers, and they suffer more cross noise among different directions. However, they have a role, albeit a fairly minor one currently.

1.2.3 Proximity Sensors

The introduction of proximity sensors in the 1950s was indeed a major step forward in the monitoring of rotating machinery. For the first time, engineers were able to follow the motion of a rotor directly rather than inferring information from data gathered at the stator.

Donald Bently developed the proximity transducer for use on rotating machines. It is important to note that these instruments measure displacement rather than velocity or acceleration. There are two main types: the eddy current type (as introduced by Bently) and the capacitive probe.

While seemingly similar, these two operate on quite different principles and consequently have rather distinct applications. Whilst eddy current probes rely on the magnetic field between the instrument and the target, the capacitive probe examines the electric field and consequently there are differences in both sensitivities and performance.

The eddy current probe is more sensitive to variations in the material properties of the rotor and consequently there will often be an "electrical run-out" (or baseline variation) observed, although this is usually very small. These probes also need calibration with respect to a particular rotor. This is not the case with capacitive instruments which are much less sensitive to rotor (target) material.

Against this, eddy current probes have much higher tolerance to hostile environments and higher temperature. They tend to have a higher power usage which can be an issue in some vacuum installations.

It is true to say that the introduction of the proximity transducer represented a major step forward in measurement and hence understanding of rotating machinery.

1.2.4 Linear Voltage Differential Transformer (LVDT)

These devices have a much less significant role in machinery monitoring but they can provide useful insight on occasions. The instrument is simply a variable resistor (in a structure similar to a bicycle tire pump) in which the motion of the target moves the contact varying the resistance of a measurement circuit. Owing to the mechanical system involved, these instruments are only useful at very low frequencies (1 Hz or less). They can be helpful, for instance, in evaluating the influence of slow pipework vibration/motion. An interesting case arose some years ago: during the investigation of problems on a feed pump, an LVDT was mounted to measure the relative motion of the pump casing to the rotor. Note that on this pump the bearings were mounted on separate pedestals. As the rotor accelerated to full speed from rest, the relative vertical position changed by 40 μm. Initially, this seemed very surprising but the explanation soon became clear. As the pump's body comprised a rigid cylinder with a hole cut in the top (connected to a flexible pipework system), a substantial force was effectively pushing the pump's body downward as a pressure was developed. This insight proved valuable in resolving the plant problems, and no other instrumentation would have been so convenient.

1.2.5 Strain Gauges

In the context of studies on machinery, strain gauges are rather specialized and are used for specific investigations rather than as a standard tool. One case in which strain gauge was invaluable is discussed in Section 9.4. In more general structural dynamics, the strain gauge has been in use for many years and is the fundamental sensing element for many types of sensors, including pressure sensors, load cells, torque sensors, position sensors, etc. The majority of strain gauges are foil types, available in a wide choice of shapes and sizes to suit a variety of applications.

They consist of a pattern of resistive foil which is mounted on a backing material. They operate on the principle that, as the foil is subjected to strain, the resistance of the foil changes in a defined way. The strain gauge is connected into a Wheatstone bridge circuit with a combination of four active gauges (full bridge), two gauges (half bridge), or less commonly a single gauge (quarter bridge). In the half and quarter circuits, the bridge is completed with precision resistors.

Although installation is a somewhat demanding process, strain gauges have benefits including lightness, wide frequency response, and temperature stability. A problem with respect to use on rotors is recovering the data. This is most conveniently achieved with a telemetry system as described in Chapter 9.

1.2.6 Acoustic Emission Sensors

Acoustic emission (AE) is the transmission of elastic waves in a material generated by some change such as growth of a crack. High frequencies arise from local resonance of the wideband spectrum generated by the breaking of molecular bonds and these are usually in the region of 25 kHz to 1 MHz. Joseph Kaiser, in 1950, was the first researcher to record that engineering materials in general emit low-amplitude, high-frequency clicks when stressed, which we now call AEs. Kaiser's most fundamental finding was that material does not emit unless the applied stress exceeds any previously applied stresses. This is known as the "Kaiser effect."

The sensors are piezo-electric high-frequency accelerometers. In many cases their frequency response is less smooth than those of standard accelerometers, but in many instances, this is of little importance. This is because many AE investigations aim to identify the activity just by counting so-called events. It is only in recent years that investigators have been examining frequency content and its implications. This trend may well dictate some changes in the emphasis.

1.2.7 Storage of Data

It was, at one time, common practice to make analogue recordings of plant data and store on magnetic tape for subsequent analysis. Whilst there are occasions where an analogue record can be useful, in the vast majority of cases, recordings are now made digitally. It is important to appreciate, however, that in making a digital record, some important decisions have to be made. The most obvious is the sampling frequency which must be at least double the maximum signal frequency of interest.

Another point to be considered is the possibility of leakage (which is discussed more in Chapter 2). On the assumption that the data at some point will undergo Fourier transformation, it is important that higher frequencies in the data are not allowed to yield ghost "aliased" frequencies in the range of interest. If it is suspected that any high frequencies are present in the signal, appropriate antialiasing filters must be applied. The issues related to sampling and aliasing are discussed in Chapter 2.

1.3 Mathematical Models

Many of the principles throughout this text are explained and illustrated with mathematical models, in some cases following analytic expressions and in others a numerical model. Some readers may find this surprising, as in many cases, a machine operator may not have precise data with which to formulate such models: this will be so particularly in specifying fine clearances which will change with time as the machine components wear, but this does not invalidate the need or the role for theoretical models – in many ways, it emphasizes the role. The aim of the model is to enhance and extend operational experience: to a degree, the actual numbers are secondary and what really matters is the extent to which the different parameters influence each other. This type of sensitivity is extremely valuable in gaining an understanding of the machine's operation. The sentiment, which may surprise some readers, is completely aligned with the theme presented here; although mathematical models and computing (and hence numbers) are used extensively throughout the text, the basic motivation is to elicit understanding of the underlying physical processes as opposed to any precise prediction. It is far more important to use models to examine interdependencies of various parameters and to understand the way in which physical changes (or uncertainties) can/will influence the measured behavior than for a precise prediction of behavior. While the onset of problems can often be identified using purely statistical or empirical approaches, a full diagnosis and localization of the fault will almost invariably require some form of model. This should not be surprising as the model is simply a means of expressing knowledge of the operation of a system.

Such an approach will not always be appropriate. For small items of plant which are readily interchangeable, a suitable approach would be to remove the item from service pending a general overhaul. For high capital plant, however, this approach is rarely feasible and it is in such cases that fault diagnosis, as a means to minimizing maintenance/repair time, is imperative: this is where it becomes vital to glean all the information possible from the data available.

The need to gain insight into a machine's operational behaviour raises another interesting contrast to which we will return several times, namely the distinction between data and information. This distinction is discussed more quantitatively in Chapter 7 in the context of elucidating the influence of foundations on the dynamics of machines, an effect which is particularly important with very large machines such as turbo-generators.

For a given machine the various measured parameters will be linked and one of the aims of mathematical modeling must be to elucidate these links with a view of understanding the fluctuations which occur in machines running under nominally steady-state conditions. To some extent, steady-state is a figment of the engineer's imagination: a machine may be operating at a steady (average) speed and load, but in reality, there will always be some short-term fluctuations; understanding these varia-tions will often yield significant insight into a machine's operation. These fluctuations may arise from the system's control or may be an indicator of a fault and it is important to make the distinction. One example for this sort of phenomenon is rubbing as discussed in Chapter 6. Gearbox oscilla-tions on a pumping system are discussed in Chapter 5, together with a detailed case study in Chapter 9. In many cases a full understanding of the connections between variables is beyond our current analysis methods, principally due to a lack of detailed knowledge of machine's parameters such as subsoil properties beneath the machines that are influencing the foundation flexibility. Indeed, there is ample statistical evidence that, at coastal power plants, the state of the tide has a substantial influence on vibration levels of machines. Whilst the mechanism of this interaction is understood, modeling techniques have not reached the level of quantita-tive predictions of such effects. In such cases artificial neural networks (ANNs) can yield insight, effectively forming a nonlinear curve fit to the measured data. Chapter 8 gives some examples for this type of study: such approaches, while they do not have an acceptable end in themselves, may well aid in progress toward more complete models.

As focus changes to the second question, the diagnosis stage, significantly more detail of data (or at least the analysis of that data) together with some input of the physical processes of the machine's operation, is essential. A prime requirement for any detailed study is the time development of the signal. (Here we refer primarily to vibration, pressure, or voltage signals; temperature variations will normally be very much slower.) Often these signals will be analyzed in terms of frequency and, in particular, their relationship with the shaft rotation speed. Often this relationship will give a good indicator as to the nature of the problem and will be the first stage in the diagnosis. Although some small items of plant may not warrant a thorough diagnosis of any fault, the situation on a high capital item, such as a turbo-alternator, is very different. The cost, in terms of lost revenue of one of these large machines, can be up to £0.5M per day. Because of this high revenue cost, it is very important to minimize the time required to rectify the fault and return the unit to service and this means that fault diagnosis and localization can be extremely cost effective. Some of the techniques available for this important task are reviewed in subsequent chapters.

The third question posed relating to the plant's lifetime is rather more difficult and is, as mentioned, still at the research and development stage; nevertheless, useful predictions can be made with care. This can be extremely beneficial commercially, for instance, by deferring an emergency outage and completing necessary repairs within a scheduled maintenance period. To achieve, this requires a combination of understanding the fault, predicting its probable progress, modeling the influence on the machine's behavior and carefully tracking this against monitored parameters until the scheduled removal from service.

1.4 Machine Classification

The range of rotating machines is immense and any attempt at categorization may appear at first sight, daunting. The masses of rotors span over a range of about eight orders of magnitude. Whilst small machines may have a rotor with a mass of a few grams, some large machines have rotating elements weighing several hundred thousand kilograms. Similarly, the range of speed is immense: while wind turbines typically rotate at around 1/3 revolutions per second (rev/s), turbo-molecular pumps run around 2,000 rev/s and micro-turbines, currently being developed, operate over 1 million rev/min.

Given this wide variation, classification of rotating machine is not a simple matter, but it is necessary to progress with the basic task of this book, namely the elucidation and clarification of some of the main techniques for the interpretation of vibration signatures. Some of the key determining parameters are mass of the rotor, load on bearings, rotor speed, types of bearings, and the duty of the machine. The mass of the rotor, in the case of a machine with a horizontal rotor, will play a dominant role in determining the load on the bearings and indeed, this will dictate the type of bearings to be used in a machine. The choice of bearing type is a crucial design decision and although it is anticipated that for the most part readers of the present text will be operators rather than designers, for the interpretation of vibration characteristics, it is important to appreciate the behavior of the main types of bearing.

1.4.1 Bearing Types

The most basic form of bearing is simply a brass bushing through which the shaft passes, involving a small air-filled clearance. This has the advantage of simplicity (and cheapness) and is suitable for small light machines. In its simplest form, the bearing relies on hydro-elastic effects from the rotation to generate lift forces, but more sophisticated versions employ some pressurization of the air. Another variant of the air bearing incorporates a flexible foil within the clearance and this supports the rotor at low speed; as the rotor speed increases, the hydro-elastic force dominates.

As the loading increases, it may become necessary to use a rolling element bearing. These bearings are considerably more complicated, but they provide a robust support giving good stiffness properties but it is in only small damping terms. As the name implies these bearings have a series of balls or rollers between the rotor and the stator and hence all contact between the mating surfaces is essentially rolling rather than sliding. The minor component of sliding is accounted for in the lubricant. The rolling components are mutually located in a cage which itself rotates at a fraction (up to half) of shaft speed. However, this relatively complex construction means that a series of frequency components will be generated including cage frequency, ball spin frequency, and, of course, the shaft frequency (as discussed in more detail in Chapter 6). The relative importance of these will vary with the condition of the bearing. Some very large machines, particularly low speed such as wind turbines, are supported on rolling element bearings.

At higher speeds, however, the majority of heavy machines are supported on oil journal bearings in which rotor and stator are separated by a film of oil. In hydrodynamic bearings (the over whelming majority) pressure in the oil film is generated from viscous stresses arising from the rotation of the shaft. Although the system is in essence much simpler than the rolling element bearing, the behavior of the oil film is speed dependent and nonlinear. In the plane bearing, subsynchronous instabilities can arise but, where appropriate, a modified type of bearing incorporating tilting pads can be used to overcome the problem. In subsequent chapters, there is considerable discussion of phenomena arising with oil journal bearings. Their popularity arises from their high load capacity but an additional desirable feature is the contribution to system damping. In large turbo-generators, the oil films account for about 99% of the total system damping.

Whilst these represent the most prominent forms of bearing, other types are gaining acceptance for particular types of application. The hydrostatic bearing is a type of oil film bearing but, in this case, the pressure is supplied by an external pump rather than arising from the rotatory motion. The provision of an external pump clearly adds complexity to the system, but this gives the significant advantage of a degree of control over the dynamic properties of the bearing. These units also have greater ability to operate at low speeds as compared to hydrodynamic units.

The ability to control the dynamic behavior of a machine is currently an active research area. Whilst hydrostatic bearings have a potential role here, much attention has been focused on active magnetic bearings (AMBs). In these units, the rotor is levitated in a magnetic field whose strength is controlled via a feedback loop using the measured shaft displacement as the controlling parameter. Hence the rotor, bearing, and feedback loop become a single entity and must be analyzed as a system. AMBs have great potential but do have high cost. Because there is no contact and no fluid shearing (apart from air), friction is virtually zero and consequently these bearings are suitable for very high-speed applications. Damping can be introduced by adjusting the control algorithm. This is one of the greatest advantages of AMBs – the introduction of adaptive forces. The bearings are limited in load capacity, although there have been some applications in large compressors in gas pipelines. The fact that no lubricating oil supply is required presents a very significant advantage, particularly in remote locations. This is the motivation in recent development trends, aiming at completely electrical aircraft and ships, but it is likely to take some time before these efforts reach fruition. The different types of bearing are discussed in Chapter 6, Section 2.

1.4.2 The Rotor

Turning now to the rotor, a marked trend toward flexible rotors is clear over the last half century. It is important to determine whether a rotor is essentially rigid or flexible, as this can have a marked influence on machine behavior and will prove crucial in assessing monitored vibration signals. Prior to World War II, all rotors were rigid and it was only with increased understanding that it was deemed feasible to operate machine with flexible rotors. Large alternators are now almost invariably flexible and an increasing number of smaller machines come into this category. Chapter 4 discusses the salient distinction between rigid and flexible rotors. In some respects, the differences are modest, but these distinctions can give rise to some marked changes in behavior owing to the presence of higher modes and stress variations within the rotating elements.

1.5 Considerations for a Monitoring Scheme

Let us examine the logic which determines the type of monitoring system suitable for a given application. The prime question will be the capital value of the asset; if it is a low value asset which is readily replaceable, there may be little value in installing a complex, expensive system (the issue of costs will be addressed a little later), rather the requirement is mainly to detect an incipient failure and ensure that a spare unit is readily available. The requirement is at stage one, in terms of the classification scheme outlined above: purely statistical approaches can usually be applied to this problem and this is often as simple as specifying that the vibration level must remain below a certain threshold level. A marginally more complicated system may consider a change in the shaft orbit, a problem which is discussed in Chapter 2.

The task of devising an appropriate monitoring scheme is complex and will depend on a number of operational parameters together with the range of anticipated fault conditions. Having decided to monitor this item of plant, an important issue is the frequency of monitoring – a very important issue which often carries major resource implications. To make an appropriate decision, the operator must consider the range of faults which occur on the type of plant in question and in particular the timescale on which the faults develop from inception to any threat to machine operation or integrity. The interval chosen for monitoring must be less than the development time of the faults or clearly a damaging fault could be completely missed by the system. On the other hand, monitoring too often generates excessive data which then requires adequate systems and staff resources to make use of the data. This process of formulating the appropriate system has been formalized into a methodology called failure modes and effects analysis (FMEA). A discussion of FMEA is outside the scope of this volume, but it has been widely discussed elsewhere.

For a slowly developing fault, or a combination of faults, intermittent checking on a daily or even weekly basis may well suffice and in such cases it is likely that the required data can be logged on by staff, either manually or increasingly with hand-held monitors which can then feed data centrally. It is unlikely that in such a case the financial and other costs in implementing a dedicated computerized system could be warranted. The overriding consideration in the design of monitoring systems must be the cost effectiveness, but the accurate evaluation of this is by no means a trivial task.

If particular faults on plant develop more rapidly, then some automation of the process becomes necessary or at least desirable. However, this involves a hidden cost: not only is there a cost involved in gathering the data, but also a not insignificant resource requirement in evaluating the data.

For high-value plant, the situation is rather different. In this case there will rarely be an option of simply replacing the plant item and the monitoring system must address (at least partially) the issue of fault diagnosis with a view to optimizing the repair strategy and minimizing the so-called down time. The requirement for a diagnostic capability almost inevitably implies a need for examining the frequency content of the signal and hence the need to record data at least several times per revolution of the shaft (actually this is governed by Shannon's sampling theorem, requiring that the sampling frequency is at least double that of the signal). Clearly, such rapid sampling calls for a computerized system and with a profusion of data, appropriate strategies are needed to extract meaningful information which can make a real contribution to plant's operation. Of course, there is still the decision to be made as whether the parameters are monitored continuously or at periodic intervals, but we need not dwell on this here.

Clearly, it is unrealistic to produce an accurate model of all machinery operations, although the later chapters report significant progress in recent years towards this aim. Like many technical topics, one is forced into a regime of diminishing returns in which progress becomes limited by the restricted marginal benefit. It is appropriate to examine in general terms the realistic goals for any monitoring system. To make a meaningful contribution to plant operation, a monitoring system is needed to provide full data in the event of an incipient fault, but this implies that the overall volume of data produced is likely to be enormous and it becomes essential to devise some automated system to provide an initial filtering of the raw data. The direct automated diagnostics is somewhat beyond current capabilities and hence the practical aim for the system is to alert the engineer (i.e. a human expert) of potential difficulties.

In preparing this text, efforts have been made to minimize reliance on mathematical analysis. However, as the guiding theme is to gain an in-depth understanding of machine behavior, some mathematics is inevitable but in all cases, this is basically expressions of Newton's laws coupled with some degree of Fourier analysis. Whilst there is an element of nonlinear behavior in most systems, this is relatively minor in the majority of cases. Indeed, it is often the onset of this nonlinearity, which is the vital clue to establish a change. In all the discussions presented, the mathematical details are secondary to the theme: the central focus is the physical source of phenomena. The purpose of any theoretical or numerical model is to relate the observed behavior to the processes within the plant item. Alternatively, this may be expressed as the conversion of data into information and it is worth considering this in a little more detail. If, for example, an operator has a vibration reading of a given amplitude and phase, this in itself is extremely difficult, if not impossible, to interpret. But there is a variety of ways in which this data can be rendered useful. Perhaps the most basic of these techniques is the formation of trends over time – and the period involved will depend on how a fault (or rather a phenomenon) develops. The challenge is then to relate the change in behavior to some changes in the condition. Clearly this task may be made easier with complete records but only rarely will it be a trivial exercise.

The relationship between monitored behavior and internal processes is often complex, and in this situation, it is necessary to utilize all information available. Clearly, measured data contains information; this is not to say that the data is absolutely correct, indeed, it will inevitably incorporate noise and possibly some systematic errors, but it will contain some information. In the same way, a model of the machine, either analytical or numerical will reflect some of the physics of the real machine and while containing some limitations, if used in the correct manner, it can be used to give additional insight. Chapter 7 discusses the incorporation of foundation effects using this type of "blended" approach, but more generally the resolution of machine problems often requires the use of information from a number of different sources.

The present state of knowledge in this field is far from complete, and in the next few years, there are prospects for significant advances. Over recent years, it has become possible to formulate reasonably accurate theoretical models to replicate the dynamic behavior of practical machines. This has resulted partly from enhancement in our basic understanding and partly due to the advancement in computing technology. Currently the main features of behavior are reflected in models, but with more recent identification techniques it is becoming feasible to gain an insight into the more subtle distinctions between nominally identical machines. This is significant, as this will make it possible to more fully utilize on-load data of machine operation. It has been a dilemma for many years that all plant operators have a vast quantity of data of their machines operating on-load, yet the information content of these records is relatively low because of the lack of interpretative capability. Only gross changes could be interpreted with any degree of confidence. A significant component of the variation was the inter-machine variation which is now better understood than in previous years.

1.6 Outline of the Text

In monitoring and diagnosing problems on rotating plant, a wide variety of parameters are assembled and gathered. Vibration data is commonly utilized but, because of its rapidly varying nature, the sheer volume of data leads to a question of presentation. This is not a trivial matter because different formats of presentation bring out different features of the behavior and it is frequently necessary to consider a number of different representations to gain a general overview of the phenomena. Chapter 2 discusses a wide range of the presentation of vibration phenomena and illustrates how various modes of presentation may be used to bring out different aspects of a machine's behavior.

Chapter 3 discusses the fundamentals of machine modeling. After examining the need for theoretical models, the chapter outlines the FE method: it is fortunate that one of the simplest FEs is also one of the most important in the representation of rotating machines. Having established the methodology, an explanation is given for the various ways in which data may be usefully presented. Often, this is slightly different to the approaches presented in Chapter 2 because a different range of information is made available. At the end of the chapter, we discuss the application of mode shapes and perturbation theory to enhance the understanding gained from models.

Chapters 4, 5, and 6 give an extensive discussion of some of the more common faults on rotating machinery. It is not feasible to cover all the possibilities in a single text, but the range covered here is representative. The rationale for this discussion is that it is simply not feasible to diagnose a fault without some idea of what the fault would look like, in terms of external parameters. Note that this is a distinguishing feature between detection and diagnosis. At what we might call level 1 of monitoring, for the detection of a fault, it is quite possible and often vital, to make a decision without any further knowledge. Such a simple system, usually based purely on statistical models or even just some threshold criteria, can trigger the removal of a machine from service or a reduction in load. However, as discussed above, to proceed to make a diagnosis, some insight into the physics of the situation and some form of model is needed. Chapter 4 discusses rotor imbalance, the single most prevalent fault in rotating machinery. The chapter begins with a detailed discussion of the differences between rigid and flexible rotors, as this has a number of important consequences and the methods used for correcting imbalances must be appropriate to the rotor class. After outlining the principal approach to rotor balance, the discussion turns to rotor bends. As discussed in Chapter 4, it is important that bends and imbalance are not confused, even though they both give rise to synchronous vibration.

The survey of machine faults is continued in Chapter 5 with a discussion of misalignment, cracked rotors, torsion, and rotor rubbing. These are complicated topics and some are still areas of active research. They are, however, important practical problems on real machines as well as being of substantial academic interest. The phenomena associated with misalignment remain incompletely understood, even though it ranks second only to imbalance as a major problem on rotating machines. Later in the chapter, torsion is discussed and as a particular important example of this, a discussion of the dynamics of gears is given; this is another case in which clearances and wear patterns can play a profound part in determining both performance and dynamic characteristics of the system.

Chapter 6 focuses on the influences of clearances in various types of rotating machines. There are many fine clearances occurring within machines, usually filled with either working fluid or lubricant. The effect that these clearances have on the dynamics of the machine is typified by the case of a large centrifugal pump and this is discussed at some length. The important issue is that the internal neck-rings act as subsidiary bearings and also have a profound effect on the internal pressure and flow distribution within the unit, with consequent influence on the unit's overall performance. Clearances at blade tips in turbines, on the other hand, give rise to the so-called Alford's force. Actually, the behavior of a shaft within an annular clearance is complex and a discussion of the range of phenomena, and the ensuing machine faults are described.

Chapter 7 is a little more specialized than the others and focuses on recent work on the problems of taking into account the dynamics of the supporting structure. On large machines in particular, such as turbo-alternators, the support can have a very significant influence on the dynamic properties of a machine and there are a number of factors which render conventional modeling of these structures virtually impossible, the reasons for which are set out in the chapter. In principle of course, all supporting structures will have some influence, the difference with these very large machine is that it is impractical to make their supports sufficiently stiff to minimize the effect. Because the supporting structures are complicated, it is impractical to obtain an adequate model of its influence on the machine simply by FE methods. Such an approach has been attempted by various authors over the last few decades, for example Lees and Simpson (1983) and Pons (1986). The overall conclusion is that, although these models give general trends which can be useful by highlighting problems at the design stage, they are insufficiently accurate for use in monitoring and diagnosis of operational plant. This is because each support has a significant number of complex joints which are extremely difficult to measure, even if complete data were available, but this is rarely if ever the case. The overall problem is illustrated by the fact that nominally identical units can have appreciably different dynamic conditions. Chapter 7 outlines the recent progress in this area, showing how a dynamic model can be enhanced or updated in the light of measured data. These developments offer the prospect of significant progress in the coming years, as models may be sufficiently refined to study the variations in nominally steady-state conditions. Current methods only make very cursory use of this source of data.

Chapter 8 considers some of the more modern methods of treating data. These include ANNs and kernel density estimation. Both of these approaches provide a route to establish connection between data; they may both be regarded as nonlinear fitting methods but this can be invaluable in establishing relationships and thereby understanding. In this chapter, we also explore the methods for dealing with strongly nonstationary signals, all being a part of the general field of time–frequency analyses. Some of these are still at the research stage but practice over the next few years will do much to evaluate their usefulness. Kernel density estimation, wavelet transform, Huang–Hilbert transforms, and expert systems are just some of the procedures discussed in this section, in addition to more model-based approaches.

Emphasis is given on the potential of blending deterministic and more statistical based approaches as a means of gaining the greatest insight into the operation of a piece of machinery. Indeed, this is the constant theme of the book – the constant effort to use available data to gain insight. In doing this, measured plant data and the predictions of theoretical or numerical models can contribute significantly to overall understanding.

Chapter 9 describes some case studies, taken from both the author's experience and the literature. The five examples given cover a range of machine faults and in most of the cases, a number of techniques are combined to arrive at a satisfactory solution to the problem. This is entirely realistic, as most practical problems demand some degree of ingenuity for their solution.

Finally, Chapter 10 is slightly more speculative than other parts of the text. An outline is given of a range of techniques currently available for both data analysis and model improvement. It is anticipated that a steady improvement in data analysis, perhaps in tandem with enhanced computing capabilities, will continue for some years and at the present time, it is not totally clear what the form of monitoring will be in decades to come. Nevertheless, with due trepidation, some predictions are offered as to schemes which may become feasible (and cost effective) in years to come. The culmination of developments in condition monitoring, diagnosis, and control may combine to form what might be called a "smart machine." That is, a machine which will detect and diagnose an incipient fault, then apply a corrective force which will compensate for, or at least mitigate, the effects of the fault. There are several strands to the research in this area and although it is still a little away from realization, the work is at a stage where the progress may be rapid. A brief overview is given in Chapter 10 of the recent developments in this area.

Pending the fruition of these developments, work on the monitoring and diagnosis of problems in rotating machinery is a role that is vital across a wide range of key industries. It is hoped the techniques and fault descriptions presented here will aid that work.

1.7 Software

Throughout this text, the importance of a physical understanding of processes within a machine has been emphasized and almost invariably this understanding requires some form of mathematical model. In some cases, these take an analytical form, but more usually the complexity of real machines demands a software model. The work presented here uses the toolbox developed by Friswell *et al.* (2010), which is freely available at www.rotordynamics.info. This set of routines has been supplemented by a set of user interfaces and some other scripts for various common tasks are given. These are available on the website. No other toolboxes are required and readers with access to MATLAB 6 or later should have no difficulty in running any of the codes. Other readers might gain access by using the free SCILAB package. The real benefit of the numerical models is not specific answers to numerical case, but rather the ease of varying parameters and investigating the consequences of various scenarios. Readers are encouraged to experiment with the software as this offers a powerful learning experience.

References

Childs, D., 1993, *Turbomachinery Rotordynamics: Phenomena, Modelling and Analysis*, Wiley, New York.

Friswell, M.I., Penny, J.E.T., Garvey, S.D. and Lees, A.W., 2010, *Dynamics of Rotating Machines*, Cambridge University Press, New York, NY.

Inman, D.J., 2008, *Engineering Vibration*, Third Edition, Prentice-Hall, Upper Saddle River, NJ.

Lees, A.W. and Simpson, I.C., 1983, Dynamics of turbo-alternator foundations, *Institution of Mechanical. Engineers Conference*, London, paper C6/83, February.

Pons, A., 1986, Experimental and numerical analysis on a large nuclear steam turbo-generator, *IFToMM Conference 'Rotordynamics'*, Tokyo.

Price, E.D., Lees, A.W. and Friswell, M.I., 2005, Detection of severe sliding and pitting fatigue wear regimes through the use of broadband acoustic emission, *Proceedings of the Institution of Mechanical Engineers, J-Journal of Engineering Tribology*, 219 (J2), pp. 85–98.

Sikorska, J.Z. and Mba, D., 2008, Challenges and obstacles in the application of acoustic emission to process machinery, *Proceedings of the Institution of Mechanical Engineers, Part E, Journal of Process Engineering*, 222, pp. 1–19.

2

Data Presentation

2.1 Introduction

The process of resolving problems on rotating plant, or simply monitoring performance, is in essence made up of two complimentary strands: firstly, obtaining plant data and secondly interpreting it, either on the basis of experience or by the use of some form of theoretical model. In either case, the engineer is implicitly (or explicitly) fitting the measured data to a model, or to his/her understanding of the machine's operation, to infer the internal parameters of the machine being studied.

In this chapter, consideration is given to some of the basic forms of the measured data in the study of rotating machinery. In some respects, this is similar to the study of structural dynamics but there is an important complicating factor; the variation of both the system and the forcing with shaft rotational speed. This is an obvious pre-requisite to the material which follows in subsequent chapters.

We note that there are two fundamental distinctions between a structure and a rotating machine, as far as the dynamic behavior is concerned. Firstly, the dynamic properties of a rotating shaft depend on the rotational speed owing to gyroscopic terms and, in many cases, the property of the bearings. Secondly, the shaft rotation provides a potential energy source which implies that under certain conditions, vibrations can grow – a condition of instability. Both these issues will be addressed briefly in this chapter and further more extensively in Chapters 4–6. Prior to the detailed examination however, these issues have some influence on the present discussion.

The introduction of shaft rotation means that there is a wide variety of ways in which machine vibration data may be presented and in this chapter some of the main methods are presented. It is important to note that there is no ideal approach for all occasions: different presentation approaches bring out different features in seeking to understand plant behavior.

2.2 Presentation Formats

Although mathematical analysis of systems is discussed in Chapter 3, it is helpful to briefly mention the form of dynamic equations here. Any rotating machine can be described by an equation of the form

$$\mathbf{K}x + \mathbf{C}\dot{x} + \mathbf{G}\Omega\dot{x} + \mathbf{M}\ddot{x} = \mathbf{F}(t) \tag{2.1}$$

The evaluation of the terms in this equation can be a little involved, but here attention is focused on the structure of the equation. On the left-hand side, in addition to the stiffness, damping, and mass terms ($\mathbf{K}$, $\mathbf{C}$, and $\mathbf{M}$, respectively) there is a gyroscopic term $\mathbf{G}$. This is discussed in some detail in Chapter 3, but the essential point is that it arises from the conservation of angular momentum and an important consequence is that the dynamic behavior of a rotor is dependent on the speed of rotation of the shaft. An extensive discussion of the gyroscopic term is given by Friswell *et al.* (2010).

In addition to gyroscopic effects discussed above, bearing properties often vary with shaft speed in a pronounced way. Because both the natural frequencies and mode shape of a rotor vary with speed, the presentation of plant vibration data is somewhat more complicated than that for a nonrotating structure.

As mentioned in the introduction, there are two types of data measured on a typical rotating machine: a) data from a steady running condition and b) transient speed plots (run-ups and rundowns). On many types of plant, the former is plentiful and forms an important part of condition monitoring systems, but the transient data is extremely rich in information, but at a cost! The cost is in terms of difficulty of analysis.

2.2.1 Time and Frequency

The fundamental problem is that the dynamic properties of a rotating machine are functions of the rotation speed and indeed often functions of other parameters such as pressure or load. We consider first the dependence on speed, as this is (almost) invariably the key parameter. This variation with speed is of course one of the factors distinguishing a rotating machine from any general structure and the variations may arise from the behavior of bearings and seals as well as the effects of gyroscopic couples. The reader may note that gyroscopic terms tend to give rise to relatively small corrections for systems such as that shown in Figure 2.1 in which the rotor mass is between bearings. However, for situations where there is an appreciable overhang as is often the case with fans, as shown in Figure 2.2, gyroscopic terms may become very significant or even dominant.

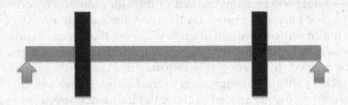

FIGURE 2.1
Basic Rotor System.

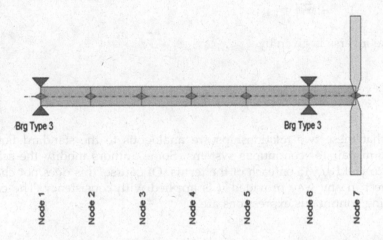

FIGURE 2.2
Layout of an Overhung Fan.

At one time vibration data would be recorded on magnetic tape for subsequent analysis. In essence, this was an analogue process but the more modern approach is to directly digitize data and store on a disc or other medium. It should be recognized, however, that any digitization process involves decisions related to the form of the data. The sampling rate must be chosen to adequately capture the machine's behavior. This choice will be governed by some assessment of the anticipated outcomes and considerations of aliasing and leakage problems (see e.g. Bossley *et al.*, 1999). There are two viable strategies for this; the first is to sample as rapidly as possible then to resample in order to make evaluation viable. The alternative, and in some ways preferable, approach is to lock sampling to shaft rotation and sample at a number of times per shaft revolution. This strategy automatically eliminates leakage (a term explained below) as a problem for most investigations. The essential point however is that important decisions have been made at this stage.

The stored data will be a time record of the vibration and on occasions this is in itself a key part of decisions on the machine's condition. More usually, however, the main focus of attention will be on the spectral content of the time records as this provides a route by which forcing and response frequencies can be related. In essence, this requires the use of the discrete Fourier transform (DFT), the digital equivalent of the Fourier transform. Given a set of measurements, x_r, evenly spaced in time, the frequency spectrum is given by X_k and these are related by the DFT which is defined as

$$X_k = \sum_{r=0}^{n-1} X_r e^{j2\pi kr/n} \qquad\qquad k = 0, 1, 2, ..., n-1 \qquad (2.2)$$

and the inverse is given by

$$X_r = \frac{1}{n} \sum_{k=0}^{n-1} X_k e^{j2\pi kr/n} \qquad\qquad r = 0, 1, 2, ..., n-1 \qquad (2.3)$$

Note that these two relationships are analogous to the standard Fourier transform pair for continuous systems. Some authors modify the scaling factor to yield $1/\sqrt{2\pi}$ on each of the terms. Of course, this does not change the effect in any way provided it is applied with consistency. The corresponding continuous expressions are

$$x(t) = \frac{1}{2\pi} \int_{-\infty}^{\infty} X(\omega) e^{j\omega t} d\omega \qquad (2.4)$$

$$X(\omega) = \int_{-\infty}^{\infty} x(t) e^{-j\omega t} dt \qquad (2.5)$$

The calculation of the DFT involves n^2 summations and a much faster algorithm, the fast Fourier transform (FFT), was reported by Cooley and Tukey (1965). This reduces the number of summations required to $n/2\log_2 n$. For large n, this represents a very significant reduction and this algorithm has gained virtually universal acceptance.

The FFT algorithm, together with some subsequent refinements, has become a cornerstone of vibration analysis and signal processing in general. In the case of rotating machines, the shaft rotation speed is crucial and the expression of data in the frequency domain clarifies the relationships of various components.

2.2.2 Waterfall Plots

In the analysis of structures, it is usual to present the spectrum of vibration by performing an FFT on the vibration signal and this gives considerable insight into both the structure and the exciting forces. The same logic applies for rotating machinery but there is an important distinction: in this case, both the forces and the structural properties are functions of shaft rotation speed and so a spectrum is required for each shaft speed. The difficulties in doing this are discussed a little later, but the result is a so-called waterfall plot, an example of which is shown in Figure 2.3.

The scale labeled "number" can be rotor speed or time, or indeed other parameters such as load. The waterfall plot, although a fairly crude device, offers a comprehensive overview of the system dynamics and in many instances, it will form a prelude to a more detailed study. It is interesting to make some observations on this particular plot. The data was taken from a large laboratory rig at Swansea University, which is shown in Figure 2.4. From this simple plot, two features are immediately obvious. In addition to the main diagonal line showing synchronous excitation arising from imbalance, there are a number of other lines appearing corresponding to excitation at harmonics of shaft speed. These undesirable features must arise from nonlinearity in the system and this was quickly

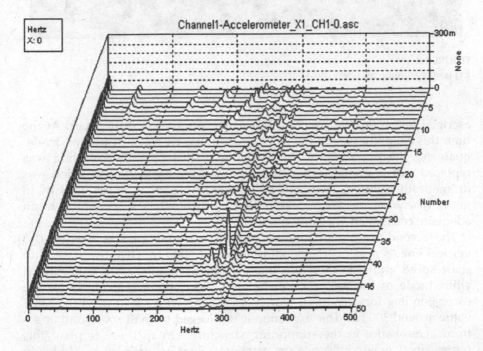

FIGURE 2.3
Sample Waterfall Plot.

Reproduced with permission from the Institution of Mechanical Engineers. Published as part of Lees *et al.* (2004).

FIGURE 2.4
Experimental Rig – Swansea University.

ascribed to the flexible coupling between the motor and the shaft. At the time this coupling was of the single membrane type which proved inadequate for the purpose. On the evidence of this plot, the coupling was replaced with a double membrane type and the harmonic excitation was dramatically reduced. There will be unavoidable sources of nonlinearity in many machines, but an understanding of their effect is a prelude to an adequate correlation with any model.

The second feature of interest on this waterfall plot is the (nearly) vertical line of peaks corresponding to about 300 Hz. This decreases with shaft speed and it is therefore likely to be associated with a backward whirl mode of the rotor. Somewhat less clear, there is evidence of the corresponding forward mode slowly increasing with speed from the same value around 300 Hz. This excitation is not locked to shaft speed and arises from a resonance in the supporting structure. On real power plant this represents a major influence on machine vibration properties. Modeling the foundation is in principle straightforward; the (major) difficulty is in having the appropriate data to use as each individual foundation has its unique properties. Much lower down in frequency, there is a resonance at

48 Hz which corresponds to a critical speed of the system and which is largely independent of speed.

The details are not particularly important but the key point here is that so much information may be gleaned from a waterfall plot of a single channel. Added information may be presented through similar plots from other channels of data. Recent work on this problem is discussed in Chapter 7. The interested reader may also refer to a recent review (Lees *et al.*, 2009).

2.2.3 Scatter Plots (or Carpet Plots)

Just as the waterfall plot provides a simple "broad brush" overview of behavior over a range of speeds, the so-called scatter plot is a useful tool for viewing on-load data at constant speed. An important part of vibration monitoring concerns the analysis of data while the machine is operating, very often at a nominally constant speed. Whilst this is a regime in which there is inherently less information available from the data, since it is the state of normal operation, there is a vast amount of data available at essentially no cost. However, the vast data resource contains limited information and it is important to extract the limited information which is accessible.

It is informative to commence our discussion by posing the question "Why does the vibration behavior vary over time when the rotational speed is fixed?" The precise answer to this will depend on the type of machine being studied but for the sake of illustration we will discuss the case of a large turbine generator. The variation of the magnitude of vibration on the front bearing of the LP turbine of a large turbo-alternator is shown in Figure 2.5, but this only tells part of the story. Nevertheless, it can be seen that the variations are not insignificant.

This figure shows the amplitude of the vibration velocity at two minute intervals over a period (about 11 days). For the vast majority of this period, the machine was operating at a nominally constant speed of 3,000 rev/min. Despite this "steady operation" the vibration amplitude shows considerable variations and this raises the following two issues:

a) Can any insight be gained from theses variations over time?

b) Can the data be sensibly used for condition monitoring purposes?

The naïve answer to the second question is clearly "yes" since if the vibration undergoes a large change, this would clearly invoke some investigation as to the cause so we repose this question as "How small a change in vibration levels can be considered as significant?"

First, however, we address the more fundamental question of why these variations arise. The answer is rather complex and the details are still the subject of active research work.

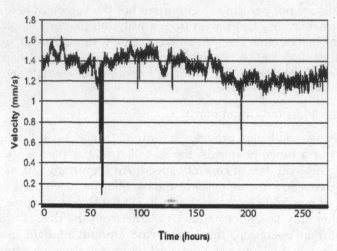

FIGURE 2.5
Bearing Vibration over an Extended Period.

Some of the variations which influence the machine are as follows:

a) ambient temperature
b) water level in the condenser
c) steam temperature
d) steam pressure
e) variations in power supplied
f) variations in reactive power
g) rotor and stator cooling operation
h) rotor current settings

and many more: for each of the data points shown, 54 pseudo-static parameters are recorded. Of course, not all of these are equally important. A method for deciding the importance of different parameters is introduced next. For instance, at coastal power stations, the state of the tide has been observed to have some influence over the dynamics, no doubt by effecting changes in the water table. For other types of machine, the list of factors will be different, but there will invariably be features which give rise to variations during nominally constant operation. A different perspective is gained if we replot this data showing the real part against the imaginary part. The result is Figure 2.6. This is a *scatter* or *carpet plot*. The

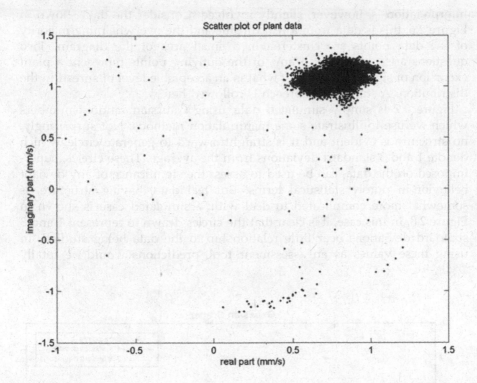

FIGURE 2.6
Typical Carpet Plot.

measured vibration velocity at each of the intervals is plotted on the Argand diagram, thereby representing both magnitude and phase. Note that ideally, under constant running conditions, all the points should coalesce, but because of fluctuations there is also some spread. The problem is to assess whether this spread is within normal bounds.

This can be simply a plot of the time data of the motion in two orthogonal directions. The real and imaginary parts of the synchronous term could also be plotted in this way but this complication is unnecessary. The simple plot gives an overview of long-term variations.

The precise mechanism of the various interactions is rather beyond the current state of machine modeling, although it is hoped that recent developments will lead to progress in this area. For the present, however, statistical approaches are used to gain some insight into on-load data. This plot shows some interesting structure, but the general properties are first discussed using a simulated example.

One advantage of using a scatter plot is that, despite its simplicity, it gives information on the complete vibration signal, in two directions.

Interpretation is, however, slightly involved. Consider the data shown in Figure 2.6: this is data from a real turbine and the overwhelming majority of the data points are clustered in a small area of the diagram. Two questions arise: a) which, if any, of the outlying points represent a plant excursion of significance and b) what is an acceptable level of spread in the distribution. A statistical approach is followed here.

Figure 2.7 is simply simulated data using Gaussian random numbers which we use to illustrate some manipulation methods. Not surprisingly, no structure is evident and it is straightforward to generate circles which denote 1 and 3 standard deviations from the average. These circles, super-imposed on the data, can be used to assess the significance of any deviant behavior in purely statistical terms. But real data, having structure, is somewhat more complicated to deal with: a simulated case is shown in Figure 2.8. In this case, it is clear that the circles drawn to represent 1 and 3 standard deviations bear little relationship to the data being studied. In using these values as an assessment tool, predictions would be totally

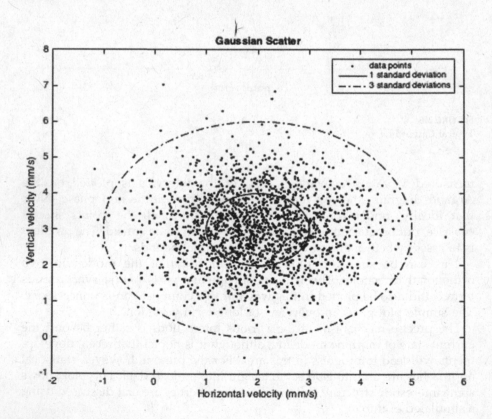

FIGURE 2.7
Scatter Plot of 2D Gaussian Noise.

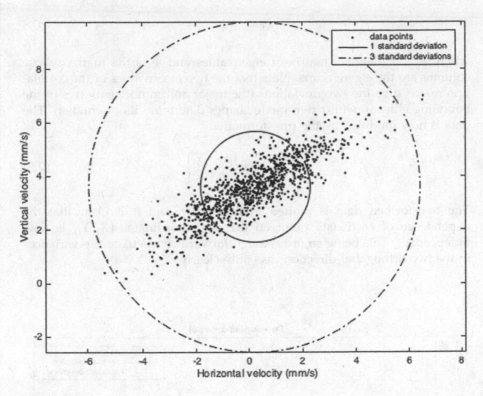

FIGURE 2.8
Simulated Correlated System.

inconsistent, sometimes being unduly restrictive, whilst others being too lax. The problem now is to determine a way in which the collection of data can be adequately represented. In Figure 2.8, the use of circles to define the spread of the data is clearly inappropriate, not only do the two directions scale differently but there is also an issue with the orientation of the bulk of the data.

To resolve these issues, consider the set of data points x_i, y_i. The first step is to take averages $\bar{x}, \bar{y}$ and use these to calculate the variance matrix **A** as

$$\mathbf{A} = \frac{1}{N} \begin{bmatrix} \sum_{i=1}^{i=N} (x_i - \bar{x})(x_i - \bar{x}) & \sum_{i=1}^{i=N} (y_i - \bar{y})(x_i - \bar{x}) \\ \sum_{i=1}^{i=N} (x_i - \bar{x})(y_i - \bar{y}) & \sum_{i=1}^{i=N} (y_i - \bar{y})(y_i - \bar{y}) \end{bmatrix} \quad (2.6)$$

This matrix is now used to resolve the data into principal components, essentially transforming the data into a new coordinate system. The key step in this is the solution of the following eigenvalue problem:

$$\mathbf{A}\Phi = \Lambda\Phi \tag{2.7}$$

where Λ is a diagonal matrix of eigenvalues and Φ is the matrix whose columns are the eigenvectors. Note that the two eigenvalues of the correlation matrix give the two deviations (the major and minor semi-axes of the bounding ellipse), whilst two mode shapes determine its orientation. The data is now modified by the transformation

$$\begin{Bmatrix} X_i \\ Y_i \end{Bmatrix} = [\Phi] \begin{Bmatrix} x_i - \bar{x} \\ y_i - \bar{y} \end{Bmatrix} \tag{2.8}$$

The transformed data is plotted in Figure 2.9, and it is clear that the dependence of each data point on the new coordinates (X_i, Y_i) is now independent. This being so, it is straightforward to calculate the variances in the two orthogonal directions as (since clearly, $\bar{X} = \bar{Y} = 0$)

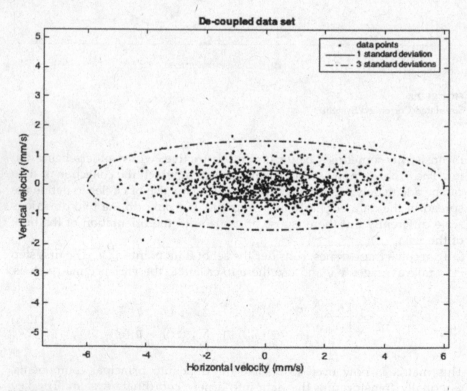

FIGURE 2.9
Plot of Orthogonalized Data.

$$\sigma_x^2 = \sum_{i=1}^{i=N} \frac{X_i^2}{N} \qquad \sigma^2{}_y = \sum_{i=1}^{i=N} \frac{Y_i^2}{N} \tag{2.9}$$

and using this information, ellipses may be drawn around the data showing the extent of variability at one and three standard deviations. Note that the two variances shown above are just the two eigenvalues of the matrix **A**. Transforming everything back to the original coordinates, the result is Figure 2.10. The important point to note here is that the ellipses representing deviations from the norm are now far more representative of the data set. Hence, any judgment as to whether or not a change has taken place is far more accurate using these curves. The circles of Figure 2.8 would yield totally inconsistent warnings.

The prediction and understanding of the subtle variations revealed by these plots are beyond the scope of current modeling techniques, although the recent developments discussed in Chapter 7 offer some prospect of

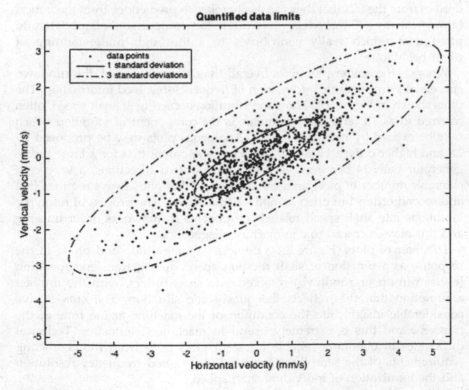

Quantified data limits

· data points
— 1 standard deviation
—·—· 3 standard deviations

Vertical velocity (mm/s)

Horizontal velocity (mm/s)

FIGURE 2.10
Analyzed Scatter Plot.

advances in this area. A significant difficulty has been that, in practice, no two machines are identical and the changes due to damage are often commensurate with the differences between nominally identical machines. One possible way forward in the understanding of "steady load" conditions is the use of artificial neural networks and other techniques which are somewhat loosely termed "artificial intelligence." In some ways, this might be considered to be somewhat against the theme of this book where physically based modeling is advocated in the quest to understand dynamic behavior. However, physics-based and statistically based approaches are not necessarily opposing methods and a brief outline is given in Chapter 8.

2.2.4 Order Tracking in Transient Operation

In most, if not all, turbo-machinery, the predominant frequency of excitation is related to the rotational speed of the shaft. This implies that during transient operation, that is, during run-up to operating speed or coast down to rest, a broad range of frequencies is "excited" and hence a comprehensive assessment of the machine's dynamic properties is available from a careful treatment of the monitored data. Much of the power of rundown curves derives from the fact that they can easily relate to predictions from theoretical (almost always FE) models: it is often the comparison of models and plant data which really contributes to a thorough understanding of plant behaviour.

Whereas the scatter plot gives overall trends over a period, the rundown curve gives time-specific information of frequency-resolved information. The plots shown in Figure 2.11 show the vibration occurring at shaft speed, often referred to as 1X (and this, of course, is the component of vibration which may be excited by rotor imbalance). Analogous plots may be produced by 2X and higher orders. However, it becomes apparent that for a large turbo-generator with 14 bearings, each vibrating in two directions, a very considerable number of plots are generated given that one always requires first and second orders but often 3X and 4X in addition. The process of resolving vibrations into shaft speed related components is known as order tracking and this plays a crucial role in machine diagnostics.

This pair of plots (Figure 2.11) depicts the magnitude and phase of the response as a function of shaft rotation speed during transient operation (either run-up or rundown) resolved into shaft orders, with the implicit assumption that this reflects the quasi-static situation. The curves give considerable insight into the condition of the machine at the time of the transient and this is extremely useful in machine diagnostics. Technical difficulties in obtaining reliable transient plots arise from the conflicting requirements of the sampling time needed for good frequency resolution and the localization of individual shaft speed.

Both amplitude and phase show significant variation with shaft speed and Figure 2.11 shows the synchronous components. Similar plots can be

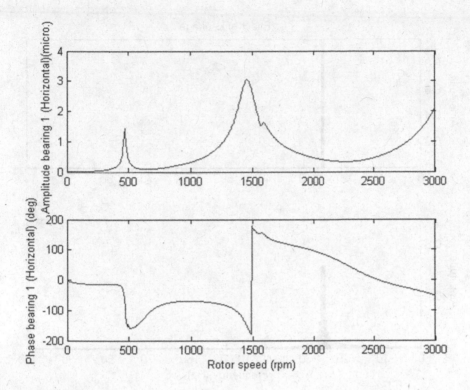

FIGURE 2.11
Typical Rundown Plot.

made for the higher orders, the twice per revolution being particularly useful. The problem arises of how to calculate the frequency response since a straightforward Fourier transform assumes that the system is time invariant. The simplest, and still the most commonly used, technique is the short-term Fourier transform (STFT), but two difficulties arise as we now illustrate. The terminology short-time Fourier transform is also used for this approach. The first difficulty is the problem of treating the data in a meaningful way to give an accurate reflection in the frequency domain. The second issue is to relate this transient behavior to a quasi-static situation by reflecting on how rapidly the machine can respond.

2.2.4.1 Short-Term Fourier Transform

Figure 2.12 shows part of the FFT of a sinusoidal signal at 25 Hz over 2 sec. Notice there is a single line in the FFT at 25 Hz. Also shown in the figure is the same signal but taken over 2.02 seconds and the FFT looks

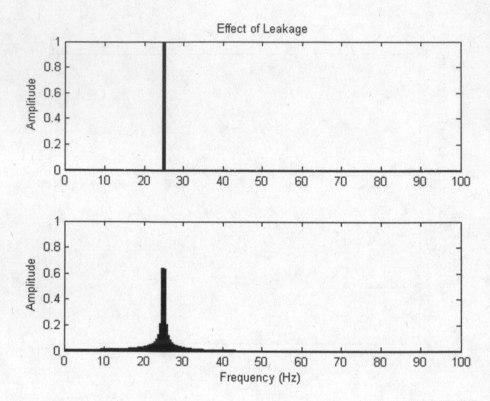

FIGURE 2.12
A Comparison of the Spectra over 2 and 2.02 Sec.

rather different: although the sampling time has been longer, an inferior result has been obtained because of the phenomena known as leakage. This occurs because in taking an FFT of a signal over a finite time interval, the signal is implicitly repeated throughout all time to create a signal of infinite length and this gives rise to inaccuracies if the beginning and end of the measure signal do not join continuously. Another way of saying this is that the measured sample must contain an integer number of cycles. In rotating machinery, the way around this problem that is normally used is to fix the data acquisition with relation to shaft position rather than time. This is highly relevant in that in most cases the principal concern is to relate the vibration signal to rotation speed rather than an absolute frequency reference. This locking can be achieved either in the data acquisition hardware or in software by digitally resampling. Before proceeding we note some of the salient features of Figure 2.12. The resolution is, of course, set by the overall sample period and since this changes by only 1%, no difference in resolution is apparent. The major difference in the second case is that the time span does not represent an integral number of cycles. The graph of this case shows that the peak is now close to (but not equal to) 25 Hz and its magnitude has decreased to a considerable degree. On the other hand, the value of the FFT at other frequencies has increased. The distribution has broadened: the "signal" which should all be at 25 Hz has "leaked" into neighboring frequency bands.

If a structure is being studied, analysis is quite straightforward – one simply samples the motion over a period chosen to give the required frequency resolution from

$$\Delta f = \frac{1}{T} \tag{2.10}$$

with frequency expressed in Hertz and T (in seconds), the period over which the data is sampled. Equation 2.10 has a simple physical meaning; the maximum resolution which can be observed corresponds to a difference of a single cycle over the period monitored.

With machinery in transient operation the situation is more difficult in that during the interval T the speed of the machine, and consequently the excitation will change. Hence, the choice of the sampling period will sometimes be a compromise. Let us assume that a machine runs down to rest exponentially so that the speed, Ω, is given by

$$\Omega = \Omega_0 e^{-at} \tag{2.11}$$

This expression is only approximately true, but it will suffice for the current discussion. So we may write

$$\frac{d\Omega}{dt} = -a\Omega \tag{2.12}$$

In this context, it is rather more meaningful to express the sampling interval in terms of the numbers of shaft rotations, N, then the decrease in speed during this interval is given by

$$\Delta\Omega \approx -\,a\Omega T = -a\Omega\frac{2\pi N}{\Omega} = -2\pi a N \tag{2.13}$$

The frequency resolution (in Hertz) is given by

$$\Delta f = \frac{1}{T} = \frac{\Omega}{2\pi N} \tag{2.14}$$

Hence, Equations 2.13 and 2.14 yield two conditions which tend to conflict: Equation 2.13 expresses the fact that given a long sampling time will allow the shaft to decelerate, hence degrading the frequency resolution, while on the other hand 2.14 shows the need for a long sample period to maximize the frequency resolution (i.e. to minimize Δf).

Putting these two conditions together the best resolution possible is achieved when

$$\Delta f_1 = \frac{\Delta\Omega}{2\pi} = -\alpha N = \Delta f_2 = \frac{\Omega}{2\pi N} \tag{2.15}$$

$$\therefore N = \sqrt{\frac{\Omega}{2\pi\alpha}} \tag{2.16}$$

Using this value, we see that

$$\Delta f = \sqrt{\frac{\alpha\Omega}{2\pi}} \tag{2.17}$$

So that as the speed decreases, the frequency resolution improves: the vibrational properties at very low rotor speeds are of less significance as any forces arising are low. Equation 2.17 also shows that for high values of α, the slew rate, frequency resolution is poor and this makes conventional rundown methods inaccurate. For applications involving high rates of change of rotational speed, a rather different method has been developed which involves the use of Kalman filters. This approach is somewhat more involved and is discussed in the next subsection (2.2.4.2).

To illustrate the method in Equations 2.10–2.17, consider a gas turbine used for power generation: the graph of rotor speed as a function of time is shown in Figure 2.13.

Clearly this curve is not exponential, but it is not too far away and an estimate for the effective α value is readily obtained by considering this curve, in particular the initial deceleration rate, in this case about 6 rev/min/sec.

To effectively assess the transient behavior, the operator has a dilemma: in order to maximize the frequency resolution, a long time block is required, but an excessive period will bring errors due to the machine's speed variation. Therefore, we seek an optimum compromise and rather than consider time explicitly, consider the block in terms of a number of complete shaft rotations (this automatically resolves any possible leakage problems).

If we sample over N cycles, with a time frame T, the resolution is given by

$$res = \frac{1}{T} = \frac{\Omega}{2\pi N}$$

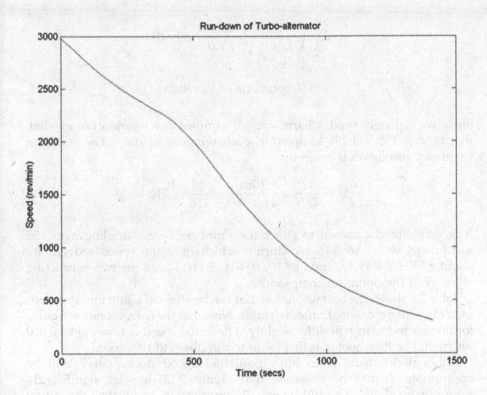

FIGURE 2.13
Speed Decay Curve.

But in this time the shaft slows by

$$\Delta\Omega = -\alpha\Omega T = 2\alpha\pi N \text{ (in radians/sec)}$$

$$\frac{\Delta\Omega}{2\pi} = \Delta f$$

Hence, from Equation 2.15

$$\frac{\Omega}{2\pi N} = \alpha N$$

The value of α is effectively given in the question: α is just the initial deceleration divided by the initial speed (in consistent units), i.e. $\alpha = \dfrac{6}{3,000} = \dfrac{1}{500}$. Using this yields

$$\therefore N = \sqrt{\frac{\Omega}{2\pi\alpha}} = \sqrt{\frac{50}{1/500}} = 50\sqrt{10}$$

$$\Rightarrow N \approx 158$$

$$= 158 \text{ rotations of the shaft.}$$

Since we typically need 4 harmonics, 8 samples per rotation are needed, that is, $8 \times 158 = 1{,}264$ samples in each window of data. The resulting frequency resolution is given by

$$\Delta f = \frac{\Omega}{2\pi N} = \frac{100\pi}{2\pi \times 158} = \frac{50}{158} \approx \frac{1}{3}\,\text{Hz}$$

It is very straightforward to check the numbers here: sampling over 158 revolutions will take 3.16 sec during which time the speed reduces by about $6 \times 3.16 = 19$ rev/min or $19/60 \approx 1/3$ Hz; hence the two conflicting effect yield the optimum compromise.

Table 2.1 shows the best resolution that can be attained using this approach for a range of exponential rundown rates. Note that the corresponding figures for linear rundowns will differ slightly. The initial speed is taken to be 3,000 rev/min. The time quoted is that taken to run down to 10% speed.

Although in many cases an exponential speed decay curve will be appropriate, it may be observed that Figure 2.13 deviates significantly from this ideal and it would be equally appropriate to idealize this curve as linear. Exactly the same logical argument may be applied to estimate the appropriate rundown parameters. Denoting the speed S in rev/sec (as it makes things a bit clearer than rev/min), then $S = \Omega/2\pi$ and

$$\frac{dS}{dt} = -\frac{2}{60}$$

TABLE 2.1

Rundown (from 3,000 rev/min) Frequency Resolution

Time	α	N	Δf
	(sec)	(revs)	(Hz)
10	0.23	15	3.4
50	0.046	33	1.5
100	0.023	47	1.1
500	0.0046	104	0.48
1000	0.0023	147	0.34

The time interval for N cycles is (initially) $N/50$ sec and in this time the shaft speed will reduce by $\Delta S = -\dfrac{2}{60} \times \dfrac{N}{50}$. Consequently, the optimum sampling occurs when

$$\Delta f = \frac{50}{N} = \frac{2}{60} \times \frac{N}{50}$$

or

$$N = 50\sqrt{30}$$

Note that this figure is not substantially different from the value used earlier and it may be inferred that the precise detail of the rundown rate is not a very sensitive determinant of the optimum sampling. A discussion of the process of order tracking is given by Fyfe and Munck (1997).

For machines with higher slew rates, there is an alternative means of analyzing rundown behavior known as the Vold–Kalman method, described in the following section.

2.2.4.2 The Vold–Kalman Method

In the previous section, the analysis of machine transient operation is discussed and it is shown how the study of run-up and rundowns often yields significant insight into a machine's condition. In other words, it is information rich. There is a contradiction inherent in the analysis of transient phenomena with essentially steady-state methods based on Fourier methods. This places a limitation on the frequency resolution which can be obtained and, perhaps not surprisingly, the resolution decreases as the speed of the transient increases. For machines which undergo rapid transients, an alternative approach is needed and a suitable methodology was developed in the 1990s. In fact, there are two distinct problems associated with rapidly varying machines: firstly, the machine itself will respond dynamically rather than in a quasi-static manner, and this issue is addressed in Section 2.2.4.3. The second issue is the difficulty in obtaining meaningful frequency resolution: this is discussed here.

The approach was first reported by Vold and Leuridan (1995) although there have been a number of developments since then. The method does not involve Fourier transforms, thereby avoiding some of the inherent difficulties of resolution. Nevertheless, the approach does involve very significant computing demands. There are several variations of this approach all yielding subtly different properties but they all seek to fit order tracked components to measured data without resorting to Fourier methods. This is achieved be minimizing a combination of measurement errors and so-called structural errors.

To begin the analysis, we observe that the vibration signal may be expressed as

$$y(t) = \sum x_k(t) + \eta(t) \qquad (2.18)$$

In this equation, y represents the observed time data and x_k is a sinusoidal function of time at a set of angular frequencies ω_k, often some multiple of shaft speed ω. Note that the shaft speed ω is itself a function of time, and η is the so-called nuisance component: this includes that part of the signal which is not related to shaft speed as well as any noise contributions. Of course, this requires some constraint on the variation of the signal which gives the so-called structural equation. In the original version of the filter, an equation was required to, in effect, specify that the components were slowly varying terms and this was achieved using a structural equation.

In this formulation of the method, it was assumed that the signal follows a sinusoidal variation and hence, using standard trigonometric identities

$$\sin(\omega k\Delta t) + \sin(\omega[k-2]\Delta t) = 2\cos(\omega t)\sin(\omega[k-1]\Delta t) \qquad (2.19)$$

Hence, the so-called structural equation may be written as

$$x(n\Delta t) - 2\cos(\omega\Delta t)x([n-1]\Delta t) + x([n-2]\Delta t) = \varepsilon(t) \qquad (2.20)$$

where the term on the right-hand side is recognition of the fact that the signal is not strictly sinusoidal. Before proceeding with the analysis it is worth considering Equation 2.20 a little further: Note that the speed term ω is varying in time, a fact which is the whole basis of our problem and hence a more precise expression for the structural equation would be

$$x(n\Delta t) - 2\cos\left(\int_0^{[n-1]\Delta t} \omega(t)dt\right)x([n-1]\Delta t) + x([n-2]\Delta t) = \varepsilon(t) \qquad (2.21)$$

The order tracked signal can now be estimated by minimizing the total error for a given ratio $r = \eta/\varepsilon$. This is known as angular-velocity approach and is readily formulated as a dynamic online filter (see e.g. Pan and Lin, 2006).

The alternative approach is the so-called angular-displacement approach which is used here to illustrate the overall method. Note that there are a number of closely related methods which have subtly different results. Consider now machine transient over a period which is divided into uniform time steps. The angular speed ω is clearly a function of time. The total angular rotation at the nth time step may be written as

$$\Theta(n\Delta t) = \int_0^{n\Delta t} \omega(t)dt \qquad (2.22)$$

Using this notation, the measured data at the nth point may be expressed as

$$y(n) = x(n)e^{j\Theta(n)} + \eta(n) \qquad (2.23)$$

$x(n)$ here represents the (complex) amplitude of the wave $e^{j\Theta(n)}$, while $\eta(n)$ is a "noise" term. In this formulation, x is automatically a sinusoid and hence a separate equation of the type above is not required. Note that for normal diagnostics, $x(n)$ is the term of interest. Equation 2.23 is conveniently recast in the form

$$\{\eta\} = \{y\} - [C]\{x\} \qquad (2.24)$$

where

$$[C] = \begin{bmatrix} e^{j\Theta(1)} & 0 & \vdots & 0 \\ 0 & e^{j\Theta(2)} & \vdots & 0 \\ \cdots & \cdots & \ddots & \cdots \\ 0 & 0 & \vdots & e^{j\Theta(n)} \end{bmatrix}$$

The square of the error vector norm is given by

$$\{\eta\}^H\{\eta\} = \left(\{y\}^T - \{x\}^H[C]^H\right)\left(\{y\} - [C]\{x\}\right) \qquad (2.25)$$

However, we are not yet in a position to complete the analysis of the transient data. To achieve this goal, some assessment must be given as to how rapidly the wave amplitude can change and this is done via the so-called structural equation. For example, using a two-pole filter representation gives

$$x(n) - 2x(n+1) + x(n+2) = \varepsilon(n) \qquad (2.26)$$

Noting that use of a higher-order filter is equivalent to allowing more rapid changes in the response carrier amplitude. This equation is essentially specifying that the amplitude of the components is slowly varying. Equation 2.26 may be conveniently written as

$$[A]\{x\} = \{\varepsilon\} \qquad (2.27)$$

For the two-pole case,

$$
\begin{bmatrix}
1 & -2 & 1 & 0 & \cdots & 0 & 0 & 0 \\
0 & 1 & -2 & 1 & \cdots & 0 & 0 & 0 \\
\cdots & \cdots & \cdots & \cdots & \cdots & \cdots & \cdots & \cdots \\
0 & 0 & 0 & 0 & \cdots & 1 & -2 & 1
\end{bmatrix}
\begin{Bmatrix}
x(1) \\ x(2) \\ \cdots \\ x(N)
\end{Bmatrix}
=
\begin{Bmatrix}
\varepsilon(1) \\ \varepsilon(2) \\ \cdots \\ \varepsilon(N-2)
\end{Bmatrix}
\tag{2.28}
$$

The square of the error term can now be expressed as

$$
\{\varepsilon\}^T\{\varepsilon\} = \{x\}^T[\mathbf{A}]^T[\mathbf{A}]\{x\}
\tag{2.29}
$$

Hence in this approach, there are two distinct error terms to be minimized and the weighting, r, between these two, largely dictates the nature of the solution obtained. The first term refers to the magnitude of measurement noise while the second determines the rate at which the response envelope can change. The global solution is obtained by minimizing

$$
J = r^2\{\varepsilon\}^T\{\varepsilon\} + \{\eta\}^T\{\eta\}
\tag{2.30}
$$

Expanding this expression gives

$$
J = r^2\{x\}^T[\mathbf{A}]^T[\mathbf{A}]\{x\} + \left(\{y\}^T - \{x\}^H[\mathbf{C}]^H\right)\left(\{y\} - [\mathbf{C}]\{x\}\right)
\tag{2.31}
$$

This weighted penalty function can be easily minimized as

$$
\frac{\partial J}{\partial x^H} = \left(r^2[\mathbf{A}]^T[\mathbf{A}] + [\mathbf{E}]\right)\{x\} - [\mathbf{C}]^H\{y\} = 0
\tag{2.32}
$$

Leading to a solution

$$
\{x\} = \left(r^2[\mathbf{A}]^T[\mathbf{A}] + [\mathbf{E}]\right)^{-1}[\mathbf{C}]^H\{y\}
\tag{2.33}
$$

where

$$
[\mathbf{E}] = [\mathbf{C}]^H[\mathbf{C}]
\tag{2.34}
$$

It is important to appreciate that this approach circumvents the difficulty of resolution which are inherent in Fourier approaches by fitting directly to harmonics of the shaft speed and in this respect it has something in common with Prony analysis. The fundamental limitation on the "accuracy" of the approach arise from the assumption made on the maximum rate of change of the vibration levels as reflected in the choice of r, the weighting factor in Equation 2.33. This is actually quite a complicated issue which depends on the rate at which the shaft is accelerated or decelerated and the system damping.

As a machine changes speed the vibration levels at any one time will not equate to the levels corresponding to the instantaneous speed because the system does not have time to saturate. Because of this, the example shown here in somewhat idealistic insofar as immediate response to forcing is assumed: this does have the virtue, however, of providing a known target for our calculations. Figure 2.14 shows the ideal frequency response function of our test rotor.

For a more realistic model, the transient profile will be dependent on the rate at which the speed is changed. This can be seen from a comparison of the observed vibration levels as our machine is rundown from 3,000 rev/min in 100, 10, and 2 sec, respectively. These are shown in Figures 2.15–2.17.

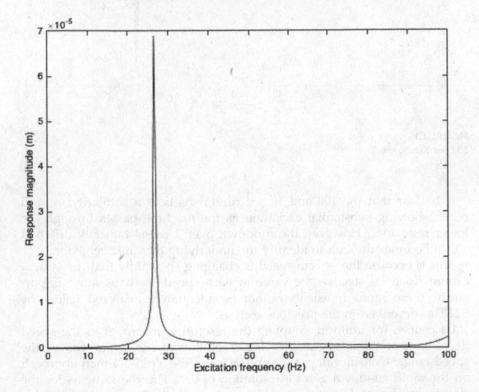

FIGURE 2.14
The Exact Frequency Response.

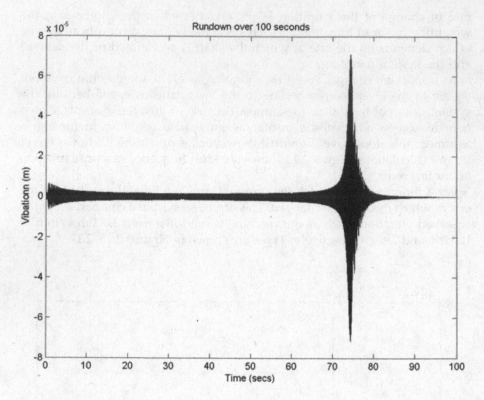

FIGURE 2.15
100 Sec Rundown.

It is clear that the 100 and 10 sec rundowns bear a similarity to each other, showing substantial excitation as the machine passes through the lower resonance. However, the rundown over 2 sec is radically different and it becomes difficult to identify the underlying dynamic behavior.

This is because the system speed is changing so rapidly that the system cannot reach its steady-state value at each speed. Perhaps more importantly, these rapid transients cannot be adequately analyzed using the STFT as described in the previous section.

Of course, for uniform sampling the resolution improves as the speed decreases but these figures give the resolution at the upper end of the speed range. Consider now the application of the Vold–Kalman approach on the rapid rundown gives a resolution of 0.34 Hz (the same as for the slowest rundown). Given the sampling rate used, the Nyquist frequency is 100 Hz. Tůma (2005) gives an expression relating the weighting factors used: for a two-pole filter

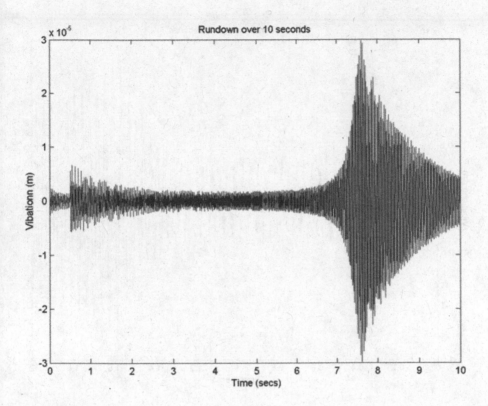

FIGURE 2.16
Faster Rundown.

$$r = \sqrt{\frac{\sqrt{2}-2}{6 - 8\cos(\pi\Delta f) + 2\cos(2\pi\Delta f)}} \approx \frac{0.00652097315}{\Delta f^2} \qquad (2.35)$$

In this case, $\Delta f = 0.34/100$ and hence $r = 564$. This yields the appropriate frequency resolution, but the price paid for this is reduced smoothing of signal noise. Without detailed prior knowledge of the system this compromise cannot be avoided. Figure 2.18 shows the synchronous excitation calculated with weighting terms of 564, 100, and 50, respectively, using the rapid rundown data.

The most rapid transient shown here, in Figure 2.17, is at first sight confusing and no clear evidence of the critical speeds, or indeed shaft orders are immediately apparent. Only very poor resolution can be obtained using FFT techniques. Nevertheless, application of the Vold–Kalman approach can yield valuable information.

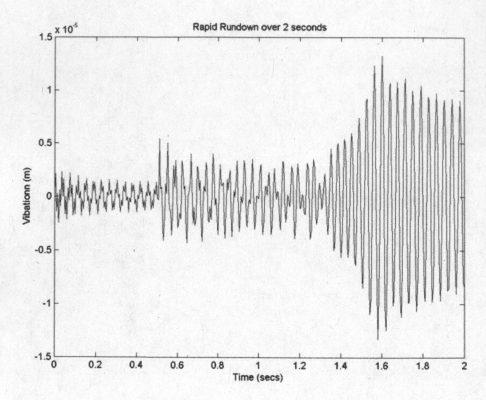

FIGURE 2.17
Very Rapid Rundown.

Figure 2.19 shows a waterfall plot in which the synchronous vibration component is plotted against shaft speed and weight factor and it is immediately apparent that two critical speeds have influence in this speed range.

It is worthy of note that the shape of the response derived is dependent on the value chosen for the weighting factor. As discussed in the preceding pages the weighting factor expresses the ratio of the errors in the amplitude variation and the signal noise in the overall "target" which is optimized. Therefore, a low value of the weighting permits rapid transients in amplitude of response, whereas it is clear that at higher values, the peak at about 90 Hz is considerably suppressed. In this case, a second-order filter has been applied.

At this point, it is important to note some remarks concerning the amplitude of the responses close to resonance. Although it is often of only minor importance in problem resolution, in the case of slow transients, the

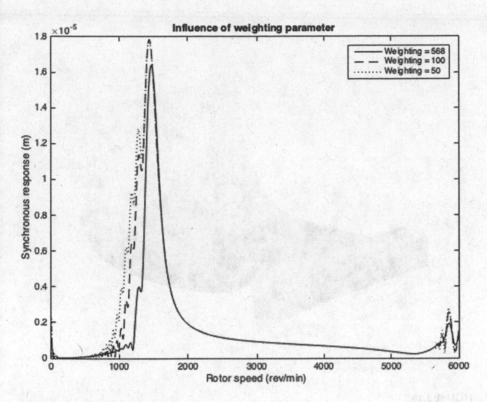

FIGURE 2.18
Comparison of Weighting Effects.

response to a given imbalance is dependent on mode shape and damping only. With rapid transients, however, the rate of change of rotor speed has also a marked influence on the peak response. This is because any system takes some time to reach its full steady-state response to sinusoidal excitation, a fact which is frequently used to advantage in the deliberate rapid transit through critical speeds. This is an important topic and the interested reader is referred to the discussion given by Friswell *et al.* (2010).

In conclusion, it should be noted that the original Vold–Kalman method has now developed into several closely related but distinct approaches. All have a common basis in philosophy and differ only in the detailed treatment of variables and the way in which the computed orders are allowed to vary. This gives rise to some differences in resulting bandwidth and consequently, some experience is needed to arrive at optimum results.

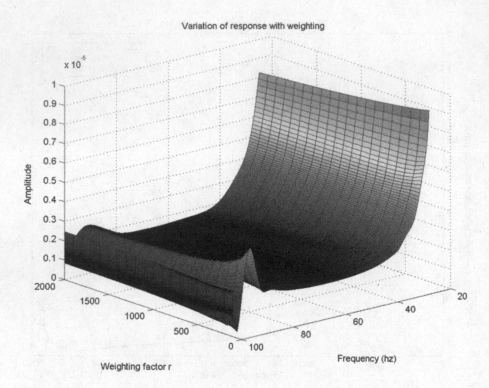

FIGURE 2.19
Variation of the Vold–Kalman Method.

2.2.4.3 System Response

At this stage, we turn attention to another aspect of the interpretation of transient vibrational signals: clearly, this has to be considered before any sound judgements can be made on the condition of a piece of machinery. To gain some insight into the rate at which a machine can reach the response levels of steady running, we consider two cases: the first is the response to a step (constant) force, whilst the second case is the initiation of a sinusoidal excitation.

The transient response of a single degree of freedom system to a unit step change is readily calculated (using Laplace transforms) as

$$c(t) = 1 - \frac{1}{\beta}e^{-\zeta\omega_n t}\sin(\omega_n\beta t + \theta) \tag{2.36}$$

where $\beta = \sqrt{1-\zeta^2}$ and $\theta = \tan^{-1}(\beta/\zeta)$. ω_n is the (undamped) natural frequency and ζ is the damping ratio (i.e., the fraction of critical damping).

Figure 2.20 illustrates the response to a unit force for various values of damping. Notice that there is a characteristic time of the response, $\tau = \dfrac{1}{\zeta \omega_n}$.

The same factor occurs if we now consider the effect of an oscillator starting an oscillator from rest. Clearly, it will take time for the response to develop to the steady-state response. The equation of motion can be expressed as

$$m\ddot{y} + c\dot{y} + ky = me\omega^2 \cos \omega t \qquad (2.37)$$

where e is the mass eccentricity. This may be rewritten in the usual form as

$$\ddot{y} + 2\zeta \omega_n \dot{y} + \omega^2_n y = e\omega^2 \cos \omega t \qquad (2.38)$$

The solution is readily written in the form

$$y = \frac{me\omega^2 \cos \omega t}{\sqrt{(\omega^2_n - \omega^2)^2 + 4\zeta^2 \omega^2 \omega^2_n}} + e^{-\zeta \omega_n t}(A \sin \omega_d t + B \cos \omega_d t) \qquad (2.39)$$

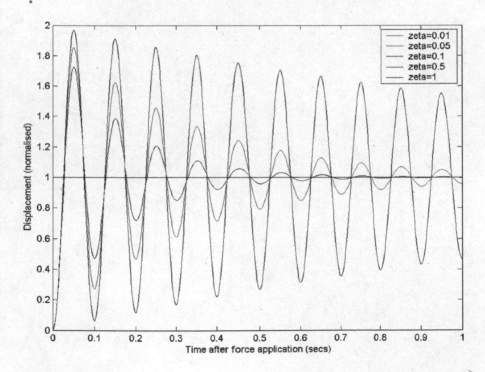

FIGURE 2.20
Response to a Step Force.

where the damped natural frequency is given by

$$\omega_d = \omega_n \sqrt{1 - \zeta^2}$$

An example of how the response converges to the steady-state value is shown in Figure 2.21.

In practice, the difference between the damped and undamped natural frequencies will rarely be of concern. It is now a straightforward, if tedious task to determine the constants A, B to give the complete solution. Clearly, the approach to the steady-state solution is governed by the time scale $1/(\zeta \omega_n)$.

If we consider now the time required for the response to be within 1% of the steady-state value, then

$$\zeta \omega_n t = 2\pi \zeta m = log_e(100)$$

where m is the number of complete cycles. For a typical example, consider 2% damping in a system with $\omega_n = 80\pi$. This gives a value for m of 37 cycles

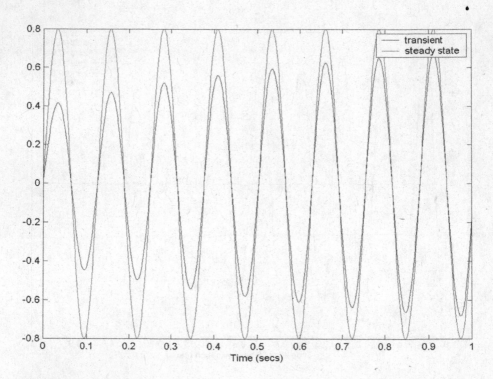

FIGURE 2.21
Convergence to Steady State with $\omega_n = 50$, $\xi = 0.025$.

which at 50 cycles/sec represents 0.75 sec for the system to saturate. Provided the machine in question does not reduce speed significantly over this period of time, quasi-static analysis is appropriate. It is worth noting that an undamped system will never settle to the steady-state value. To analyze the general system in the general case, resort must be made to consider the general response in the time domain.

At this stage, it will be clear that for the analysis of rapid transients, there are two quite distinct difficulties. Firstly, there is the technical problem of obtaining an adequate time–frequency description and this is discussed in Section 2.2.4.1, but the quite separate issue is how the system responds to transient forcing.

The general response is described by

$$h(t - \tau) = \frac{e^{-\zeta \omega_n (t - \tau)}}{m \omega_d} \sin \omega_d (t - \tau) \quad \text{for } t > \tau \tag{2.40}$$

The displacement of the rotor may be written as

$$y(t) = \int_0^t h(t - \tau) \left[e\omega^2 \cos \omega\tau + e \frac{d\omega}{dt} \sin \omega\tau \right] d\tau \tag{2.41}$$

Using the convolution theorem, the Fourier transform of this equation can be readily formed (although this may involve substantial computing – the precise requirements will be examined a little later). Knowing the Fourier transforms Y and F, the function H is readily calculated. This implies that the difficulties of what may be termed "unsaturated response" have been overcome and this is indeed the case because full use has been made of the machines dynamic response via the impulse response function. In practice, most analysts assume an approximation to saturated conditions as this simplifies computation considerably. Given a validated model, a simulation of a defined transient can be simulated although the marginal improvement in accuracy means this is rarely justified.

2.2.4.4 Comparison of Methods

There is considerable interest in interpreting transient behavior because it yields considerable insight into the structural properties of a machine. This is because a broad spectrum of frequencies are excited during the process. Figure 2.22 shows the analyses of three rundown from 6,000 rev/min in times of 2, 10, and 100 sec, respectively. Figure 2.22a shows results using the Vold–Kalman approach, while those shown in Figure 2.22b are from STFT analysis. There are some points to note from this comparison. All traces do identify the principal resonance with varying resolution, but note that the amplitude of the processed signal reduces and the period of the

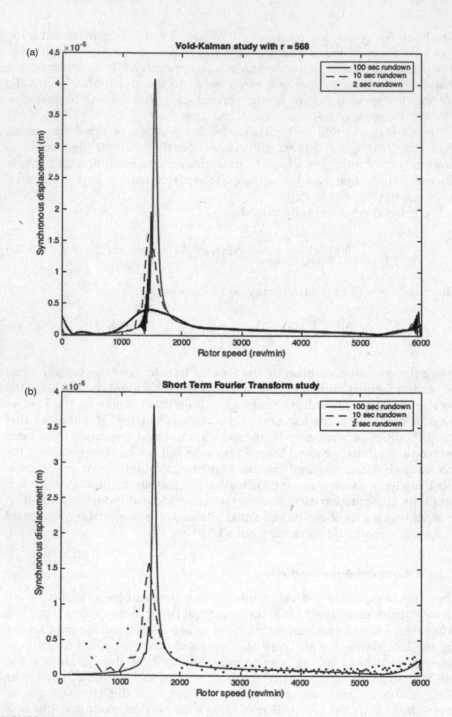

FIGURE 2.22
A Comparison of Transients and Methods.

transient reduces. This is due to the time response of the machine. The Vold-Kalman approach does tend to give more satisfactory results in the case of rapid transients, but there remains some debate concerning the optimum weighting factor to use.

2.2.5 Shaft Orbits

The effective evaluation of machinery inevitably involves a wide variety of techniques, both sophisticated and reasonably simple. An important maxim of the analyst must be "never neglect the obvious." An extremely simple measure of plant performance is a plot of an orbit of either the shaft or bearing motion (as both can offer valuable insights). This is simply a matter of plotting x against y over one, or preferably a series of shaft cycles.

Orbits can take a number of forms and we do not attempt a comprehensive discussion here. Figure 2.23 gives four possible orbits and these can help give considerable insight into the operation of a machine. Like most techniques in machine diagnosis, the shape of the orbit will rarely give conclusive evidence of some machine fault but it will very often provide an important part of a collection of interlocking pieces of evidence. So, what can be inferred from the orbits shown in Figure 2.23. An important point to note is that all four figures are records at a fixed shaft rotation speed. Figure 2.23a shows a purely circular orbit and this may suggest that the machine is symmetric, in the sense that it is equally stiff in the two directions orthogonal to the axis of the shaft. To be conclusive about this, one would need to check the orbit at some other shaft speed; a symmetric machine will show circular orbit at all speeds. The orbit also reveals that the forcing is harmonic and not predominantly in one direction.

In Figure 2.23b, the orbit has changed to an elliptical form, and this shows clearly that the machine is not equally stiff in the two orthogonal directions. It would be tempting to suggest that the machine is more flexible in the vertical direction, but more evidence is needed. At first sight, this may be a little surprising, but given that the stiffness is asymmetric, natural frequencies are different for vertical and horizontal directions. It is possible that the speed chosen happens to be close to the vertical resonance. Information at other speeds would be needed to reach conclusions on the relative stiffness in the two directions.

Figure 2.23c again shows an elliptic orbit, but the ellipse is tilted. This suggests some cross coupling between the x and y directions. This could arise in the supporting structure, the bearings (particularly oil film journal type) or gyroscopic effects, but the latter tends to be significant only in overhung systems. Finally, the last orbit, Figure 2.23d shows a marked change in that an internal loop has appeared. This implies the system has some nonlinearity or, less often, some external forcing at a frequency other than that corresponding to the operating speed. Common sources of nonlinearity are oil bearings,

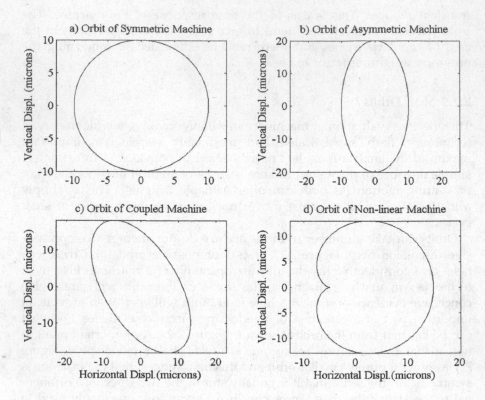

FIGURE 2.23
Some Common Orbit Shapes.

shaft rubs, gearboxes, and (less often, fortunately) shaft cracking. These features will be discussed in greater depth in Chapter 5.

2.2.6 Polar Plots

Plant engineers frequently use the polar plot to assess the performance of a machine over a range of rotations speeds. Rather than plotting data in terms of the real and imaginary parts as in the scatter plots, plotting is in terms of magnitude and phase although, of course, the two representations are entirely equivalent. Recordings are made at a series of rotational speeds, but the plot is not extended over a long period of time as was the case with scatter diagrams. The simple reason for this is the attempt to encapsulate different aspects of the dynamic properties. Figure 2.24 shows an example of a polar plot.

The points on the curve shown represent the magnitude and phase of vibration (at a given measurement point on the machine) for different shaft

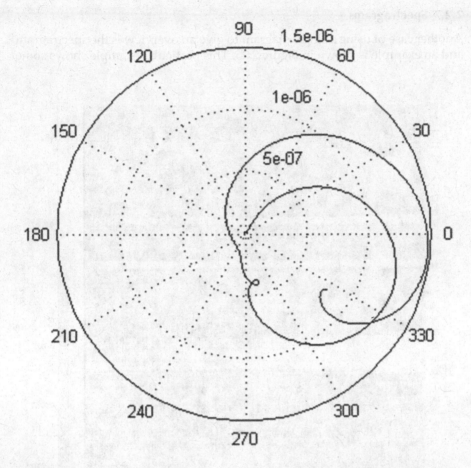

FIGURE 2.24
Typical Polar Plot.

speeds. Clearly, this implies we are referring to synchronous vibration components. From the curve shown, we immediately see the variation of response with frequency throughout a transient operation (either run-up or rundown). Each point on the curve may be associated with an operational speed. Any critical speed close to running speed would normally give rise to substantial phase changes.

The information on the polar plot is precisely the same as that on first-order rundown plot (transient response): it is just presented in a compact form. This is a perennial problem – the presentation of the vast amount of data which emerges from rotating machinery. This plot is taken from a gas turbine undergoing a rapid run-up to speed. The three circular shapes (i.e., two large diameter circles and a single small one) are indicative of resonances.

2.2.7 Spectrograms

Another case of using a single diagram to give an overview is the spectragram and an example is shown in Figure 2.25. This particular example shows some

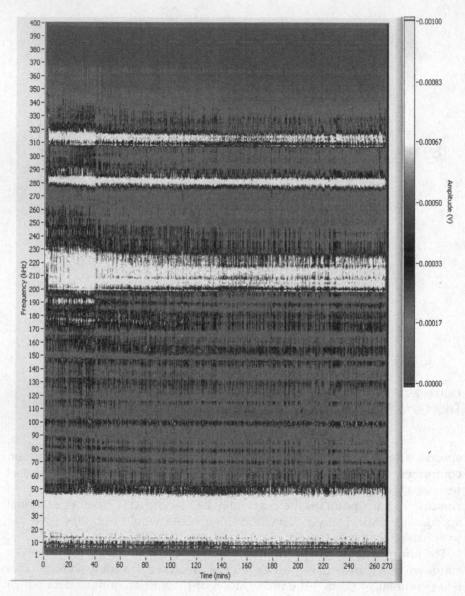

FIGURE 2.25
Spectragram Showing AE Data.

Reproduced with the permission of the author from Quiney (2011), unpublished.

acoustic emission (AE) hence the high frequencies shown, but exactly the same technique may be applied with good effect to vibration data. This plot is closely related to the waterfall plot, with a series of FFT plots arranged for each time frame. The heights at each point are normally denoted by color but in this instance it is shown in shades of grey. Clearly, the same time resolution problems, as discussed elsewhere, must be taken into consideration, in preparing such a plot. This is a case of time–frequency analysis and the wider problem is discussed in Chapter 8. It is another example of the presentation of a significant volume of information on a single plot. Whilst for many purposes there may be a need to focus on more specific details, it can be extremely helpful in defining the context of an investigation.

2.3 Comparison with Calculations

Section 2.2 discusses some of the more common ways of presenting data and an outline is given there of some of the physical principles involved in the interpretation of the measured signal. The important step of course is the use of this data to infer the condition of the machine by detecting the physical processes which give rise to the measured parameters. Expressed another way, the engineer is inferring information on internal parameter from external measurements: furthermore, the measurements will be subject to some errors and noise. This step is far from simple and is sometimes not accorded the attention it deserves.

On rare occasions one may be able to measure precisely a parameter of interest, but more usually information is inferred from the generality of some other data. This process of interpreting data relies on a variety of sources including knowledge of similar machines, records of previous operation, maintenance records, and mathematical models, and in the last few decades, the importance of mathematical models in this interpretation has increased dramatically. Of course, this is no substitute for records where available, but the key benefit of an adequate model is the ability to probe questions such as "what if?". The word "adequate" applied to a model is an interesting concept which is explored elsewhere in this text, particularly in Chapter 7, in which measured data is used to enhance the model.

It is worth emphasizing that the model, of whatever form, should be regarded as a tool to increase one's overall understanding of the machine in question. Since it incorporates our (albeit limited) physical understanding of the operation, it should be used to add to the overall information available. Just as measurements are subject to some uncertainty, so is the model of whatever form.

Many of these models will be based on the finite element method which is discussed in detail as applied to rotor dynamics by Friswell *et al.* (2010). A more general discussion of the approach is given by Irons (1987) among many other books on the topic. Irons does, however, give a very appropriate overview in his preface "Computing is for understanding, not numbers." This is a sentiment with which the present author is in total agreement.

2.4 Detection and Diagnosis Process

In monitoring and diagnosis of plant problems, there is considerable variety and each instance will have particular features. Although the discipline cannot be reduced to a routine or checklist, some common themes do emerge as useful patterns. Comprehensive records are, of course, absolutely invaluable and with modern data storage technologies, this is easily implemented. Section 2.2.2 outlines some of the simple statistical approaches that may be applied to long-term data and these methodologies may give the first indication of some plant "excursion" that warrants investigation. Experience may well dictate machine shutdown under certain conditions. In general, however, the statistical approach will just highlight an area or event to be examined.

To take matters further usually (but not always) requires a study of the data in the frequency domain enabling the engineer to relate phenomena to the operation of the machine. The most common faults occurring in turbo-machines are discussed in Chapters 4–6, but here some rough outlines are offered. If excitation is at synchronous shaft speed, the most common problem is imbalance, although (less commonly) it may also indicate a bent rotor. These two effects can only be distinguished by further examination as outlined in Chapter 4. Higher-order excitation usually indicates some form of nonlinearity, possibly a rotor crack or misalignment. In either case, further investigation will be needed probably involving the analysis of machine transients.

The presence of components of vibration which are not multiples of shaft speed normally indicates some nonlinear behavior of the system. If it is below half rotor speed, then, in machines with oil bearings, oil whirl may be suspected and this is another consequence of misalignment or the incorrect loadings of bearings. Rubbing between rotor and stator may also give rise to nonsynchronous excitation and a wide range of phenomena.

Hence, the techniques presented in this chapter offer a "first filter" to indicate changes in a machine's behaviour. To positively identify the nature of an event requires further insight into possible problems together with further analysis of the data.

2.5 Concluding Remarks

This chapter describes a number of approaches for the examination of vibration data. Sometimes these methods may be sufficient to resolve a difficulty, but very often it is necessary to consider in more detail the mechanics of the plant and this step invariably requires some form of model. In Chapter 3, the principal methods of modeling and analysis are discussed and later chapters describe features of a range of plant faults.

Problems

2.1 Explain what is meant by "order tracking" and why it is an important concept for the monitoring of rotating machines.

On shutdown a large boiler feed pump runs down from 6,000 rpm at an initial rate of 120 rpm/sec. The response up to four times shaft speed is required.

　i. How would you sample to obtain maximum resolution, describing block size, and sampling rate?

　ii. What is the resulting frequency resolution?

2.2 Assuming an exponential rundown rate in question 2.1, how does the resolution vary during the rundown? If the rundown rate is linear, how is the resolution changed?

2.3 A signal is known to be $y = \sin(50\pi t) + 3\cos(100\pi t)$. Using 200 sampling points, obtain the FFTs using sample lengths of 0.5, 1, 2, and 4 sec. Comment on the results.

2.4 A signal has the form $y = sin(50\pi t) + 4\cos(55\pi t)$. Determine appropriate sampling rates and the number of samples to obtain adequate separation of the two components. Obtain the FFT and discuss your results.

2.5 Explain the following data presentation formats and describe their application in condition monitoring:

　i. Orbits

　ii. Waterfall plots

　iii. Rundown plots

2.6 A scatter diagram contains 1,000 data points which are described by the parameters x_i and y_i. It has been established that

$$\sum_{i=1}^{1000} (x_i - \bar{x})^2 = 73,570 \qquad \sum_{i=1}^{1000} (x_i - \bar{x})(y_i - \bar{y}) = -23,600$$

$$\sum_{i=1}^{1000} (y_i - \bar{y})^2 = 53,860$$

where $\bar{x} = -3$, $\bar{y} = 2$ are mean values. Derive the size and orientation of the ellipse which defines the standard deviation of this set of data. What is the probability of observing a point on the diagram within the interval $x = 7 \pm 1$, $y = 2 \pm \frac{1}{2}$?

2.7   A machine·can be represented at a uniform rotor 0.3 m diameter and 4 m between bearings each of which has constant stiffness of 5×10^7 N/m and damping 12,000 Nsec/m. There are two discs 1 m inboard of each bearing and these two discs are 0.5 m diameter and 0.05 m thick. The first disc has an imbalance 0f 0.0003 Kg m. The rotor is accelerated steadily from rest to 3,000 rev/min in 30 sec. Calculate the response and compare with the observed behavior on a slow run-up. Using the results, apply the Vold–Kalman technique to obtain a rundown profile.

References

Bossley, K.M., McKendrick, R.J., Harris, C.J. and Mercer, C., 1999, Hybrid computer order tracking, *Mechanical Systems & Signal Processing*, 13(4), pp. 627–641.

Cooley, J.W. and Tukey, J.W., 1965, An algorithm for the calculation of complex Fourier series, *Mathematics of Computation*, 19, pp. 297–301.

Friswell, M.I., Penny, J.E.T., Garvey, S.D. and Lees, A.W., 2010, *Dynamics of Rotating Machines*, Cambridge University Press, New York.

Fyfe, K.R. and Munck, E.D.S., 1997, Analysis of computed order tracking, *Mechanical Systems & Signal Processing*, 11(2), pp. 187–205.

Irons, B.M., 1987, *Numerical Methods in Engineering: Numbers Are Fun*, John Wiley, Oxford, UK.

Lees, A.W., Price, E.D. and Friswell, M.I., 2004, Identification of rotor dynamic machinery – A laboratory trial, *Institution of Mechanicsl Engineers Conference on Vibrations in Rotating Machinery*, pp. 373–382, Swansea, UK.

Lees, A.W., Sinha, J.K. and Friswell, M.I., 2009, Model based identification of rotating machines, *Mechanical Systems & Signal Processing*, 23(6), pp. 1884–1893.

Pan, M.C. and Lin, Y.F., 2006, Further exploration of Vold-Kalman filtering order tracking with shaft speed information – I: Theoretical part, numerical implementation and parameter investigations, *Mechanical Systems & Signal Processing*, 20(4), pp. 1134–1154.

Quiney, Z.A., 2011, Use of acoustic emission for bearing monitoring, Ph.D. Thesis, Swansea University.

Tůma, R., 2005, The passband width of the Vold-Kalman order tracking filter, *Proceedings of Scientific Works at VŠB-TU Ostrava, Mechanical Engineering Series* year. LI, 2005. č. 2, paper no. 1485, pp. 149–154.

Vold, H. and Leuridan, J., 1995, High resolution order tracking at extreme slew rates using Kalman tracking filters, *Shock and Vibration*, 2, pp. 507–515.

3

Modeling and Analysis

3.1 Introduction

In this chapter, consideration is given to some of the basic methods of analysis in the study of rotating machinery. In some respects, this is similar to the study of structural dynamics but there is an important complicating factor; the variation of both the system and the forcing with shaft rotational speed. This is an obvious prerequisite to the material which follows in subsequent chapters.

We note that there are two fundamental distinctions between a structure and a rotating machine, as far as the dynamic behavior is concerned. Firstly, the dynamic properties of a rotating shaft depend on the rotational speed owing to gyroscopic terms and, in many cases, the properties of the bearings. Secondly, the shaft rotation provides an energy source which implies that under certain conditions, vibrations can grow – a condition of instability. Both these issues will be addressed in this chapter but the issue of instability is discussed more fully in Chapter 5.

3.2 Need for Models

The objective of any monitoring scheme on any kind of equipment is the rapid identification of any trend in performance. The overall problem can be approached at different levels and one could devise a very simple system which equates increased vibration to a general deterioration and at some point, removes the plant from service. However, with sophisticated plant it is usually much more desirable to use the monitored data to gain an understanding of the machines operation and use this knowledge to guide maintenance and operations. This almost invariably requires some form of mathematical model to aid understanding, and in most instances, for all except the simplest of machines, this will mean a finite element model (FEM).

Basically, understanding of the machine's behavior is gained by comparing the measured vibration levels with those obtained from the model. Of course, this requires some care. A full discussion of the FEM is beyond the scope of the present text but many books cover the basis of the approach (see e.g. the classic text by Zienkievicz *et al.*, 2005). The method as it applies to rotating machinery is discussed by Friswell *et al.* (2010), and a very brief outline is given in the following section.

3.3 Modeling Approaches

Why is this section included in a text on condition monitoring? Some engineers would regard modeling and plant measurements as two quite disparate disciplines, but if the effort is focused on maximizing the understanding of the internal processes of plant, then the two approaches are in general inextricably linked and are equally essential for a comprehensive status overview. A model without measurements is somewhat divorced from reality, whereas measurements in themselves do not give a view of the internal operation. In establishing the methodology of the FEM, although consideration of a bending beam is among the simplest of examples, it is also one of the most important.

The rotor of many machines can be modeled as a rotating beam onto which one or more discs are attached. Some machines required more sophisticated models, particularly those in which significant distortion of the cross section takes place. Examples of this are in a minority and are discussed by Friswell *et al.* (2010).

3.3.1 Beam Models

FEM can be viewed as an extension to the Rayleigh–Ritz approach to the analysis of structures. In this approach, a trial function is taken for the body's displacement, and then the kinetic and potential energies can readily be calculated. The key point in seeking to represent motion in terms of trial function is the dramatic reduction in the number of degrees of freedom from a very large (effectively infinite) number to the parameters of the trial functions. Whereas in the classical Rayleigh–Ritz approach only a few trial parameters were considered, in FEM considerably more are allowed. Fortunately, a considerable number of rotating machines can be adequately modeled using a simple linear element and so the principles of the FEM are illustrated for this case. Let us consider the steps in modeling a simple machine, the rotor of which can be represented as a beam mounted on bearings at each end. The first step is to divide the rotor into sections, called elements; in principle, we can choose the number of these elements, but this is a very important issue to which we will return. For the sake of illustration, simply divide the rotor into two elements. At the end of each element, there is a node – which may be regarded as a point at which the displacement is determined. The rotating assembly of many, but not all, rotating machines can be accurately represented by beams and hence our attention is now focused on the modeling of beams. The simplest description of a bending beam is that given by Euler. We consider a beam of rectangular cross section (dimensions $a \times b$) and length L, which is considerably greater than either a or b. A rectangular cross section is used here for the sake of clarity but the transformation to cylindrical coordinates to deal with circular shafts is straightforward. As the beam bends it is clear that there is a curved surface which can be drawn through the beam on which there is no extension or compression: this is called the neutral axis and in a uniform beam it would lie along the center line of the beam. Points within the beam above this surface will be subject to tensile stresses, whilst those below are subject to compression as shown in Figure 3.1. An important assumption of the Euler model is that as the beam bends, cross sections remain plane and furthermore they are at right angles to the neutral axis. In other words, there is no shear deformation.

At some point, distance x from the neutral axis, the bending may be taken to have a local radius of curvature R. The situation is shown in Figure 3.1 with the beam lying along the z-axis and it is clear that all points

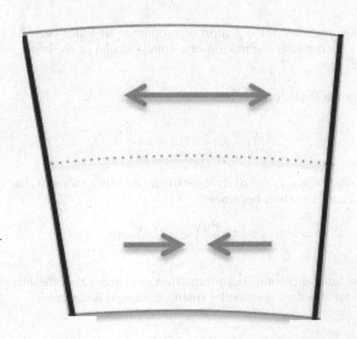

FIGURE 3.1
Strains during Bending.

above the neutral axis are in tension while points below the neutral axis are in compression. Then the strain is given by

$$\varepsilon = \frac{x}{R} \tag{3.1}$$

And the corresponding stress is given

$$\sigma = E\frac{x}{R} \tag{3.2}$$

Summing moments about the neutral axis gives

$$M_r = \int E\frac{x^2}{R}\,dx\,dy = \frac{EI}{R} \tag{3.3}$$

where I is the second moment of area and E is Young's modulus. The radius of curvature R is the inverse of the second derivative of the bending deflection and hence

$$M_r = EI\frac{\partial^2 u}{\partial z^2} \tag{3.4}$$

Using this expression, the equation of motion of an Euler beam is readily derived. If we consider the motion of a small portion of the beam between z and $z+\delta z$, then

taking P as the shear force

$$P(z + \delta z) - P(z) = \rho A\frac{d^2 u}{dt^2} \tag{3.5}$$

Since the shear force is the derivative of the bending moment, the Euler–Bernoulli beam equation becomes

$$\frac{\partial^2}{\partial z^2}\left(EI\frac{\partial^2 u}{\partial z^2}\right) - \rho A\frac{d^2 u}{dt^2} = 0 \tag{3.6}$$

This is the Euler–Bernoulli beam equation, and from this the nth natural frequency for this case is given by (using the usual notation)

$$\omega_n = \frac{n^2\pi^2}{L^2}\sqrt{\frac{EI}{\rho A}} \tag{3.7}$$

3.3.2 Finite Element Method

It is important to recall that the FEM approach is an application of the Rayleigh–Ritz variational method, albeit a highly sophisticated extension. Just as with all such approaches, in calculating natural frequencies, accuracy improves when the trial function is given sufficient freedom to faithfully represent the shape of the modes of the machine. This implies in general that the more elements in the model the better the result, but of course, a significant penalty is paid for this in terms of computing effort. In practice, one must also ask what accuracy is required. In dealing with the resolution of problems on real machines, the analyst is usually presented with the problem of reconciling measured data with a theoretical model in an attempt to gain understanding and insight. Clearly, the accuracy must be sufficient to aid this process, but there is a point at which improved precision is largely academic.

In fact, the idealization of a given rotor is the area where the analyst's skill is most important. There are two major and sometimes conflicting assessments to be made. On the one hand, the model must reflect the geometry of the machine, with appropriate changes of section, but consideration is also needed as to the number of elements required to give the appropriate measure of confidence. It is shown that these two factors are sometimes reconciled by using an approach known as "node condensation" whereby the geometry is modeled in considerable detail but then the resulting very large model is reduced in size prior to the vibration calculations. We first consider the problem of modeling a uniform beam which is pinned at each end. Initially, a FEM of an Euler beam is considered since an analytic solution is readily available. The corrections due to shear and rotary inertia are considered subsequently.

3.3.2.1 Element Formulation

For this illustration, the complete beam is taken to comprise two elements as shown in Figure 3.2. Consider one of the beam elements with nodes at each end lying along the z-axis. At each node it is assumed that the displacement u, transverse to its axis, and the slope du/dz are nodal parameters. Assume that the transverse displacement within the element is given by

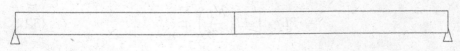

FIGURE 3.2
Simple Rotor on Two Bearings.

$$u = a_0 + a_1 z + a_2 z^2 + a_3 z^3 \tag{3.8}$$

Because the parameters a_n are not readily related to the global system, they should be eliminated and this can be done using nodal variables. Note here, that for this element moving in a single direction, there are now only four degrees of freedom, corresponding to the parameters a_0, a_1, a_2 and a_3. We may form expressions for the four nodal variables and write in vector and matrix notation for an element of length l_e

$$\begin{Bmatrix} u_1 \\ \left(\dfrac{du}{dz}\right)_1 \\ u_2 \\ \left(\dfrac{du}{dz}\right)_2 \end{Bmatrix} = \begin{bmatrix} 1 & 0 & 0 & 0 \\ 0 & 1 & 0 & 0 \\ 1 & l_e & l_e^2 & l_e^3 \\ 0 & 1 & 2l_e & 3l_e^2 \end{bmatrix} \begin{Bmatrix} a_0 \\ a_1 \\ a_2 \\ a_3 \end{Bmatrix} \tag{3.9}$$

The subscripted values on the left-hand side of the equation refer to values at the respective nodes. From this equation, we may readily write the vector of the $\{a\}$ parameters in the form

$$\{a\} = [C]^{-1}\{v\} \tag{3.10}$$

where $\{v\}^T = \left\{ u_1 \quad \left(\dfrac{du}{dz}\right)_1 \quad u_2 \quad \left(\dfrac{du}{dz}\right)_2 \right\}$

and it is readily shown that

$$[\mathbf{C}]^{-1} = \begin{bmatrix} 1 & 0 & 0 & 0 \\ 0 & 1 & 0 & 0 \\ -3/l_e^2 & -2/l_e & 3/l_e^2 & -1/l_e \\ 2/l_e^3 & 1/l_e^2 & -2/l_e^3 & 1/l_e^2 \end{bmatrix} \tag{3.11}$$

To calculate the potential energy (*PE*) in the element during deformation in the prescribed form, we may write

$$PE = \int_0^L EI \left(\frac{\partial^2 u}{\partial z^2}\right)^2 dz \tag{3.12}$$

which may be expressed as

$$PE = EI \int_0^l (2a_2 + 6a_3 z)^2 dz \tag{3.13}$$

However, because we wish to eliminate the $\{a\}$ parameters, it is far more convenient to express the energy in matrix terms; hence

$$PE = EI \int_0^{l_e} \{a\}^T \begin{Bmatrix} 0 \\ 0 \\ 2 \\ 6z \end{Bmatrix} \{0 \quad 0 \quad 2 \quad 6z\}\{a\}dz \tag{3.14}$$

but since we need to express this energy in terms of the nodal variables, the $\{a\}$ parameters are eliminated and we obtain from Equation 3.10

$$PE = EI \int_0^{l_e} \{v\}^T [\mathbf{C}]^{-T} \begin{Bmatrix} 0 \\ 0 \\ 2 \\ 6z \end{Bmatrix} \{0 \quad 0 \quad 2 \quad 6z\}[\mathbf{C}]^{-1}\{v\}dz \tag{3.15}$$

After completing the integrals and multiplications, an expression for the *PE* of the element is obtained. This becomes

$$PE = \{v\}^T [\mathbf{K}]\{v\}$$

where

$$\mathbf{K} = \frac{2EI}{l_e^3} \begin{bmatrix} 6 & 3l_e & -6 & 3l_e \\ 3l_e & 2l_e^2 & -3l_e & l_e^2 \\ -6 & -3l_e & 6 & -3l_e \\ 3l_e & l_e^2 & -3l_e & 2l_e^2 \end{bmatrix} \tag{3.16}$$

The mass matrix is calculated by considering the kinetic energy (*KE*). In this case, we obtain

$$KE = \int_0^{l_e} \{v\}^T [\mathbf{C}]^{-T} \{1 \quad z \quad z^2 \quad z^3\} \begin{Bmatrix} 1 \\ z \\ z^2 \\ z^3 \end{Bmatrix} [\mathbf{C}]^{-1}\{v\}dz \tag{3.17}$$

On completing the calculations, the expression for *KE* is given by

$$KE = \omega^2 \{v\}^T [\mathbf{M}] \{v\}$$

where

$$\mathbf{M} = \frac{ml_e}{420} \begin{bmatrix} 156 & 22l_e & 54 & -13l_e \\ 22l_e & 4l_e^2 & 13l_e & -3l_e^2 \\ 54 & 13l_e & 156 & -22l_e \\ -13l_e & -3l_e^2 & -22l_e & 4l_e^2 \end{bmatrix} \qquad (3.18)$$

where m is the mass per unit length of the beam.

However, there is one further step in deriving the element matrices. So far, we have analyzed the motion in a single direction, whereas in most problems on machinery we must consider at least two orthogonal motions: this is easily accomplished and the element matrices become 8×8 rather than 4×4. Using the Euler–Bernoulli formulation for an axisymmetric rotor, the element mass matrix becomes

$$\mathbf{M_e} = \frac{\rho_e A_e l_e}{420} \begin{bmatrix} 156 & 0 & 0 & 22l_e & 54 & 0 & 0 & -13l_e \\ 0 & 156 & -22l_e & 0 & 0 & 54 & -13l_e & 0 \\ 0 & -22l_e & 4l_e^2 & 0 & 0 & -13l_e & -3l_e^2 & 0 \\ 22l_e & 0 & 0 & 4l_e^2 & 13l_e & 0 & 0 & -3l_e^2 \\ 54 & 0 & 0 & 13l_e & 156 & 0 & 0 & -22l \\ 0 & 54 & -13l_e & 0 & 0 & 156 & 22l_e & 0 \\ 0 & -13l_e & -3l_e^2 & 0 & 0 & 22l_e & 4l_e^2 & 0 \\ -13l_e & 0 & 0 & -3l_e^2 & -22l_e & 0 & 0 & 4l_e^2 \end{bmatrix}$$

$$(3.19)$$

The element stiffness matrix becomes

$$\mathbf{K_e} = \frac{E_e I_e}{l_e^3} \begin{bmatrix} 12 & 0 & 0 & 6l_e & -12 & 0 & 0 & 6l_e \\ 0 & 12 & -6l_e & 0 & 0 & -12 & -6l_e & 0 \\ 0 & -6l_e & 4l_e^2 & 0 & 0 & 6l_e & 2l_e^2 & 0 \\ 6l_e & 0 & 0 & 4l_e^2 & -6l_e & 0 & 0 & 2l_e^2 \\ -12 & 0 & 0 & -6l_e & 12 & 0 & 0 & -6l_e \\ 0 & -12 & 6l_e & 0 & 0 & 12 & 6l_e & 0 \\ 0 & -6l_e & 2l_e^2 & 0 & 0 & 6l_e & 4l_e^2 & 0 \\ 6l_e & 0 & 0 & 2l_e^2 & -6l_e & 0 & 0 & 4l_e^2 \end{bmatrix} \qquad (3.20)$$

The simple element derived here is based on Euler–Bernoulli beam theory. Whilst this is acceptable for the modeling of slender rotors (with a high length/diameter ratio), it is often necessary to base the model on the more sophisticated Timoshenko beam theory which includes shear and rotary inertia effects. After a slightly more complicated derivation, the stiffness becomes

$$\mathbf{K} = \frac{EI}{(1 + \Phi_e)l_e^2} \begin{bmatrix} 12 & 6l_e & -12 & 6l_e \\ 6l_e & l_e^2(4 + \Phi_e) & -6l_e & l_e^2(2 - \Phi_e) \\ -12 & -6l_e & 12 & -6l_e \\ 6l_e & l_e^2(2 - \Phi_e) & -6l_e & l_e^2(4 + \Phi_e) \end{bmatrix} \quad (3.21)$$

where

$$\Phi_e = \frac{12EI}{\kappa GA_e l_e^2}$$

and κ is the shear constant, G is the shear modulus and A_e is the cross-sectional area.

The derivation of the mass matrix is rather, but it produces two terms which are each 4×4 matrices representing the mass and rotary inertia, respectively. Full details can be found in Chapter 4 of Friswell *et al.* (2010). The mass matrix for the element is now given as

$$\mathbf{M_e} = \frac{\rho_e A_e l_e}{840(1 + \Phi_e)^2} \begin{bmatrix} m_1 & m_2 & m_3 & m_4 \\ m_2 & m_5 & -m_4 & m_6 \\ m_3 & -m_4 & m_1 & -m_2 \\ m_4 & m_6 & -m_2 & m_5 \end{bmatrix} +$$

$$\frac{\rho_e I_e}{30(1 + \Phi_e)^2 l_e} \begin{bmatrix} m_7 & m_8 & -m_7 & m_8 \\ m_8 & m_9 & -m_8 & m_{10} \\ -m_7 & -m_8 & m_7 & -m_8 \\ m_8 & m_{10} & -m_8 & m_9 \end{bmatrix} \quad (3.22)$$

where

$$m_1 = 312 + 588\Phi_e + 288\Phi_e^2, \qquad m_6 = -(6 + 14\Phi_e + 7\Phi_e^2)l_e^2,$$
$$m_2 = (44 + 77\Phi_e + 35\Phi_e^2)l_e, \qquad m_7 = 36,$$
$$m_3 = 108 + 252\Phi_e + 140\Phi_e^2, \qquad m_8 = (3 - 15\Phi_e)l_e,$$
$$m_4 = -(26 + 63\Phi_e + 35\Phi_e^2)l_e, \qquad m_9 = (4 + 15\Phi_e + 10\Phi_e^2)l_e^2,$$
$$m_5 = (8 + 14\Phi_e + 7\Phi_e^2)l_e^2, \qquad m_{10} = -(1 + 5\Phi_e - 5\Phi_e^2)l_e^2.$$

The second part of the mass matrix represents the effect of rotary inertia. It should be emphasized that although the derivation of the Timoshenko element is somewhat more complicated than that of the Euler element, there is no added difficulty in use. This is discussed in Section 3.3.3.

3.3.2.2 Matrix Assembly

Both components of the energy of the element have now been written in terms of the nodal coordinates and then the elements may then be assembled into structures. To illustrate this, consider the beam comprising two elements in Figure 3.2. If the vibration is in one plane only and there are now three nodes with six degrees of freedom. Denoting elements from the nth element as $(k_{ij})_n$, then the assembled stiffness matrix for the complete beam will be

$$
\mathbf{K} = \begin{bmatrix}
(k_{11})_1 & (k_{12})_1 & (k_{13})_1 & (k_{14})_1 & 0 & 0 \\
(k_{21})_1 & (k_{22})_1 & (k_{23})_1 & (k_{24})_1 & 0 & 0 \\
(k_{31})_1 & (k_{32})_1 & (k_{33})_1 + (k_{11})_2 & (k_{34})_1 + (k_{12})_2 & (k_{13})_2 & (k_{14})_2 \\
(k_{41})_1 & (k_{42})_1 & (k_{43})_1 + (k_{21})_2 & (k_{44})_1 + (k_{22})_2 & (k_{23})_2 & (k_{24})_2 \\
0 & 0 & (k_{31})_2 & (k_{32})_2 & (k_{33})_2 & (k_{34})_2 \\
0 & 0 & (k_{41})_2 & (k_{42})_2 & (k_{43})_2 & (k_{44})_2
\end{bmatrix}
$$

$$(3.23)$$

and the overall mass matrix is made up in the same manner. Thus, we can express the energy and therefore the force distribution in terms of the nodal parameters of the full structure.

For more complicated elements types, the mathematical approach is a little different but the physical basis of the argument remains the same.

The system matrices are assembled following the pattern shown in Equation 3.23, adding in terms for masses on the shaft and bearing parameters. The equation of motion for the system can be given as

$$\mathbf{Kx} + \mathbf{C\dot{x}} + \mathbf{G}\Omega\mathbf{\dot{x}} + \mathbf{M\ddot{x}} = \mathbf{F}(t) \qquad (3.24)$$

Two extra terms have been added to this equation: $\mathbf{C}$ represents the damping while $\mathbf{G}$ is the gyroscopic term. The damping is sometimes difficult to quantify accurately, but arises in machines, mainly from the bearings. Damping arising in the rotor itself demands a rather different treatment and this can cause instabilities (see Section 5.6). The gyroscopic term arises from consideration of angular momentum conservation and the matrix $\mathbf{G}$ is skew-symmetric: the important point to note is that the presence of this term makes the dynamic properties of the system dependent on the rotational speed of the shaft. The importance of this term clearly depends on the geometry of the shaft under consideration; in practical machines, gyroscopic terms are rarely important for cases in which the dominant inertias are between the bearings, but for overhung rotors they become a crucial consideration.

It is worthwhile reflecting on what has been achieved by representing the machine in this manner. Despite considerable underlying algebra, the physical importance of the process is quite straightforward. By imposing a fixed functional relationship within each element (Equation 3.8), the original system with an infinite number of degrees of freedom has been reduced to one with a finite number of degrees of freedom (equal in total to four times the number of nodes). The degree to which the motion of the model accurately represents the behavior of the real system is determined by the ability of the model to represent the true deflection and this will be determined by the number of elements in the model. In principle, clearly the more elements the better, but this approach leads to heavy computational overheads so it is important to gain some insight into the way in which model refinement is reflected in accuracy. To illustrate the point, we consider Example 3.1.

Example 3.1: A circular shaft with a diameter of 10 mm is supported on bearings at either end, 0.5 m apart. Determine the first four natural frequencies using 1, 2, 4, and 8 Euler–Bernoulli finite elements.

Solution
Since the model is using Euler–Bernoulli beam theory, an exact solution is available. This is given by

$$\omega_n = n^2 \pi^2 \sqrt{\frac{EI}{ml^4}}$$

The results of the FE calculations are shown in Table 3.1. As shown in the table, the lower-order modes are predicted well with few elements but the higher modes require a much finer FE mesh. This is simply a manifestation of the need to represent the mode's deflection shape.

Figure 3.3 shows the model of this simple rotor with 1, 2, 4, and 8 elements, respectively.

There are several points which are clear from this fairly simple set of calculations:

a) The FE-derived natural frequency is in all cases higher than the exact value. This is always the case with this type of element (which is said to be conforming meaning that there is continuity of the

TABLE 3.1

Values of Natural Frequencies (Hz) for Simply Supported Beam

No of Elements	1	2	4	8	Exact
Mode 1	90.5	81.8	81.5	81.5	81.5
Mode 2	414.6	361.9	327.3	326.1	326.0
Mode 3	–	909.5	746.9	734.5	733.5
Mode 4	–	1,658	1,447	1,309	1,304

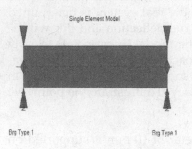

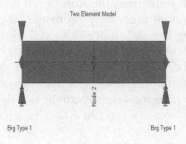

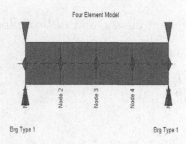

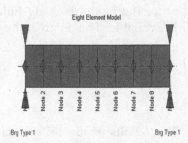

FIGURE 3.3
Rotor Modeling with FE.

displacement and slope between adjoining elements). The reason for this is clear physically: the model is more restricted that the true plant and consequently is stiffer.

b) The accuracy increases with the number of elements but decreases with mode number.

c) A model with only two elements gives the first mode accurately but shows considerable error for the second mode.

Having set out the basic framework for the modeling of rotating machines, we now consider the evaluation of measured data. Often modeling and the plant data are considered as quite separate entities, but throughout the present text it is proposed that they should be intimately related: experience on the plant should inform the modeling whilst predictions of the model aid understanding of the plant. The two are inseparable. Although this approach is suitable for the study of an ideal uniform beam, it becomes more clumsy to apply this approach to a beam of nonuniform cross section. For the study of most realistic analyses, the most convenient approach is with the FEM.

3.3.3 Modeling Choices

Of course, the examples so far all use the Euler beam theory which neglects shear and rotary inertia. The use of this type of element is just one of the choices that the engineer must make on how to represent the system. The main decisions to be made are as follows:

a) Is a beam model appropriate?
b) Is shear deformation and/or rotary inertia important (depending on the rotor's aspect ratio and the modes considered)?
c) The number of elements used.
d) Other factors such as bearings, seals, and the distribution of mass.

Timoshenko's beam theory includes both shear and rotary inertia and Table 3.2 shows the difference these terms make for a circular cross section using 16 elements. This table is expressed in terms of the normalized frequency, f/Ξ_0 where

$$\Xi_0 = \frac{L^2}{\pi^2} \sqrt{\frac{\rho A}{EI}} \qquad (3.25)$$

The table shows the normalized frequency or a number of aspect ratios. Note how the significance of the extra terms increases for higher mode numbers.

While Timoshenko's model is superior to that of Euler for thick beams, it is not exact and as the ratio of diameter-to-length increases the whole idealization of beams breaks down and the only adequate model would be a full three-dimensional solid model. Fortunately, this is rarely necessary for the analysis of rotating machines. Although the derivation is a little involved, this element is available in all commercial software and it is no more difficult to use than the Euler version. It is therefore advisable to use this as standard in rotor calculations except, of course, where more sophisticated models are required.

TABLE 3.2

Influence of Aspect Ratio on Normalized Frequency Calculations

Mode	Ratios (D/L)				Euler Solution
	0.025	0.05	0.075	0.1	
1	0.9992	0.9970	0.9933	0.9882	1
2	3.9881	3.9531	3.8971	3.8235	4
3	8.9414	8.7717	8.5136	8.1947	9
4	15.8213	15.3161	14.5915	13.7567	16

3.4 Analysis Methods

In the previous chapter, a description is given of the main ways of presenting measured data from rotating plant. There is a considerable quantity of data available in many instances and sometimes a rather confusing scenario is portrayed: it is often necessary to view data in a number of different ways in order to digest the significance. In any case of fault diagnosis, the essential feature is to correlate the observation with an understanding of the plant operation and this means some theoretical model. In many instances, there is not sufficient information to provide an accurate model, but this does not negate the usefulness of the modeling approach in gaining understanding. Chapter 7 discusses recent techniques for the correlation of models with plant measurements and in particular we discuss the derivation of the properties of supporting structures which is a common difficulty, particularly in the case of very large machines. Once established, the model can produce a range of presentations of results. In some cases, these replicate the measured plots discussed in the previous section, but since there is greater access to data, some other equally informative plots are available.

The development of FEM has transformed the analysis, design, and understanding of rotating machinery. It provides a general formulation to gain insight into the dynamics of machinery.

3.4.1 Imbalance Response

Mass unbalance is the most prevalent fault in turbo-machinery so it is natural to begin our discussion with the assessment of an unbalanced shaft. This topic, arguably the most important in the study of rotating machinery, is discussed in detail in Chapter 4; it is, however, convenient to give a short summary at this point in order to utilize some of the basic ideas in subsequent discussions. The general equation of motion Equation 3.24 may be rewritten to give the response at a shaft speed corresponding to Ω radians per second. The frequency of vibration arising from imbalance is $\omega = \pm\Omega$, the rotor speed is sinusoidal. The amplitude of the response is given by the equation

$$Kx + jC\Omega x + jG\Omega^2 x - M\Omega^2 x = M\Omega^2 e \qquad (3.26)$$

The important point to note here is that the shaft itself is providing the forcing term which gives rise to vibration. The solution to this may be conveniently computed as

$$\mathbf{x} = \left[\mathbf{K} + j\mathbf{C}\Omega + j\mathbf{G}\Omega^2 - \mathbf{M}\Omega^2\right]^{-1}\mathbf{M}\Omega^2 e \qquad (3.27)$$

This expression can be used to calculate and plot a rundown curve given a vector of mass eccentricities. It should be emphasized that this calculation is quasistatic in the sense that it strictly simulates a rundown taken over an infinite length of time. However, the difference caused by this should be fairly small.

Before leaving Equation 3.27 we note that for even moderate size models over a frequency range, this represents a substantial computation. For reasons of computational efficiency, the calculation is normally carried out in terms of modal parameters. This approach is explained in Section 3.5.1

3.4.2 Campbell Diagram

Following the philosophy for nonrotating structures, the first step here is to consider natural frequencies, but with the added complication that these natural frequencies will be a function of rotational speed. A plot of the natural frequencies as a function of shaft speed is the Campbell diagram, an example of which is shown in Figure 3.4. The curved lines emanating near the origin arise from modes controlled by the fluid bearings which have speed-dependent properties. The modes at 9, 27, and 58 Hz all have two components corresponding to forward and backward modes. In this instance, the splitting is only slight, but there is more discussion of this phenomenon in Chapter 4.

For a given rotation speed Ω, the natural frequencies are given by the eigenvalues of

$$\mathbf{K}x + j\mathbf{C}\omega x + j\mathbf{G}\Omega\omega x - \mathbf{M}\omega^2 x = 0 \tag{3.28}$$

In this case, Ω and ω are distinct quantities, the former being the rotation speed while the latter is the frequency of vibration. Because the exciting forces are usually locked to the shaft rotation, lines along which the frequency is in a fixed relationship to the rotation speed are of particular importance and the intersections of these lines and the natural frequency lines are called critical speeds. For imbalance response $\omega = \pm\Omega$. These critical speeds are of crucial importance in both the design and operation of rotating machinery. But even this does not give the whole picture of what is a very complicated story: in particular, the presence of a critical speed on a Campbell diagram does not necessarily imply a problem, because the particular resonance may not be excited to any significant degree. A resonance may not be excited owing to heavy damping, inappropriate distribution of forces or geometric effects; a very important example of this occurs in the case of reverse modes which cannot be excited by imbalance in the case of a machine mounted on symmetrical supports. This is discussed further in Chapter 4. While the Campbell

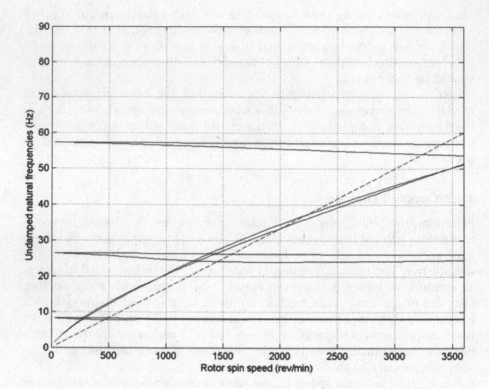

FIGURE 3.4
Campbell Diagram.

diagram is invaluable as a diagnostic aid, owing to the lack of information on damping it is incomplete and we seek to supplement our system knowledge.

The four plots shown in Figure 3.5a–d give an overview of some of the varieties seen in Campbell diagrams. The four cases shown above all relate to a simple fan with a large impeller mounted on the shaft between rolling element bearings. Because of the large impeller, gyroscopic moments play a part in some of the modes depending on the angular deflection of the mode in question. In Figure 3.5a, since the impeller is centrally mounted, the first mode is just a "bounce" motion; hence, there is no gyroscopic splitting. There is splitting in the second mode. Note that in this case there are only two critical speeds (one forward and one reverse). In Figure 3.5b, the supporting structure has been given different stiffness in the x and y direction and this can be seen by the frequency splits at zero shaft speed. In Figure 3.5c and d the axial location of the impeller has been changed and is off center. This means that even the lowest mode of vibration will involve some rotation and hence there is gyroscopic splitting of the first

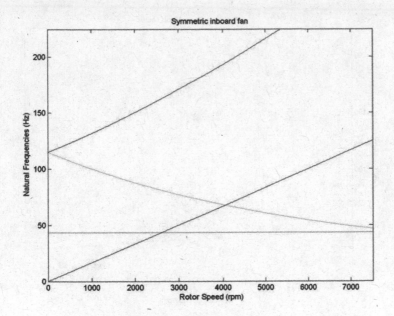

FIGURE 3.5A
Campbell Diagrams for Various Geometries. a Central Disc, Symmetric Support.

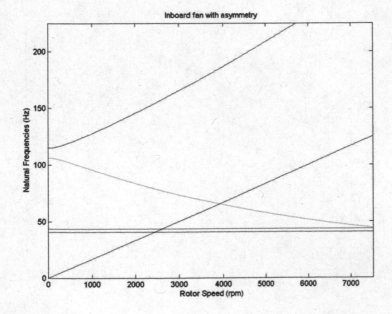

FIGURE 3.5B
Central Disc, Asymmetric Support.

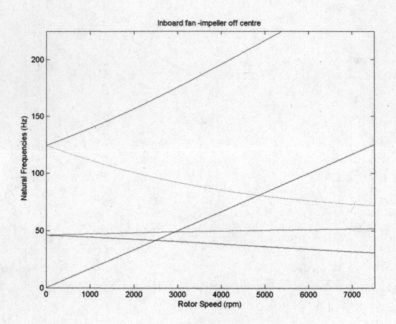

FIGURE 3.5C
Offset Disc, Symmetric Support.

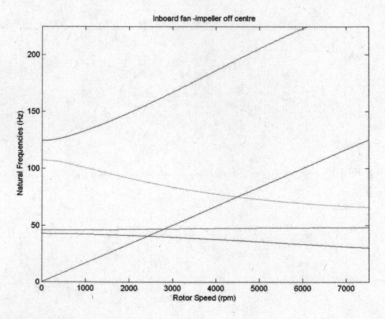

FIGURE 3.5D
Offset Disc, Asymmetric Support.

mode. Note also that with simple bearings (i.e. not oil film type) it is easily shown that forward modes increase in frequency with rotation speed whereas reverse modes decrease in frequency. Whilst this gives some insight, it is not the complete story because if the supporting structure and bearings are symmetric, no reverse mode (i.e. $\omega = -\Omega$) can be excited by imbalance. Of course, these modes are present and can be excited by external forces.

3.4.3 Analysis of Damped Systems

In the discussion so far, the evaluation of rotor response as a function of rotor speed has been demonstrated for simple cases and a more extensive survey of the topic is given in Chapter 4. Although the forced response is often the most valuable aspect of analysis, understanding the natural frequencies and mode shapes will often add significantly to the engineer's insight into machine operation. In analyzing the free vibration, the analyst seeks the damping coefficient of each mode in addition to the natural frequencies and mode shapes.

At first sight this presents a problem, as there are now three matrices to deal with in the analysis of the form

$$\mathbf{K}x + \mathbf{C}\dot{x} + \mathbf{M}\ddot{x} = 0 \tag{3.29}$$

A method is required to obtain natural frequencies, mode shapes, and damping coefficients with the three matrices $\mathbf{K}$, $\mathbf{M}$, and $\mathbf{C}$ involved in the calculation. The gyroscopic term is neglected here for the sake of clarity, but its inclusion is straightforward. This problem is most conveniently solved by expressing Equation 3.29 in terms of so-called state space. To do this, we regard x and $\dot{x}$ as two separate variables describing position and velocity, respectively. This means that there must be an extra equation expressing the fact that velocity is the time derivative of position and so the second-order Equation 3.29, becomes two first-order equations

$$\mathbf{K}x + \mathbf{C}\dot{x} + \mathbf{M}\frac{d\dot{x}}{dt} = 0 \tag{3.30}$$

$$\dot{x} = \frac{dx}{dt} \tag{3.31}$$

These two equations may now be combined into a single one of the form

$$\begin{bmatrix} \mathbf{K} & 0 \\ 0 & -\mathbf{M} \end{bmatrix} \begin{Bmatrix} x \\ \dot{x} \end{Bmatrix} + \begin{bmatrix} \mathbf{C} & \mathbf{M} \\ \mathbf{M} & 0 \end{bmatrix} \frac{d}{dt} \begin{Bmatrix} x \\ \dot{x} \end{Bmatrix} = 0 \tag{3.32}$$

This is of the form

$$\mathbf{Aq} + B\dot{\mathbf{q}} = 0 \tag{3.33}$$

The unknown vector is now $\{\mathbf{x} \quad \dot{\mathbf{x}}\}^T$ which clearly has twice the dimension of the original problem. Note also that this analysis gives a direct calculation of natural frequency, rather than the natural frequency squared obtained in the undamped second-order calculation. In general, the natural frequencies will be complex, of the form $\omega_k = \alpha_k + j\beta_k$, and it is worth reflecting on the physical meaning of this. Recall that an equation such as Equation 3.33 is solved by assuming a form $\mathbf{q} = \mathbf{q}_0 e^{\lambda t}$ and after substitution, this leads to the eigenvalue equation

$$\mathbf{Aq}_0 + \lambda \mathbf{Bq}_0 = 0 \tag{3.34}$$

If there are n degrees of freedom in the original (second-order) problem, there will be $2n$ solutions to the eigenvalue problem of Equation 3.30, each comprising an eigenvalue, λ, and corresponding eigenvector, $\mathbf{q}_0$ The kth solution to the free vibration Equation 3.33 becomes

$$q_k = q_{0k} e^{\alpha_k t} (A \sin \beta_k t + B \cos \beta_k t) \tag{3.35}$$

It is clear from this that provided α is negative, any disturbance will decay in time but if α is greater than zero, disturbance will grow and the system will be unstable. A, B are just constants chosen to fit the initial conditions. By comparison with a single degree of freedom system, it is easily shown that the kth natural frequency ω_{nk} and the damping coefficient ζ_k are given by

$$\omega_{nk} = \sqrt{\alpha_k^2 + \beta_k^2} \tag{3.36}$$

and

$$\zeta_k = \frac{\alpha_k}{\sqrt{\alpha_k^2 + \beta_k^2}} \tag{3.37}$$

It is clear from these two relationships that β_k gives the damped natural frequency and that

$$\omega_{dk} = \beta_k = \omega_{nk}\sqrt{1 - \zeta_k^2} \tag{3.38}$$

To clarify these issues, consider the equation of motion for the free vibration of a single degree of freedom system. This is described by

$$kx + c\dot{x} + m\ddot{x} = 0 \tag{3.39}$$

Assuming the trial form $x = x_0 e^{st}$ leads to the equation

$$(k + cs + ms^2)x_0 e^{st} = 0 \tag{3.40}$$

From this the two roots are given as

$$s = \frac{-c \pm \sqrt{c^2 - 4km}}{2m} = \frac{-c}{2m} \pm \sqrt{\left(\frac{c}{2m}\right)^2 - (k)^2} \tag{3.41}$$

Setting $\zeta = \frac{c}{2m}\sqrt{\frac{m}{k}}$ and taking a factor of (-1) out of the square root gives

$$s = -\zeta\omega_n \pm j\omega_n\sqrt{1 - \zeta^2} \tag{3.42}$$

Note that when $\zeta = 1$, the imaginary part of s, corresponding to the oscillatory component, vanishes and the solution becomes real. Under this condition the system is critically damped and all oscillation ceases. As an example, Figure 3.6 uses the same model of an alternator and turbine on four bearings and shows the damping coefficient, ζ_k for the first few natural frequencies, ω_{nk} for the first few modes. For this particular model, variation with rotor speed is not significant, but could be readily included in a 3D plot.

In a multidegree of freedom system, this applies to individual modes; although in many rotor systems overall damping may be modest, but specific modes may be critically damped or overdamped. This type of behavior is particularly prevalent in systems with oil-journal bearings (which usually account for an overwhelming proportion of the system damping).

There is one further issue related to system damping and the resulting dynamic behavior. In the analysis of structures, it is frequently assumed that the damping matrix is a linear combination of the stiffness and mass matrices, that is, $\mathbf{C} = \alpha\mathbf{K} + \beta\mathbf{M}$.

This is the so-called classical or proportional model of damping. Various authors have sought some justification for such a model but in truth there is none. The reason this model is used is that it leads to substantial mathematical simplification and for modest damping at least, the answers are reasonable. This is so because, in many cases, the damping is only important in parts of the frequency range close to a resonance. In many

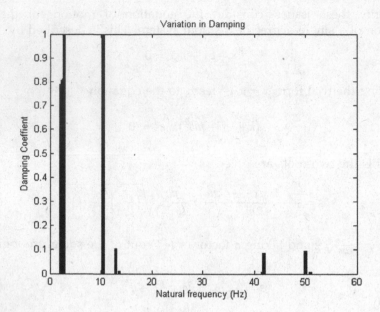

FIGURE 3.6
Damping Coefficients for Different Modes.

rotating machines however, the damping is dominated by the bearings and this has some interesting consequences.

An important feature of "proportional damping" is that the mode shapes are real and are those of the undamped system. More general damping leads to complex eigen-functions (mode shapes) and we now consider what this means physically. Suppose that the kth mode shape takes the form

$$\psi_k = A_k(x) + jB_k(x) \tag{3.43}$$

Multiplying this by the time dependence, taking the real part yields

$$y_k = Re\left((A_k(x) + jB_k(x))e^{\alpha_k \omega_{nk} t}(\cos \omega_{nd} t + j \sin \omega_{nd} t)\right) \tag{3.44}$$

Hence, for this modal contribution we get

$$y_k = e^{\alpha_k \omega_{nk} t}(A_k(x) \cos \omega_{dk} t - B_k(x) \sin \omega_{dk} t) \tag{3.45}$$

The exponential term will of course settle owing to repeated forcing but the interesting point to note is the phase of this component given by

$$\theta_k = \tan^{-1}\left(-B_k(x)/A_k(x)\right) \qquad (3.46)$$

This varies with x and so we see that the complex mode represents a travelling wave in the machine, rather than a stationary wave represented by a real mode shape.

In any case, on an actual machine the response will comprise contributions from a number of modes and although these modes may be either real or complex, their summation will produce phases that vary with location.

3.4.4 Root Locus and Stability

The previous section discusses the treatment of the damping term in the equation of motion. The role of damping is crucial in rotating machinery and this is another area where the situation is very much more complicated than the dynamics of nonrotating structures. The fundamental problem is that machines have an energy input from the rotation of the shaft and under certain circumstances, some of this energy can be transferred into the modes of vibration and the result is that minor disturbances grow in time rather than decaying. There are three potential causes of rotor instability:

a) Inappropriate bearing design or loadings
b) High damping on the rotor
c) Excessive rotor asymmetry

Within the context of fault diagnosis, the first of the alternatives will be the one most commonly met as alignment changes alter the static load on multibearing machines. This may manifest itself as a subsynchronous whirl, but this will be discussed more fully in Chapter 5. The essential point to note here is that instability is present when the real part of one of the eigenvalues becomes positive. (The implications of this are explained shortly.) In any case, the damping is clearly an import parameter as it plays a crucial role in determining the practical importance of a mode.

While a Campbell diagram is an extremely useful tool in the study of machinery, it gives incomplete information. While it is important to be aware of the existence of natural frequencies, this is only a part of the story and a mode may be either heavily damped, or may not be excited owing to the nature of the forcing function. In any event, some knowledge of the damping is clearly necessary.

In the root locus diagram, the complex frequency of each mode is plotted on the Argand plane. The term "root locus" for this plot is now widely accepted but this is a little unfortunate as it implies a (nonexistent) relationship to the analysis method of the same name in control theory. Figure 3.7 gives an example of a root locus diagram. Each line shows the trajectory of a modal frequency as shaft rotational speed varies over the range. The y-coordinate represents the frequency of oscillation while the x-coordinate gives the decay rate (and from this the damping coefficient is readily determined). Any root showing in the right-hand side of the diagram will grow with time and leads to an instability. Instability is a potential problem in machines owing to the input of energy from the rotor.

3.4.5 Overall Response

The root locus plot is a useful tool for understanding stability issues, but it is only at best a very rough guide in assessing the importance of a particular mode in the dynamics of an overall system. To arrive at

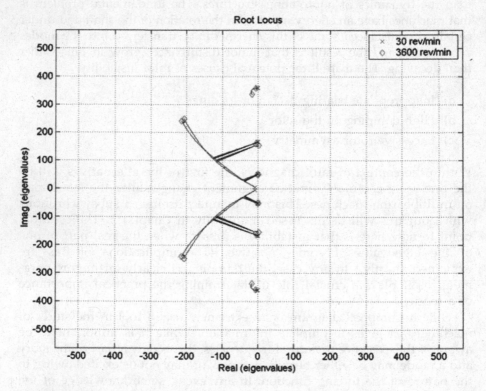

FIGURE 3.7
Root Locus Plot.

something potentially more helpful, we consider the forced response to a harmonic excitation. From Equation 3.27, it is observed that

$$\mathbf{x} = \left[\mathbf{K} + j\mathbf{C}\omega + j\mathbf{G}\Omega\omega - \mathbf{M}\omega^2\right]^{-1} f(\omega) \tag{3.47}$$

from which it can be inferred that an important parameter is

$$D = \det(\mathbf{K} + j\mathbf{C}\omega + j\mathbf{G}\Omega\omega - \mathbf{M}\omega^2) \tag{3.48}$$

Figure 3.8 shows a plot of the reciprocal of the determinant of the dynamic stiffness, as in Equation 3.48. In evaluating this, care is needed in scaling to avoid numerical difficulties.

While this cannot be regarded as a precise measure of response it does give an indication of potential problems without detailed consideration of forcing positions.

These three-dimensional plots can help give insight into the relative importance of different resonance conditions. However, it is important to

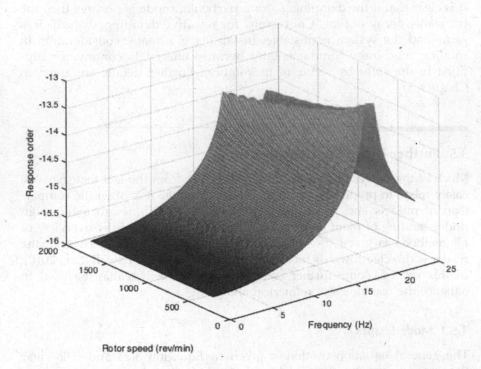

FIGURE 3.8
Plot of the Reciprocal of the Determinant (Equation 3.48).

appreciate that even this does not present the whole story: in particular, it says nothing about stability. To understand this issue, we need to reflect on the nature of second-order differential equations of the type which describe the behavior of rotating machines. In nondimensional form, we may write for a single degree of freedom system

$$\ddot{x} + 2\zeta\omega_0\dot{x} + \omega_0^2 x = F/m \qquad (3.49)$$

Let the force, F, arise from unbalance and take the form $F = me\omega^2 \cos\omega t$, then the full solution of the equation becomes

$$x = \frac{e\omega^2 \cos\omega t}{(-\omega^2 + 2j\zeta\omega\omega_0 + \omega_0^2)} + A\exp(-\zeta\omega_0 t)\cos\left(\omega_0\sqrt{1-\zeta^2}t + \phi\right) \qquad (3.50)$$

In this expression, the constants A, ϕ are determined by the initial conditions. The first term gives the steady-state response: this gives the response to the force but it tells us nothing about the stability of the system. On the other hand, the second term gives the transient response. It is clear that if the damping, ζ, for a particular mode is positive then the transients decay in time. Conversely, for negative damping, disturbances grow and the system is unstable. Instability is a major consideration in rotating machinery, fundamentally because energy is continually supplied to the rotor by virtue of its rotation. Further details are given in Chapter 5.

3.5 Further Modeling Considerations

Much of the power of rundown curves derives from the fact that they can easily relate to predictions from theoretical models: it is often the comparison of models and plant data which really contributes to a thorough understanding of plant behavior. For this reason, a very brief overview of FE methods is given in Section 3.2 but for an in-depth discussion, the reader is directed towards more detailed texts (see e.g. Friswell *et al.*, 2010). In this section, some further analysis is shown which may be used to enhance the usefulness of numerical analyses.

3.5.1 Mode Shapes

The general equation motion is given in Equation 3.24 and it is clear the motion can be described with matrices to represent the mass, stiffness, and gyroscopic terms of the system. There are several ways in which these matrices are obtained, the most common being the FEM as outlined in Section 3.3.2. It is clear from basic theory of linear equations that, under some circumstances nonzero solutions may be obtained even when there is no force acting on the system and this occurs at a natural frequency of the system. With no forces applied the equation of motion becomes (neglecting the gyroscopic term in this instance)

$$\mathbf{K}x + \mathbf{C}\dot{x} + \mathbf{M}\ddot{x} = 0 \tag{3.51}$$

Making the substitution $x = x_0 e^{j\omega t}$, we obtain the equation

$$(\mathbf{K} + j\omega\mathbf{C} - \omega^2\mathbf{M})x_0 = 0 \tag{3.52}$$

For the present, consider the case with zero damping. This equation can have a nonzero solution provided that

$$|\mathbf{K} - \omega^2\mathbf{M}| = 0 \tag{3.53}$$

This condition occurs at each of the n natural frequencies and each of these frequencies has a corresponding displacement function or mode shape u_n which satisfies the equation

$$(\mathbf{K} - \omega_n^2\mathbf{M})u_n = 0 \tag{3.54}$$

These mode shapes have some interesting and extremely useful properties which are now outlined. Denoting $\lambda_n = \omega_n^2$, then considering any two mode shapes n and m associated with two different natural frequencies

$$\begin{aligned} \mathbf{K}u_n &= \lambda_n\mathbf{M}u_n \\ \mathbf{K}u_m &= \lambda_m\mathbf{M}u_m \end{aligned} \tag{3.55}$$

Multiply the first of these equations by u_m^T, the second one by u_n^T, the result is

$$u_m^T\mathbf{K}u_n - u_n^T\mathbf{K}u_m = \lambda_m u_n^T\mathbf{M}u_m - \lambda_n u_m^T\mathbf{M}u \tag{3.56}$$

It is a straightforward exercise to show that the left-hand side of this equation is zero (a consequence of the stiffness matrix being symmetric about the diagonal). Similarly, the symmetry of the mass matrix implies that

$$u_n^T\mathbf{M}u_m = u_m^T\mathbf{M}u_n \tag{3.57}$$

Then from these two relationships we reach the very significant conclusion that

$$0 = (\lambda_n - \lambda_m)u_n^T \mathbf{M} u_m \tag{3.58}$$

Why is this significant? Because the first term on the right-hand side is clearly nonzero, the conclusion is that any two different mode shapes are orthogonal with respect to the mass matrix and this provides important techniques for the analysis of many aspects of dynamic behavior.

For example, consider the response of a general system to a force **F**. The equation of motion is given by

$$\mathbf{K}\mathbf{x} + \mathbf{C}\dot{\mathbf{x}} + \mathbf{M}\ddot{\mathbf{x}} = \mathbf{F} \tag{3.59}$$

If the forcing function is known, this equation can be integrated in time, but for any realistic model this is a very time-consuming process. Often with machinery the solution is determined in the frequency domain and is expressed as

$$\mathbf{x}(t) = \sum \mathbf{X}(\omega)e^{j\omega t} \tag{3.60}$$

Denoting the Fourier transform of $F(t)$ as $\tilde{F}(\omega)$, X is given by

$$\mathbf{X}(\omega) = \left[\mathbf{K} + j\omega\mathbf{C} - \omega^2\mathbf{M}\right]\tilde{F}(\omega) \tag{3.61}$$

However, even this approach can be very demanding in computing terms if the model is detailed and/or the response is required at a number of frequencies. A more convenient approach is to express the motion in terms of the mode shapes by writing

$$\mathbf{x}(t) = \sum_{n=1}^{N} \mathbf{u}_n q_n(t) \tag{3.62}$$

Inserting this into the equation of motion gives

$$\mathbf{K}\sum \mathbf{u_n}q_n + \mathbf{C}\sum \mathbf{u_n}\dot{q}_{nn} + \mathbf{M}\sum \mathbf{u_n}\ddot{q}_{nn} = \mathbf{F} \tag{3.63}$$

Now premultiply this equation by u_m^T, then sum over all values of n

$$\sum \mathbf{u}_m^T \mathbf{K}q_n\mathbf{u}_n + \sum \mathbf{u}_m^T \mathbf{C}\dot{q}_n\mathbf{u}_n + \sum \mathbf{u}_m^T\mathbf{M}\ddot{q}_n\mathbf{u}_n = \mathbf{u}_m^T F \tag{3.64}$$

Consider the summations on the left-hand side of Equation 3.64. From Equation 3.55, it may be seen that in the first summation, because

$\mathbf{K}u_n = \lambda_n \mathbf{M}u_n$ the only nonzero term in the first and third summations occur when $n = m$, so Equation 3.64 can be rewritten as

$$\lambda_m a_m q_m + \sum \mathbf{u}_m^T \mathbf{C} \dot{q}_n \mathbf{u}_n + a_m \ddot{q}_m = \mathbf{u}_m^T \mathbf{F} \tag{3.65}$$

In this equation $a_m = u_m^T \mathbf{M} u_m$, but because mode shapes can be scaled it is convenient to choose the scaling so that $a_m = 1$. This leaves the damping term as the remaining difficulty: a common way around this difficulty is the assumption of so-called classical damping, also known as "proportional damping" (discussed in Section 3.4.3). It is assumed that the damping matrix can be expressed as

$$\mathbf{C} = \alpha\mathbf{M} + \beta\mathbf{K} \tag{3.66}$$

With this assumption Equation 3.65 becomes

$$\lambda_m q_m + \alpha\dot{q}_m + \lambda_m\beta\dot{q}_m + \ddot{q}_m = \mathbf{u}_m^T F \tag{3.67}$$

In rotating machinery, excitation is often sinusoidal and in such cases one seeks a solution of the form
$$q_m = q_{m0}e^{j\omega t}$$

Inserting this into Equation 3.59 leads to

$$q_m = \frac{\mathbf{u}_m^T \mathbf{F}}{\omega_m^2 - \omega^2 + j\omega(\alpha + \omega_m^2\beta)} \tag{3.68}$$

This provides a very elegant approach. From Equation 3.68, we can recover the full displacement profile $x(t)$ using Equation 3.62, and this can be done over a range of frequencies with no costly matrix inversions. The one remaining difficulty to address is the validity of the classical damping assumption: in reality there is no fundamental justification for using it, but in many instances it gives a reasonable approximation to reality, so let us consider why this is the case. In sinusoidal excitation, damping only plays a significant role when the excitation frequency is close to resonance, and in Equation 3.68 we have two parameters and therefore we can model the damping at two frequencies. This is sufficient to represent the main features of many machines in practice.

It is possible, and sometimes necessary, to analyze systems with a more general form of damping. Friswell and Lees (1998) have shown how it is possible to express the general displacement in terms of the damped mode shapes which are, in general, complex (corresponding to a traveling elastic wave, rather than one which is standing). However, the mathematics involved is rather more cumbersome and hence wherever possible classical damping is assumed.

With the assumption that the damping can be regarded as proportional, the mode shapes define a natural set of coordinates which can be used to decouple the equations of motion. While this is of importance in simplifying the mathematical treatment, it is even more important in physical understanding which it imparts. To illustrate this, we take an example of an unbalanced rotor. Imbalance is discussed in greater detail in Chapter 4. Consider the rotor shown in Figure 3.9 comprising a uniform shaft which carries two discs at the two one-third span positions. The two discs have equal imbalances which are opposite in phase.

In this particular example, the critical speeds (natural frequencies) occur at 15 and 40 Hz and since the rotor is symmetric, we know that the (unnormalized) mode shapes are

$$\mathbf{u}_1 = \begin{Bmatrix} 1 \\ 1 \end{Bmatrix} \qquad \mathbf{u}_2 = \begin{Bmatrix} 1 \\ -1 \end{Bmatrix} \tag{3.69}$$

It is worth noting that in this simple case we can develop the mode shapes using physical insight. In real machines detailed calculations will often be required, but not always. Note also that significant insight may be gleaned by considering the system's symmetry.

The damping in this example has been chosen as 2% critical in each of the modes and the forcing is proportional to the square of the rotor speed, as would be the case for imbalance excitation. Figure 3.10 shows response curve for two different excitations. These curves have been calculated using only Equation 3.68, the only constants required for the calculation were the mass of the rotor (to normalize the mode shapes) and the magnitude of the forcing. In both cases, the force at the two discs is equal, but in the first case the two are in phase, whereas in the second case they are out of phase. Mathematically, we may express this by

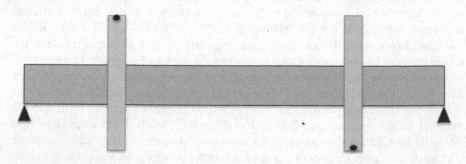

FIGURE 3.9
A Rotor with Imbalance.

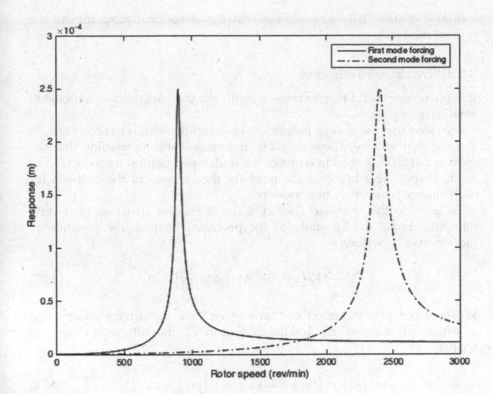

FIGURE 3.10
Imbalance Response.

$$\mathbf{F_1} = a \begin{Bmatrix} 1 \\ 1 \end{Bmatrix} \qquad\qquad \mathbf{F_2} = a \begin{Bmatrix} 1 \\ -1 \end{Bmatrix} \qquad (3.70)$$

The important point to note here is that the first force gives no contribution to the response at the second natural frequency while with the second force pattern the response at the first natural frequency is unaffected by the first mode. In other words, with the second force pattern, the system does not respond at the first natural frequency. This is important in the diagnosis of several faults and is discussed in more detail in Chapter 4. This phenomenon occurs because

$$\mathbf{u_2}^T \mathbf{F_1} = a\{1 \quad -1\}\begin{Bmatrix} 1 \\ 1 \end{Bmatrix} = 0 = \mathbf{u_1}^T \mathbf{F_2} = a\{1 \quad 1\}\begin{Bmatrix} 1 \\ -1 \end{Bmatrix} \qquad (3.71)$$

This is an example of how knowledge of the modes leads to a greater understanding of the dynamic behavior of a machine or indeed any vibrating system. It is a consequence of the orthogonal properties of the normal modes.

3.5.2 Perturbation Techniques

It is often very useful to gain some insight into the sensitivity of a model to small changes.

Provided that the change in stiffness or damping term is not too great, in a sense that will be discussed later, it is reasonable to assume that the motion can be described in terms of the undamped natural frequencies and mode shapes. This bypasses the need for the analysis of the problem for each change which may be considered.

To develop this approach, we consider a general structure neglecting damping. Using the FE study of the preceding section, the equation of motion may be written as

$$[\mathbf{K} + \Delta\mathbf{K}]\{y\} - \omega^2[\mathbf{M} + \Delta\mathbf{M}]\{y\} = 0 \tag{3.72}$$

We now consider the effect of "turning on" the perturbing effect using a parameter denoted by λ. The deflection of the nth mode may be written as

$$\{y_n\} = \{y_{0n}\} + \lambda\{y_{1n}\} + \lambda^2\{y_{2n}\} + \ldots\ldots \tag{3.73}$$

while the corresponding natural frequency is written

$$\omega_n = \omega_{0n} + \lambda\omega_{1n} + \lambda^2\omega_{2n}\ldots\ldots \tag{3.74}$$

These expansions are now inserted into (3.72) giving

$$[\mathbf{K} + \lambda\Delta\mathbf{K}]\{y_{0n} + \lambda y_{1n} + \lambda^2 y_{2n} + ..\} - \\ (\omega_{n0} + \lambda\omega_{n1} + \lambda^2\omega_{n2})^2[\mathbf{M} + \lambda\Delta\mathbf{M}]\{y_{0n} + \lambda y_{1n} + \lambda^2 y_{2n}\} = 0 \tag{3.75}$$

From this an equation for each power of λ is given, hence
for λ^0:

$$[\mathbf{K}]\{y_{0n}\} - \omega_{0n}^2[\mathbf{M}]\{y_{0n}\} = \{0\} \tag{3.76}$$

for λ^1:

$$\mathbf{K}y_{1n} + [\Delta\mathbf{K}]y_{0n} - \omega_{0n}^2\mathbf{M}y_{1n} - \Delta\mathbf{M}\omega^2{}_{0n}y_{0n} - 2\omega_{0n}\omega_{1n}\mathbf{M}y_{0n} = \{0\} \tag{3.77}$$

Equations can be written for higher powers of λ and these become progressively more complicated. However, for the present study we only require the zeroth and first power of lambda.

Equation 3.76 is simply a restatement of the problem with no perturbation, and this is conveniently solved using standard methods: hence, the frequencies ω_{0n} and the corresponding mode shapes $\{y_{0n}\}$ are known and furthermore, these mode shapes form a complete orthonormal set. The first-order changes $\{y_{1n}\}$ are expressed in terms of these known modes

$$y_{1n} = \sum a_{nm} y_{0m} \tag{3.78}$$

Inserting this expression in Equation 3.77 gives

$$[\mathbf{K} - \omega_{0n}^2 \mathbf{M}]\left\{\sum a_{nm} y_{0m}\right\} + [\Delta\mathbf{K} - \omega_{n0}^2 \Delta\mathbf{M}]y_{0n} - 2\omega_{0n}\omega_{1n}[\mathbf{M}]y_{0n} = \{0\} \tag{3.79}$$

Now taking $n = m$, the first on the left-hand side of Equation 3.79 becomes zero. The equation is now premultiplied by $\{y_{0n}\}^T$ and using the orthogonality relationships we obtain

$$y_{0n}^T[\Delta\mathbf{K} - \omega_{n0}^2 \Delta\mathbf{M}]y_{0n} - 2\omega_{0n}\omega_{1n} = \{0\} \tag{3.80}$$

This gives the first-order frequency changes as

$$\omega_{1n} = \frac{y_{0n}^T[\Delta\mathbf{K} - \omega_{0n}^2 \Delta\mathbf{M}]y_{0n}}{2\omega_{0n}} \tag{3.81}$$

It is straightforward to insert this expression back into 3.79 and assume n and deduce an equation for a_{nm} giving the changes in mode shape. The form of this expression is physically clearer by noting that the changes in frequency squared may be written as

$$\Delta\omega_n^2 = 2\omega_{0n}\omega_{1n} = y_{0n}^T[\Delta\mathbf{K} - \omega_{0n}^2 \Delta\mathbf{M}]y_{0n} \tag{3.82}$$

Before continuing the calculation, note that a reconsideration of Equation 3.79 for the case n leads to

$$a_{nm} = \frac{\{y_{0n}\}^T[\Delta\mathbf{K} - \omega_{0n}^2 \Delta\mathbf{M}]}{\omega_{0n}^2 - \omega_{0m}^2} \tag{3.83}$$

In practice, very often the main interest is in the change of natural frequencies. Equations 3.81 and 3.83 give the first-order perturbation solutions to the problem of a general system. The analysis may be extended to examine the accuracy of the calculation, by assessing the second-order contributions, but this is beyond the scope of the present discussion and is rarely used. The application of the perturbation approach is discussed by Wilkinson (1965) and Fox and Kapoor (1968) with a specific application given by Lees (1999). It is illustrated in Example 3.2.

Example 3.2: Calculate the natural frequency of a uniform beam, then with a mass of 2 tons at one quarter span. Take the span between supports as 5 m and the diameter as 0.5 m. Note that for consideration of the first mode, Euler–Bernoulli theory will suffice.

As mass is added to the rotor, its natural frequency will decrease. Considering only the variation in the frequency of the lowest mode, we denote the amplitude of the (mass normalized) mode shape at the location where the mass is added as $y_{1L/4}$, then for any mass addition the change in natural frequency is given (from Equation 3.81) as

$$\Delta\omega_1 = -\omega_1 \Delta\mathbf{M} y_{1L/4}^2$$

This equation is of course valid only for the additions of small masses. As the changes become larger, the mode shape begins to alter as shown in Equation 3.68 and the calculation becomes somewhat more complex. Nevertheless, this approach is helpful in two ways; it is helpful in making quick assessments and, perhaps more important it gives insight into parameters of significance.

Figure 3.11 shows the variation of the lowest natural frequency with the addition of a range of masses up to 2,500 kg. The frequency was recalculated fully for each case but the graph also shows the prediction of perturbation theory (to first order).

It is clear that the perturbation prediction provides a good estimate of the frequency changes, although there is a gradual deterioration in its accuracy as the magnitude of the additional mass increases. But more information can be inferred directly. For example, a direct assessment can be made about the change in natural frequency if the mass is added at the central node rather than at $L/4$. The appropriate variation can be obtained by considering the relative magnitudes of the mode shapes at quarter and midspan, respectively:

$$y_{1L/4} = 0.0114$$
$$y_{1L/2} = 0.0161$$

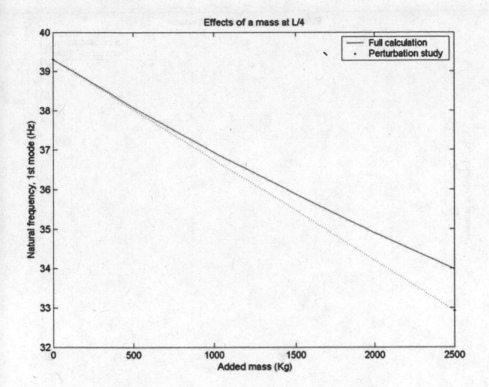

FIGURE 3.11
Comparison of "Exact" and Perturbation Solutions.

This means that (from Equation 3.81), the sensitivity of the system to a mass at midspan is r times greater than that at quarter span, where r is given by

$$r = \left(\frac{y_{1L/2}}{y_{1L/4}}\right)^2 = 1.987$$

Without further calculation, we can say that a given mass will have approximately double the effect when placed at the midspan rather than the quarter span position. It is precisely this facility to visualize the effects of change without direct calculation which makes this technique so useful. The explicit calculation of the influence at midspan is shown in Figure 3.12 together with the scale perturbation line.

It should be noted that the application of this technique enables the analyst to significantly enhance the understanding gained from a single numerical calculation.

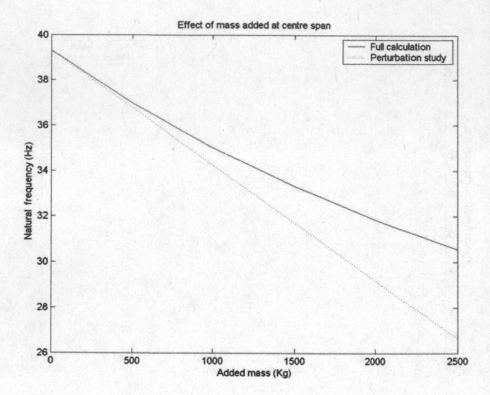

FIGURE 3.12
Comparison of Exact and Perturbed Solutions.

3.6 Summary

Data from rotating machinery is rich in information but often complex and dependent on many parameters. The effective diagnosis of plant issues demands the effective assimilation of data and the various ways in which data can be presented plays an important part in forming and overall judgement on the internal processes within a machine. This chapter presents an overview of the various ways of viewing and analyzing data, together with some of the more important modeling approaches. Almost invariably some combination of these approaches is required to deal with real plant problems. Having set out basic methods, Chapters 4–6 discuss some of the more common type of problems arising on rotating machines.

Problems

3.1 A uniform steel shaft 1.5 m in length, is supported in short rigid bearings at each end. The shaft has a diameter of 20 mm. A series of steel discs, each 1 cm thick, are available for mounting at the midspan location. Calculate the natural frequency with discs of diameter 50, 100, and 150 mm.

3.2 A shaft 1 m in length is supported at each end, on two identical bearings with a constant stiffness of 10^6 N/m. The diameter is 20 mm and it has two discs mounted 0.25 m from each bearing. Both discs have diameter 100 mm and thickness 10 mm. Using four Euler finite elements, calculate the first two natural frequencies. What happens if the number of elements is doubled?

3.3 Repeat the Problem 3.2 using Timoshenko beam elements (thereby including shear and rotary inertia effects). Comment on your answers.

3.4 Using the analytic expression for the natural frequency of a pinned–pinned Euler beam, and observing that the mode shapes are sinusoidal, calculate the first natural frequency when each of the three discs of question 3.1 are applied at the center span.

3.5 A large fan is mounted on two short rigid bearings 2 m apart. The shaft is made of steel 0.2 m in diameter and the fan impeller is overhung by 0.5 m. The impeller has a mass of 800 kg and can be regarded as a disc of diameter 1.5 m. Draw the Campbell diagram and determine the shaft speeds at which speed equals natural frequency. These are the critical speeds.

References

Fox, R.L. and Kapoor, M.L., 1968, Rates of change of eigenvalues and eigenvectors, *American Institute of Aeronautics and Astronautics Journal*, 6(12), pp. 2426–2429.

Friswell, M.I. and Lees, A.W., 1998, Resonance frequencies of viscously damped structures, *Journal of Sound & Vibration*, 217(5), pp. 950–959.

Friswell, M.I., Penny, J.E.T., Garvey, S.D. and Lees, A.W., 2010, *Dynamics of Rotating Machines*, Cambridge University Press, New York, NY.

Lees, A.W., 1999, The use of perturbation theory for complex modes, *17th International Modal Analysis Conference*, Kissimmee, FL.

Wilkinson, J.H., 1965, *The Algebraic Eigenvalue Problem*, Clarendon Press, Oxford.

Zienkievicz, O.C., Taylor, R.L. and Zhu, J.Z., 2005, *The Finite Element Method*, Sixth Edition, Elsevier Butterworth-Hienemann, Oxford.

4

Faults in Machines (1)

4.1 Introduction

The purpose of all monitoring is the understanding of the current state of the plant. In order to place in context the work presented in other chapters, it is essential to discuss the manifestation of various types of plant fault. The way in which a given machine will respond to an error type is, of course, partly dependent on the geometry of the machine question: nevertheless, several of the more important categories of fault may be described generically. Before embarking on a discussion of the more esoteric faults, it is important to examine the properties of rotor imbalance. In Chapter 3, imbalance was briefly described and used as an example to illustrate several techniques. As imbalance is the most prevalent of all rotor faults, further details are given of approaches to the identification and rectification of imbalance, but an important consideration in this is the flexibility of the rotor itself. We begin, therefore, with a discussion on the difference between rigid and flexible rotors as these issues have an important influence on the optimum approaches to rectify problems.

4.2 Definitions: Rigid and Flexible Rotors

It is important to study machine imbalance for two reasons: firstly, although it is now well understood, it remains the most common single cause of high vibration in rotating machinery. Second, the way in which local unbalances transform to give distributed vibration responses helps to gain an insight into the techniques at our disposal for the discussion of other phenomena.

The discussion here will be for a general flexible rotor – but we immediately run into a difficulty of terminology. In some older textbooks and papers, a rotor was said to be flexible if it operated above 70% of its first critical speed. For our purposes, this is a very inadequate definition. Reiger (1986) gives a far more meaningful definition: "A flexible rotor is defined as being any rotor that cannot be effectively balanced throughout its speed range by placing suitable correction weights (sic) in separate planes along its length." To clarify, we must distinguish between the rotor and the machine. The rotor is, self-evidently, just that component which rotates. So, with this distinction let us consider the simple machine shown in Figure 4.1. The mass of the rotor is 100 kg and each bearing has a stiffness of 5×10^6 N/m. This gives a resonant frequency of 50.3 Hz. Given that the rotor rotates at 2,400 rpm, the speed represents 80% of the natural frequency and so the machine may be called flexible. It will respond significantly to imbalance as discussed shortly.

However, the rotor is rigid. It moves simply as a rigid body having four degrees of freedom, two translational and two rotational. The shape of the first mode is shown as a dotted outline in Figure 4.1; it is simply a uniform translation from the equilibrium position. This mode is sometimes referred to as the "bounce" mode and can, of course, arise in the two directions orthogonal to the axis of the rotor. Note that motion in the axial direction is not excited by unbalance and torsional motion has been neglected for this case. In fact, of course there is no such thing as a truly rigid rotor; nevertheless, the concept is convenient and for a large number of machines it represents a convenient abstraction and a reasonable approximation. Friswell *et al.* (2010) carried a thorough discussion of the dynamics of a rigid rotor but the requirement here is to ascertain the validity of the concept for a given case.

Restricting attention to the plane of the page, it is clear that the rotor can only move in two ways (or combinations thereof): it can either move with equal amplitude at all points along the length or it can rotate about the central point. Any other possible motion is simply a (linear) combination of these two basic modes of vibration. Figure 4.2 shows the second mode which exists for a rigid rotor in each direction. This is often called the "rocking" mode – again it occurs in both directions.

In general, of course, the rotor will move in the two orthogonal directions and hence there will be a total of four degrees of freedom. Perhaps the most

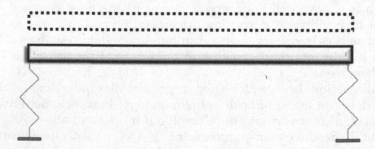

FIGURE 4.1
A Rigid Rotor on Spring Supports.

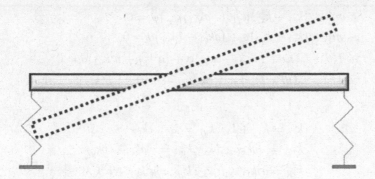

FIGURE 4.2
Second Mode ("Rocking'").

obvious way of expressing the four degrees of freedom is in terms of the displacements at the two bearings, but this choice is not unique: one could, for instance, choose the coordinates at any two points along the rotor. A less obvious notation leads to a rather more compact description of the mechanics of the system. Following Friswell *et al.* (2010), the motion of the rotor is described in terms of the motion of the center of gravity of the rotor and its rotations about the x and y axes. The equations of motion will be given by

$$m\ddot{u} + k_{x1}(u - a\psi) + k_{x2}(u + b\psi) = 0$$
$$m\ddot{v} + k_{y1}(v + a\theta) + k_{y2}(v - b\theta) = 0$$
$$I_d\ddot{\theta} + I_p\Omega\dot{\psi} + ak_{y1}(v + a\theta) - bk_{y2}(v - b\theta) = 0$$
$$I_d\ddot{\psi} - I_p\Omega\dot{\theta} + ak_{x1}(u - a\psi) - bk_{x2}(u + b\psi) = 0$$

(4.1)

where u, v represent the motions of the center of gravity in the x and y directions and θ, ψ are the rotations about the x and y axes, respectively. The center of mass is a distance a from bearing 1 and b from bearing 2. These equations are formulated simply by expressing Newton's law for both the translation and the rotations. The two terms involving $I_p\Omega$ are the gyroscopic moments which arise basically from the conservation of angular momentum. A full discussion of these moments are given by Friswell *et al.* (2010), but for many rotors, particularly in arrangements with bearing outboard of the main inertia components, the influence of these terms will be minor. They are included here for the sake of completeness as they become important in machines with overhung rotors or those with very large impellers. Simply rearranging these equations yields

$$m\ddot{u} + (k_{x1} + k_{x2})u + (-ak_{x1+bk_{x2}})\psi = 0$$
$$m\ddot{v} + (k_{y1} + k_{y2})v + (ak_{y1} - bk_{y2})\theta = 0$$
$$I_d\ddot{\theta} + I_p\Omega\dot{\psi} + (ak_{y1} - bk_{y2})v + (a^2k_{y1} + b^2k_{y2})\theta = 0 \qquad (4.2)$$
$$I_d\ddot{\psi} - I_p\Omega\dot{\theta} + (-ak_{x1} + bk_{x2})u + (a^2k_{x1} + b^2k_{x2})\psi = 0$$

Letting

$$k_{xT} = k_{x1} + k_{x2}; k_{yT} = k_{y1} + k_{y2}$$
$$k_{xC} = -ak_{x1} + bk_{x2}; k_{yC} = -ak_{y1} + bk_{y2} \qquad (4.3)$$
$$k_{xR} = a^2k_{x1} + b^2k_{x2}; k_{yR} = a^2k_{y1} + b^2k_{y2}$$

then (4.2) can be written more concisely thus

$$m\ddot{u} + k_{sT}u + k_{xC}\psi = 0$$
$$m\ddot{v} + k_{yT}v - k_{yC}\theta = 0$$
$$I_d\ddot{\theta} - I_p\Omega\dot{\psi} - k_{yC}v + k_{yR}\theta = 0 \qquad (4.4)$$
$$I_d\ddot{\psi} - I_p\Omega\dot{\theta} + k_{xC}u + k_{xR}\psi = 0$$

Of course, because of the gyroscopic terms, the motion in the u and v directions are coupled to some extent and so, in general, the four equations must be solved together to reflect the behavior of the rotor system. It is important to appreciate that here, the rotor is rigid and consequently, the motion of the two ends (or specifically, bearing locations) is determined, the dynamics of the system is completely specified. Denoting the two bearing locations as A and B and their respective motions as $y_A(t), z_A(t), y_B(t), z_b(t)$, then the motion of the center of mass (gravity) may be expressed as

$$x_C = \frac{ax_A + bx_B}{a + b} \qquad\qquad y_C = \frac{ay_A + by_B}{a + b} \qquad (4.5)$$

This is where a major distinction is encountered because this is not the case if the rotor is flexible. Flexible rotors form a major part of large machines in particular but there is a long-term trend toward their use in faster machines. Where the rotor is flexible, the relationship of the points A, B and C is more complicated and is inherently frequency dependent.

It is helpful to define the following parameters

$$f_x = x_C - \frac{x_A + x_B}{2} \qquad\qquad f_y = y_C - \frac{y_A + y_B}{2} \qquad (4.6)$$

and with these defined, a measure of the degree of flexibility in the rotor is just

$$r_f = \sqrt{f_x^2 + f_y^2} \qquad (4.7)$$

By measuring the location of the shaft at any three axial positions, a measure of this type can be defined. Figure 4.3 shows a plot of how the parameter r_f varies as the flexibility of the rotor is changed. This calculation used a uniform beam divided into four elements and flexibility was altered by varying the value of Young's modulus.

However, with less than three locations measured, no definite conclusions can be made concerning the shaft dynamics without some further assumptions or, best of all, a reliable theoretical model of the machine. Of course, this is no easy task and other chapters in this book are devoted to the derivation of such models. As mentioned elsewhere (see Chapter 7), the modeling of the rotor itself is the easiest task in the representation of a complete machine. We therefore examine the properties of the rotor in terms of its own dynamics.

The simplest case we can consider is a truly rigid rotor with a mass M and a diametral moment of inertia about its center of gravity of I_d. It is

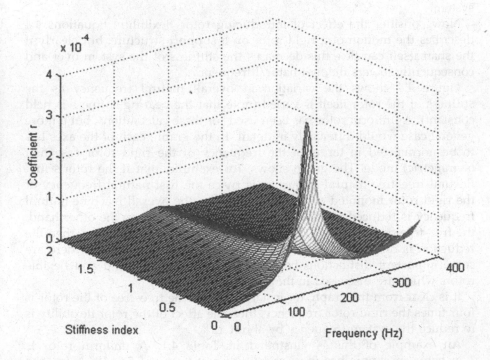

FIGURE 4.3
Variation of Modes with Rotor Stiffness.

easily shown that the two natural frequencies for a rigid rotor (neglecting gyroscopic terms) are given by

$$\omega_1 = \sqrt{\frac{k_T}{m}} \qquad \omega_2 = \sqrt{\frac{k_R}{I_d}} \qquad (4.8)$$

Consider a uniform rotor which is mounted on two idealized bearings: the dimensions and mass of the rotor are fixed, but Young's modulus is varied having the effect of changing the rotor's free–free natural frequency, or to be more specific, it's first bending (free–free) natural frequency. Note that for a real rotor this mode will be the seventh free–free mode as there will be six rigid body modes (with zero frequency) corresponding to translation along each of the three coordinates and rotations about each of the three axes. The frequency of this first flexural free–free mode provides a very effective indicator as to the behavior of the rotor in a machine. It is easily calculated from a rotor model, and since for this evaluation one is not dependent of the uncertainties of bearings and support structures, these predictions tend to be reliable. The calculated values are readily validated by suspending the rotor in slings from a crane, the applying an impact test, preferably in the horizontal direction to minimize the effect of the suspension system.

Now consider the effect of introducing rotor flexibility; Equations 4.4 describes the motion of a rigid rotor on a support structure, but clearly if the shaft itself can flex, this decreases the stiffness of the system over and consequently lowers the first natural frequency.

Figure 4.4 shows the variation in overall natural frequency as the stiffness of the rotor itself is varied (note that the bearing stiffness is held constant). A uniform rotor has been used for these calculations, but a more general case would differ only in detail. In the graph, each of the axes has to be normalized in terms of the frequency of the rigid rotor case first (symmetric) mode. The graph shows, for example, that if the rotor's first flexural free–free natural frequency is twice the first natural frequency of the rigid rotor mounted on the bearings, then the overall machine natural frequency is reduced to 68% of the rigid rotor case. If, on the other hand, the free–free frequency is 10 times that of the rigid rotor machine, then the reduction is only about 3%. Apart from changes in frequencies, there are some important distinctions between the properties of rigid and flexible rotors which are discussed in the section on balancing.

It is clear from the graph, for instance, that if the free–free of the rotor is four times the rigid rotor frequency, then the effect of the rotor flexibility is to reduce the system frequency by about 12%.

An example of this is illustrated in Table 4.1. A uniform rotor is 1 m between simple bearings, each with a stiffness of 5×10^5 N/m, the diameter is 3 cm, the density is that of steel but Young's modulus has been

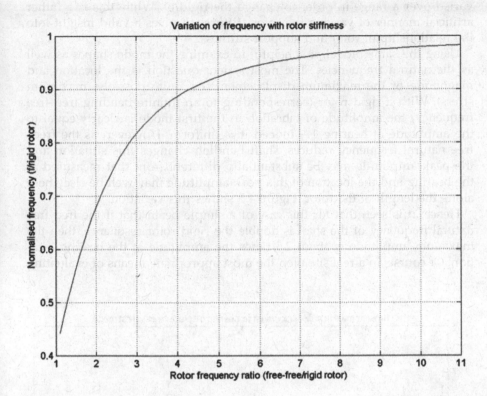

FIGURE 4.4
Influence of Rotor Stiffness.

TABLE 4.1

Variation of System Natural Frequencies

Free–Free Frequency (Hz)	Mode 1 (Hz)	Mode 2 (Hz)
67.7	27.7	88.8
117.3	42.1	106.3
151.3	48.8	110.6
179.1	52.7	112.5
203.1	55.3	113.6
224.5	57.2	114.5
244.1	58.5	114.9
262.2	59.6	115.2
279.1	60.4	115.5
∞ (Rigid)	67.8	117.4

varied over a range in order to change the rigidity. While this is a rather artificial manner of varying the rotor stiffness, it gives a valid insight into the relationship of rotor and support stiffness.

Using the same model, it is helpful to examine the mode shapes as well as the natural frequencies. The most salient question is the location and magnitude of the maximum displacement (or, quite possibly, maximum stress). With a rigid rotor (corresponding to an infinite bending free–free frequency) the amplitude of vibration in the first mode is clearly equal to the amplitude at bearing 1 – indeed it is uniform. However, as the free–free natural frequency reduces, shaft bending changes this situation and the peak amplitude may be substantially different from that measured at the bearing and the location of this peak amplitude may well be elsewhere along the length of the rotor. Figure 4.5 shows this variation.

Hence, it is seen that for this case of a simple beam that if the free–free natural frequency of the shaft is double the rigid rotor resonance, then the maximum amplitude is about 2.5 times the amplitude at the bearing location. Of course, in a real situation the most appropriate means of evaluating

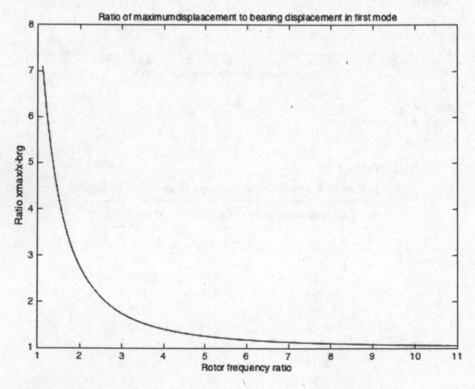

FIGURE 4.5
Modal Displacement Ratios (Peak/Bearing).

these concepts is by the use of an appropriate model: the simple case here serves to illustrate the point.

There is another very important aspect of the distinction between rigid and flexible rotors which is closely related to this discussion: a rigid rotor has precisely six free–free modes relating to the six degrees of freedom. These modes correspond to the uniform motion along the axes (two transverse modes in each direction, one torsional and one extensional). A flexible rotor by contrast has many more free–free modes. This means that the motion of a flexible rotor cannot be described in terms of its six rigid body modes, a point which has important consequences in rotor balancing, as discussed in the next section.

Many machines are designed to operate above their second critical speed and in these circumstances, it may be unclear whether the rotor is operationally rigid or flexible. Nevertheless, it is important to draw the distinction as this will dictate the form of balancing needed; are two planes sufficient or are more required? Commercial considerations dictate the absolute need to do no more than necessary. This is an important example of the many instances in which the operator must have some understanding of the machines under his or her control.

4.3 Mass Imbalance

4.3.1 General Observations

The response of rotors to mass imbalance is undoubtedly the most widely encountered problem in rotating machinery. Despite being well understood for many years it is still prevalent in a wide range of turbo-machines. The difficulties arise from the strong sensitivity of high-speed machines to relatively small manufacturing tolerances. If there are no external forces acting on the system, and gyroscopic terms are neglected for the sake of clarity, then the equation of motion for free vibration is given by

$$M\ddot{x} + C\dot{x} + Kx = 0 \tag{4.9}$$

But to understand the implications of this equation when applied to a rotating machine, careful consideration must be given to the individual terms. In the inertial term, the relevant acceleration is that of the center of mass of the rotor, rather than its geometric center, the two differing by the mass eccentricity. The issue here is that because the center of mass does not coincide with the geometric center (or, more importantly, the center of rotation), there is an acceleration imposed on the center of mass by the rotation of the shaft. The difference in location may be expressed as

$$\mathbf{x}_G = \left\{ \begin{matrix} u_G \\ v_G \end{matrix} \right\} = \left\{ \begin{matrix} u + \varepsilon \cos\phi \\ v + \varepsilon \sin\phi \end{matrix} \right\} \tag{4.10}$$

The eccentricity, ε, will usually be a vector giving the mass eccentricity at each point along the rotor. In the inertial term of Equation 4.9, $\mathbf{x}$ must be replaced with $\mathbf{x}_G$, so that the equation becomes

$$\mathbf{M\ddot{x}}_G + \mathbf{C\dot{x}} + \mathbf{Kx} = 0 \tag{4.11}$$

For constant speed of Ω radians per second, this leads to

$$[-\mathbf{M}\Omega^2 + j\mathbf{C}\Omega + \mathbf{K}]\mathbf{x} = \mathbf{M}\Omega^2\varepsilon e^{j\Omega t} \tag{4.12}$$

The steady-state solution to this equation must be of the form $\mathbf{x} = \mathbf{x}_0 e^{j\Omega t}$ and substitution of this form into Equation 4.12 enables cancellation of the sinusoidal variation to give

$$[-\mathbf{M}\Omega^2 + j\mathbf{C}\Omega + \mathbf{K}]\mathbf{x}_0 = \mathbf{M}\Omega^2\varepsilon \tag{4.13}$$

A direct solution can be expressed as

$$\mathbf{x}_0 = [-\mathbf{M}\Omega^2 + j\mathbf{C}\Omega + \mathbf{K}]^{-1}\mathbf{M}\Omega^2\varepsilon \tag{4.14}$$

Without discussing any details of the solution to the motion, three features are immediately obvious

a) The response is synchronous, that is, in steady-state operation, the vibration is at rotor speed.
b) As the speed tends to zero, the response also tends to zero.
c) As the speed increases, the vibration tends to a finite limit.

The third of these observations is perhaps worthy of note as it has a very simple physical interpretation: from Equation 4.14, it is clear that as $\Omega \to \infty$, then $\mathbf{x}_0 \to -\varepsilon$. This is simply saying that as the speed increases, the rotor will tend to rotate about its center of mass rather than the geometric center, which of course will be observed as a vibration. Indeed, on reflection this is an inevitable result as any other scenario would give rise to infinite forces.

It is instructive to extend this examination of mass imbalance by considering the modal nature of the solutions. While this is very useful in some numerical studies significantly reducing computing time, the real benefit lies in the insight it offers in understanding the nature of the phenomena.

Considerable insight into imbalance response can be gained by using some properties of eigenvalues and eigenfunctions (corresponding to natural frequencies and mode shapes). Although this requires the use of some algebra, the overall simplification is very considerable.

Provided that the damping may be regarded as "classical" or "proportional" the modes of this system form a complete orthonormal set and so the unknown response may be written as

$$x_0(z) = \sum_{n=1}^{\infty} a_n \psi_n(z) \tag{4.15}$$

Note that the so-called proportional model assumes that the damping matrix is a linear combination of the mass and stiffness matrices. As discussed in Chapter 3, there is no physical basis for this assumption but it significantly simplifies the calculation without, in many cases, significantly altering the model predictions. In fact, this assumption on damping is rarely valid: in many rotating machines, the damping will be concentrated at the bearings. However, for cases with modest damping, the assumption leads to enhanced understanding.

Inserting Equation 4.15 into 4.13, and the resulting equation is multiplied by ψ_m^T. This gives

$$\sum_{n=1}^{\infty} a_n \psi_m^T K \psi_n + j\Omega \sum_{n=1}^{\infty} a_n \psi_m^T C \psi_n - \Omega^2 \sum_{n=1}^{\infty} a_n \psi_m^T M \psi_n + = \psi_m^T M \varepsilon \tag{4.16}$$

From the orthonormality conditions (the basic properties of the mode shapes)

$$\psi_m^T K \psi_n = \omega_n^2 \delta_{nm}$$
$$\psi_m^T C \psi_n = (\alpha + \beta \omega_n^2) \delta_{nm} \tag{4.17}$$
$$\psi_m^T M \psi_n = \delta_{nm}$$

where δ_{nm} is the Kronecker delta (i.e. $\delta_{nm} = 1$ if $n = m$, 0 if $n \neq m$). Using these relationships in Equation 4.16, we see that for each value of m there is only a contribution from $n=m$, then

$$a_m (\omega_m^2 + j\Omega[\alpha + \beta \omega_m^2] - \Omega^2) = \psi_m^T M \Omega^2 \varepsilon \tag{4.18}$$

and so

$$a_m = \frac{\psi_m^T M \Omega^2 \varepsilon}{(\omega_m^2 + j\Omega[\alpha + \beta \omega_m^2] - \Omega^2)} \tag{4.19}$$

Hence, the full response can be evaluated. Rotor imbalance is an ever-present feature of rotating machinery, and the resulting synchronous vibration is characteristic of, but not uniquely related to, imbalance. As discussed in Section 4.4, a rotor bend also generates a synchronous vibration signal yet some aspects of its behavior are in marked contrast to pure imbalance.

Before proceeding it is worth reflecting on Equation 4.19. In essence, it is a rather simple expression for the modal content of the response. The essential message is that the excitation component in the mth mode is proportional to $\psi_m^T M\varepsilon$ multiplied by a fairly complicated function of shaft speed. This carries some important insights: for instance, in the case of a uniform rotor mounted on bearings of equal stiffness, an imbalance at the center span will have no influence on the second mode, because the second mode has zero amplitude at this point along the rotor. Indeed, a comparison of the relative amplitudes at different rotor speeds can often be used to locate (axially) imbalance or other faults. The concept involved is now illustrated with an example. A detailed model of a machine can offer tremendous insight into operational issues, but even limited knowledge can be used to advantage, albeit with some level of uncertainty.

Example 4.1: To illustrate this, consider the rundown profile shown in Figure 4.6. The curve shows two distinct peaks at 480 and 1,900 rev/min, although the second peak is rather damped. If it is postulated that the imbalance is dominated by a single mass, what can be said about the axial location of this imbalance? Note that the first peak is split owing to the behavior of the oil journal bearings, but despite the slightly difficult response curve, valuable assessments of plant behavior can be made.

Provided there is knowledge, even if approximate, of the mode shapes associated with the two critical speeds, then some assessment can be made. In this case, we assume that the mode shapes are those of a uniform rotor, illustrated in Figure 4.7.

From Figure 4.6 it is clear that the first peak is about five times higher than the second peak. The fact that the second peak is so significant immediately indicates that the imbalance (if there is a single imbalance) cannot be at the midspan. The damping of the two modes (as assessed by the ratio of the peak width to the resonant frequency) is rather different, and from the figure it appears that the damping of mode 2 is about three times that of the first mode. On this assumption, we may write

$$x(\omega_1) = a \sin\left(\frac{\pi z_u}{L}\right) \qquad\qquad x(\omega_2) = a \sin\left(\frac{2\pi z_u}{L}\right)$$

so that $\dfrac{x(\omega_1)}{x(\omega_2)} = 5 = \dfrac{3a \sin\left(\dfrac{\pi z_u}{L}\right)}{a \sin\left(\dfrac{2\pi z_u}{L}\right)}$

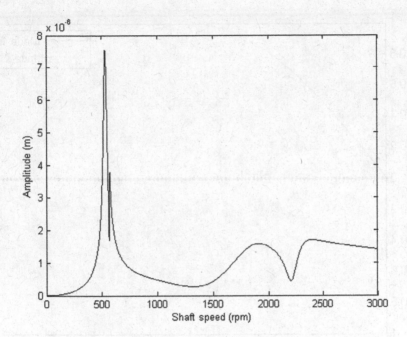

FIGURE 4.6
Transient Response Curve.

Therefore, $10 \times \sin\left(\dfrac{\pi z_u}{L}\right)\cos\left(\dfrac{\pi z_u}{L}\right) = 3 \times \sin\left(\dfrac{\pi z_u}{L}\right)$

Giving $z_u = \cos^{-1}(0.3) * L/\pi \approx 0.4L$

There is, of course, some uncertainty in the absence of some damping model, but reasonable estimates can be established using knowledge of the mode shapes.

4.3.2 Rotor Balancing

To overcome, or at least limit the influence of imbalance, a range of balancing techniques have been developed. Early work was based on rigid rotors, but in more recent years developments have focused on flexible rotors as these have assumed ever greater importance. The field was reviewed by Parkinson (1991) who quotes some 100 references, and more recent work has been surveyed by Foiles *et al.* (1998) citing a further 160 papers and indeed there are many other papers on this important subject. Although the basic methods are now reasonably well understood (at least at the research level), the large number of papers on this topic reflects the very high commercial importance. There can be little doubt that imbalance is the predominant single mechanism of concern in all rotating machinery.

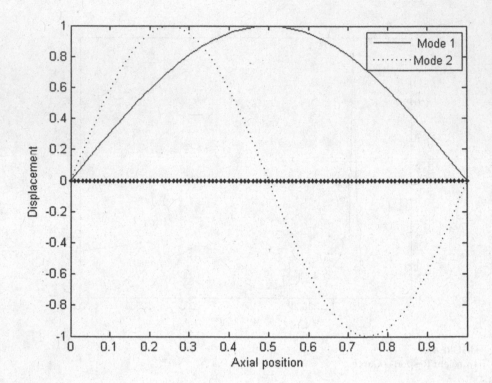

FIGURE 4.7
Mode Shape Estimates.

The approaches to balancing may be categorized into two basic approaches: the first of these investigates the influence of added masses and a fixed speed, while in the second approach (modal) balancing is performed at or near to each mode in turn. Both these approaches rely on the fundamental assumption of linearity: in practice, corrections are some-times made for nonlinearity by additional test runs thereby successively seeking to reduce the unbalance. In the case of the modal approach, how-ever, there are some rather more rigorous requirements.

In principle, all rotor balancing is based on the extrapolation of the influence of a small added mass to the full negation of any initial unbalance. In its simplest form, the process will be, within some tolerable limit, linear.

4.3.2.1 Influence Coefficient Approaches – Single Plane

The concept here is very straightforward. At a fixed running speed, the influence of an added mass on the vibration is measured in terms of both magnitude and phase. Then it is a straightforward matter to determine the required balance mass (frequently but incorrectly referred to as balance weights), but this scaling must be done in a vector sense to give the position of the mass adjustment as well as its magnitude. To efficiently balance a machine, one needs some means of identifying the angular orientation of the rotor at any given time: this might comprise a painted mark on the shaft coupled to some optical probe, or some electronic encoder system. Friswell *et al.* (2010) show that balancing can be achieved without a phase reference but this involves a rather more complex procedure and should be treated only as a last resort where no phase signal is possible. The more usual approach with phase measurement is now illustrated with a simple example.

It is appropriate to balance a machine in a single plane if only a one resonance is influencing the dynamic behavior of the machine. This will be apparent from the shape of the rundown curve which, in these cases, will have no more than one peak in, or influencing, the response curve. Single-plane balancing means that balance masses can be fitted in a single axial position along the shaft. By using trial masses, the engineer establishes the sensitivity of the system to known masses, in terms of both vibration magnitude and phase. Using this sensitivity, the magnitude and position are calculated of the mass required to counteract the initial vibration.

The operation is as follows:

a) Ensure the shaft has some form of marker which can be used to define the phase. This may take the form of a physical mark on the shaft in view of an optical probe or some electronic encoder. The important point is to ensure a phase measurement system.

b) The vibration of the machine at the test speed is measured.

c) After halting the machine, a known mass is attached to the rotor at some known angle with respect to the phase reference.

d) The machine is then up the test speed, and the vibration measured again.

e) The difference in the two vibration measurements gives the rotor's sensitivity and from this the correct mass and its required position can be calculated.

f) The machine is then stopped, the correction mass attached and the machine returned to service.

To clarify the calculation involved, let us suppose the vibration is measured with amplitude r_0 at phase ϕ_0, so that the motion in the x-direction is given by $u = r_0 \cos(\Omega t + \phi_0)$.

So, expressing this in complex form

$$u = Re\left(u_0 e^{j\phi_0}\right) \qquad (4.20)$$

The motion of the rotor is described by its equation of motion

$$ku + c\dot{u} + m\ddot{u} = \mathbf{m}\Omega^2 \varepsilon \cos(\Omega t + \phi_0) \qquad (4.21)$$

Expressing this in complex form and assuming the sinusoidal form, this may be written as

$$\left(-m\Omega^2 + jc\Omega + k\right)u = m\varepsilon\Omega^2 e^{j(\Omega t + \phi_0)} = D(\Omega)u_0 e^{j\Omega t} \qquad (4.22)$$

u_0 is, in general, complex. At this stage, the sinusoidal terms cancel to give

$$D(\Omega)u_0 = m\varepsilon\Omega^2 e^{j\phi} \qquad (4.23)$$

The point to notice here is that, although Equation 4.22 specifies the problem, it is not much help yet because all the parameters are (usually) unknown. Since the rotor speed, Ω is constant, the unknown mass, stiffness and damping may be lumped together into a single (unknown) parameter $D(\Omega)$ describing the full dynamic behavior of the machine at this particular rotor speed. $(m\varepsilon), \phi$ are the parameters which are implicitly to be determined.

A trial mass, m_1 is now fitted at angle ϕ_1, the machine is run back up to speed, and a new vibration u_1 is measured. The equation of motion can now be written as

$$D(\Omega)u_1 = m\varepsilon\Omega^2 e^{j\phi} + m_1 r_1 \Omega^2 e^{j\phi_1} \qquad (4.24)$$

Subtracting Equation 4.23 from 4.24 gives the vibration level (both amplitude and phase) resulting from the added mass and hence this yields the sensitivity of the system, resulting in the equation

$$D(\Omega)(u_1 - u_{0)} = m_1 r_1 \Omega^2 e^{j\phi_1} \qquad (4.25)$$

From which, we get

$$D(\Omega) = \frac{m_1 r_1 \Omega^2 e^{j\phi_1}}{(u_1 - u_0)} \qquad (4.26)$$

Notice that D is now fully defined. Using Equation 4.22, the original imbalance can be calculated as

$$D(\Omega)u_0 = \frac{m_1 r_1 e^{j\phi_1}\Omega^2 u_0}{(u_1 - u_0)} = m\varepsilon\Omega^2 e^{j\phi} \tag{4.27}$$

Taking the modulus gives the magnitude of the imbalance as

$$m\varepsilon = \left|\frac{m_1 r_1 u_0}{u_1 - u_0}\right| \tag{4.28}$$

and the orientation of the imbalance is given by

$$\phi_0 = tan^{-1}\left(\frac{imag(D(\Omega)u_0)}{real(D(\Omega)u_0)}\right) \tag{4.29}$$

This gives all the information required to balance the rotor. Balance is achieved by first removing the test mass, then attaching an imbalance of $m\varepsilon$ at an orientation directly opposite ϕ, that is, at $\phi + 180°$.

At this point we return to the machine described in Example 4.1, but that example sought to estimate the imbalance location, now we seek to balance the machine to run within the specified speed range.

Example 4.2: Single-Plane Balancing

The machine runs at a maximum speed of 1,500 rev/min and its rundown curve, shown in Figure 4.6, illustrates that there is only a single resonance (critical speed) influencing the running range (the split of the first peak does not invalidate this). This is an important consideration as it implies that a single-plane balancing procedure will suffice to run the machine satisfactorily at 1,500 rev/min. At full speed the machine shows a vibration on one bearing of 25 μm at a phase of 45° after the phase reference. A trial mass of 10 g is then attached to the rotor at a radius of 100 mm and a phase of 180° (i.e. opposite the phase marker). On running the machine back up to the same speed, the measured vibration was 20 μm at 30° before the phase reference. From this information, the appropriate balancing mass to completely counteract the vibration.

Expressing the initial vibration in terms of complex numbers, a complete description of the vibration can be given by a complex number involving magnitude and phase so that

$$u_0 = 25\left(\cos\frac{\pi}{4} + j\,\sin\frac{\pi}{4}\right) = 25\left(\frac{\sqrt{2}}{2} + j\frac{\sqrt{2}}{2}\right)$$

While representation in terms of magnitude and phase is valid, the Cartesian form is more convenient for the subsequent calculation. Note that the length scale can be expressed in any consistent units and in this instance are more convenient. With test mass added the amended vibration is

$$u_1 = 20\left(\cos\frac{\pi}{6} - j\,\sin\frac{\pi}{6}\right) = 20\left(\frac{\sqrt{3}}{2} - j\frac{1}{2}\right)$$

Hence, the vibration generated by the additional mass is given by

$$\Delta u = u_1 - u_0 = 10\sqrt{3} - \frac{25}{\sqrt{2}} + j\left(-10 - \frac{25}{\sqrt{2}}\right)$$

Our objective is to establish the unbalance (me) but, of course, the terms of $D(\Omega)$ are unknown; however in this case, it has been established that

$$D(\Omega)\Delta u = -0.001\Omega^2$$

This determines $D(\Omega)$ and hence the original imbalance is given by

$$m\varepsilon\Omega^2 = D(\Omega)u_0 = -\frac{0.001\Omega^2 u_0}{\Delta u}$$

This implies that the balance mass must be directly opposite this position. In the current case, this gives

$$m\varepsilon = 0.001 = \frac{25}{\sqrt{2}}(1+j) \times \frac{1}{10\sqrt{3} - \dfrac{25}{\sqrt{2}} + j\left(-10 - \dfrac{25}{\sqrt{2}}\right)}$$

which simplifies to $me = (6.4683 - j6.3035) \times 10^{-4} kg.m..$ The orientation of this imbalance is given by

$$\theta = \tan^{-1}\left(\frac{-6.3035}{6.4683}\right)$$

Then it follows that this corresponds to an imbalance of abs $(m\varepsilon) = 9.03 \times 10^{-4}$ kg m. at $-44°$. Hence, if the balance mass is to be added at a radius of 100 mm, then a mass of 9 g should be fitted at an angle of $134°$ ahead of the reference mark.

While at first sight this may appear to be an involved procedure, it is readily automated. Simply stated, the engineer is observing the effect of a trial mass, then scaling this in such a way as to negate the original vibration on the machine.

Where single-plane balancing suffices, this completes the procedure. However, if the rundown curve indicates the influence of more than a single mode, then balancing in two (or sometimes more) planes is required, but very similar logic guides the process. Recall that this single-plane technique has been employed for this example because the rundown curve (in Figure 4.6) showed that only a single resonance was influencing the machine within its running range. If this is not the case a more extensive approach must be used and the number of balance planes must equal the number of resonances influencing the rundown curve.

4.3.2.2 Two-Plane Balancing

In this section, the two-plane case is discussed as an example of multiplane balancing. It is probably the most commonly used approach but the analysis is readily extended to multiple planes. Balancing in two planes requires the use of two trial masses and measurement of the response at two distinct points along the rotor and two separate test runs at the same rotor speed. The complex vibration readings u will now be vectors having two (or more in the case of multiplane) components and the influence term D becomes a matrix. It should be emphasized that the procedure for two planes (or indeed, multi-planes) is, in principle, entirely equivalent to the single-plane approach, but inevitably the calculation is slightly more complicated.

The difference is that because there are now two balance planes, and two degrees of freedom, the system dynamic stiffness $D(\Omega)$ becomes a 2×2 matrix. Following Equation 4.25, two equations can now be written to express the dynamic behavior. Denoting our two trial imbalances as $m\varepsilon_1, m\varepsilon_2$, but noting that in this case, each of the terms is a vector. This is to reflect the fact that each of the two trials may involve masses on each of the two balance planes

$$D(\Omega)\left\{\begin{matrix} u_{11} \\ u_{12} \end{matrix}\right\} = \left\{\begin{matrix} m\varepsilon_{01}e^{j\phi_{01}} \\ m\varepsilon_{02}e^{j\phi_{02}} \end{matrix}\right\}\Omega^2 + \left\{\begin{matrix} m_{11}r_{11}e^{j\phi_{11}} \\ m_{12}r_{12}e^{j\phi_{12}} \end{matrix}\right\}\Omega^2 \tag{4.30}$$

and similarly

$$D(\Omega)\left\{\begin{matrix} u_{21} \\ u_{22} \end{matrix}\right\} = \left\{\begin{matrix} m\varepsilon_{01}e^{j\phi_{01}} \\ m\varepsilon_{02}e^{j\phi_{02}} \end{matrix}\right\}\Omega^2 + \left\{\begin{matrix} m_{21}r_{21}e^{j\phi_{21}} \\ m_{22}r_{22}e^{j\phi_{22}} \end{matrix}\right\}\Omega^2 \tag{4.31}$$

It important that the two trial cases must be linearly independent and must involve both of the balance planes. It is convenient to change notation slightly at this point and make the r terms complex, thereby absorbing the phase information into them so that

$$R_{nk} = r_{nk}e^{j\phi_{nk}} \tag{4.32}$$

Then following the same logic as in the single-plane case, using differences between each of the two trial runs and the as found condition, Equations 4.30 and 4.31, can be combined to give

$$D(\Omega) \begin{bmatrix} \Delta u_{11} & \Delta u_{21} \\ \Delta u_{12} & \Delta u_{22} \end{bmatrix} = \begin{bmatrix} m_{11} R_{11} & m_{21} R_{21} \\ m_{12} R_{12} & m_{22} R_{22} \end{bmatrix} \qquad (4.33)$$

As before, $\Delta u_{kj} = u_{kj} - u_{0j}$ represents the change introduced by the corresponding trial mass set. Equation 4.33 now encapsulates the data from the "as found" condition and the two trial runs and from this the matrix D can be determined as

$$D(\Omega) = \begin{bmatrix} m_{11} R_{11} & m_{21} R_{21} \\ m_{12} R_{12} & m_{22} R_{22} \end{bmatrix} \begin{bmatrix} \Delta u_{11} & \Delta u_{21} \\ \Delta u_{12} & \Delta u_{22} \end{bmatrix}^{-1} \qquad (4.34)$$

The imbalance on the rotor is now given by

$$\begin{Bmatrix} m\varepsilon_1 e^{j\phi_{01}} \\ m\varepsilon_2 e^{j\phi_{02}} \end{Bmatrix} = \begin{bmatrix} m_{11} R_{11} & m_{21} R_{21} \\ m_{12} R_{12} & m_{22} R_{22} \end{bmatrix} \begin{bmatrix} \Delta u_{11} & \Delta u_{21} \\ \Delta u_{12} & \Delta u_{22} \end{bmatrix}^{-1} \begin{Bmatrix} r_{01} \\ r_{02} \end{Bmatrix} \qquad (4.35)$$

This is readily expressed in terms of modulus and phase as illustrated in Example 4.3. It is important to note that this estimates the position of the effective center of mass of the rotor:

Note that to compensate for this, a balance mass must be added directly opposite to these positions (i.e., 180° from those calculated).

The balance of a machine in a speed range where there is influence from two critical speeds is now considered in Example 4.3.

Example 4.3: A two-plane balance
A rotor is known to require balancing in two planes and vibration measurements are made in the x-direction at the two bearings. The observations of the original conditions and two trial runs are shown in the table.

Table of readings for Example 4.3

	Bearing 1 Amplitude (mm)	Bearing 1 Phase (Deg)	Bearing 2 Amplitude (mm)	Bearing 2 Phase (Deg)
Original	0.4	120	0.3	240
100 g@150 mm 0°, Disc 1	0.5	140	0.35	150
100 g@150 mm 0°, Disc 2	0.45	50	0.4	300

Following the steps from Equations 4.30–4.35, the parameters are as follows.

(Note that here all measurements are kept in millimeters: any consistent set of units is acceptable.)

The as found vibration is

$$\mathbf{u_0} = \left\{ \begin{matrix} u_{01} \\ u_{02} \end{matrix} \right\} = \left\{ \begin{matrix} 0.4 \times e^{j \times 120 \times \pi/180} \\ 0.3 e^{j \times 240 \times \pi/180} \end{matrix} \right\} = \left\{ \begin{matrix} -0.2 + 0.3464j \\ -0.15 - 0.2598j \end{matrix} \right\} \text{mm}$$

On the first trial run

$$\mathbf{f_1} = \left\{ \begin{matrix} m_{11}R_{11} \\ m_{12}R_{12} \end{matrix} \right\} = \left\{ \begin{matrix} 0.1 \times 0.15 \\ 0 \end{matrix} \right\} \text{kgm}$$

and the trial response is

$$\mathbf{u_1} = \left\{ \begin{matrix} u_{11} \\ u_{12} \end{matrix} \right\} = \left\{ \begin{matrix} 0.5 \times e^{j \times 140 \times \pi/180} \\ 0.35 \times e^{j \times 150 \times \pi/180} \end{matrix} \right\} = \left\{ \begin{matrix} -0.383 + 0.321j \\ -0.303 + 0.175j \end{matrix} \right\} \text{mm}$$

Hence,

$$\Delta\mathbf{u_1} = \mathbf{u_1} - \mathbf{u_0} = \left\{ \begin{matrix} -0.183 - 0.025j \\ -0.153 + 0.4348j \end{matrix} \right\} \text{mm}$$

For the second test

$$\mathbf{f_2} = \left\{ \begin{matrix} 0 \\ 0.1 \times 0.15 \end{matrix} \right\} \text{kg m}$$

and the response is

$$\mathbf{u_2} = \left\{ \begin{matrix} u_{21} \\ u_{22} \end{matrix} \right\} = \left\{ \begin{matrix} 0.45 \times e^{j \times 50 \times \pi/180} \\ 0.4 \times e^{j \times 300 \times \pi/180} \end{matrix} \right\} = \left\{ \begin{matrix} 0.289 + 0.345j \\ 0.2 - 0.346j \end{matrix} \right\} \text{mm}$$

The difference is given by

$$\Delta\mathbf{u_2} = \mathbf{u_2} - \mathbf{u_0} = \left\{ \begin{matrix} 0.489 - 0.0017j \\ 0.350 - 0.087j \end{matrix} \right\} \text{mm}$$

Then from Equation 4.33, the imbalance is calculated as

$$\mathbf{f_0} = \begin{bmatrix} 0.015 & 0 \\ 0 & 0.015 \end{bmatrix} \begin{bmatrix} -0.183 - 0.025j & 0.489 - 0.002j \\ -0.153 + 0.435j & 0.350 - 0.087j \end{bmatrix}^{-1} \left\{ \begin{matrix} -0.2 + 0.346j \\ -0.150 - 0.250j \end{matrix} \right\} \text{kgm}$$

This simplifies to give

$$f_0 = \left\{ \begin{array}{l} -0.0192 + 0.0032j \\ -0.0135 + 0.0108j \end{array} \right\}$$

Hence,

$$|f_0| = \left\{ \begin{array}{l} 0.0195 \\ 0.0173 \end{array} \right\} \text{kgm}$$

The phases are given by

$$\theta_1 = \tan^{-1}\left(-\frac{0.0032}{0.0192}\right)$$

$$\theta_2 = \tan^{-1}\left(-\frac{0.0108}{0.0135}\right)$$

So, the corresponding phases are 171° on plane 1 and 141° on plane 2.

There are several interesting observations to be made at this point. In applying this two-plane approach, inherently this means that two degrees of freedom have been adjusted. In the case of a rigid rotor, there are only two modes (in each direction). This implies that for a rigid rotor, if it is two-plane balanced, it can be run at any speed. This is not true for a flexible rotor and hence it is important to understand this distinction and its relevance to specific plant items.

The issues involved are now illustrated by reference to the rigid rotor shown in Figure 4.8. The parameters are shown in Table 4.2, and a rundown plot is shown in Figure 4.9.

The damping and forcing terms used in the first edition have been changed to better illustrate important points. The Campbell diagram for this system is shown in Figure 4.10 which illustrates that there is a minor gyroscopic influence on this machine and the rundown plot shows peaks at 2,200 and 6,050 rpm.

Some interesting features can be seen immediately from this figure. Most importantly, the response at the two bearings seems to be almost identical and low speed and only begin to diverge above the first critical speed. This should come as no surprise as the dynamic behavior in the lower speed range will be dominated by the first mode which is symmetric with respect to the two bearings. Suppose that this machine is to be run around 2,000 rpm. This is below the first critical speed and it may be anticipated that single-plane balancing would suffice. Figure 4.11 shows the behavior of this rotor after single-plane balancing at 2,340 rpm. The balancing was

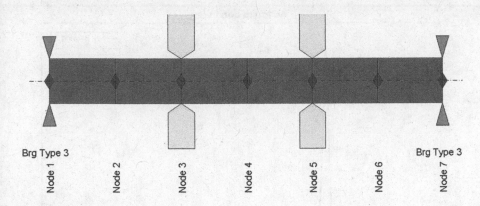

Brg Type 3 ... Node 1 ... Node 2 ... Node 3 ... Node 4 ... Node 5 ... Node 6 ... Node 7 ... Brg Type 3

FIGURE 4.8
A Rigid Rotor Example.

TABLE 4.2

Parameters of the Model

Distance between bearings (m)	1.5
Shaft diameter (m)	0.1
Bearing stiffness (N/m)	10^7
Bearing Damping (Ns/m)	4000
Disc thickness (m)	0.1
Disc diameter (m)	0.3
Material density (kg/m^3)	7800
Young's modulus (N/m^2)	2.1×10^{11}

achieved using a single trial mass and a single correction mass. It is clear that although the vibration at bearing 1 has been reduced to zero at the balancing speed of 2,340 rpm, at higher speed the amplitude has increased, but note that this takes no account of changes at bearing 2, or indeed, anywhere else on the machine.

Above the first critical speed, the vibration at a single speed and a single location can still be reduced by single-plane balancing, but the vibration elsewhere on the machine may increase. Figure 4.11 shows that at the point of measurement, the vibration can be reduced at a specific speed by single-plane balancing, but the effects at all other speeds are unpredictable without detailed knowledge of the machine's dynamic behavior. In fact, a rather more satisfactory result may be obtained even with this simple single degree

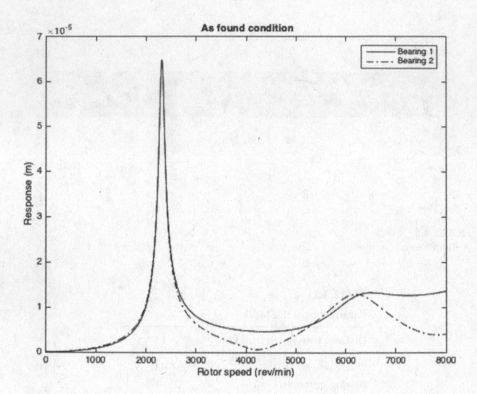

FIGURE 4.9
"As Found" Response.

of freedom approach, simply by recognizing the modal behavior of the system.

In Figure 4.11, at speeds above around 3,000 rev/min the vibration becomes dominated by the second mode, which in this case we know to be a "rocking" motion. In Figure 4.12, the calculated correction mass has been divided equally between the two discs, hence negating any chance in excitation of the second mode. The difference in behavior is striking.

It is noticeable however, that Figure 4.12 has not shown a reduction of the vibration level to zero at the balancing speed. This is because the calculations were based on a single trial mass while the correction involved two added masses: in other words, the modal content was different. The procedure may be refined by the use of a trial mass at the same position on each of the two discs and the result of this procedure is illustrated in Figure 4.13.

This approach is very similar to modal balancing which is discussed in connection with flexible rotors. In the current context, dealing with rigid rotors, two-plane balancing is, in general, the most usual method of

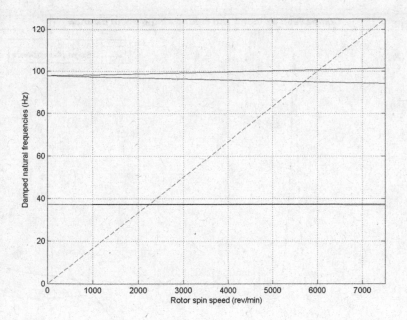

FIGURE 4.10
Campbell Diagram.

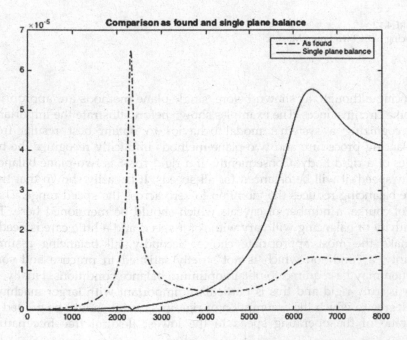

FIGURE 4.11
Before and after Balancing with a Single Mass.

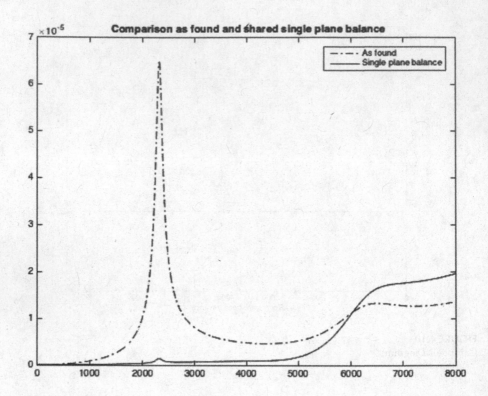

FIGURE 4.12
Balancing with Two Equal Masses.

balancing although, as shown before, single-plane methods are appropriate in some circumstances. The examples shown before illustrate the importance of recognizing a system's modal behavior to obtain best results from a balancing procedure and two-plane methods implicitly recognize the two modes of a rigid body. Consequently, if a rigid rotor is two-plane balanced at any speed, it will be balanced for all speeds. It is easily shown that two-plane balancing reduces the vibration to zero across the speed range. There are, of course, a number of caveats which should be mentioned here. The sensitivity of balancing will vary with rotor speed and a little care is needed to make the most appropriate choice. Secondly, all balancing assumes linearity, a condition which is not strictly satisfied in practice and some iteration may be required to obtain optimum balance conditions. Finally, no rotor is truly rigid and this is particularly important with larger machines: the degree to which the assumption of rigidity is valid should be guided by the ratio of the operating speed to the lowest flexural free–free natural frequency.

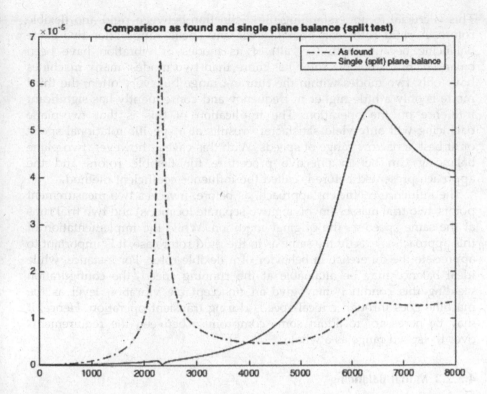

FIGURE 4.13
An Example of Balancing at a Single Speed.

Note that in Figure 4.13, the vibration is reduced to zero at a single speed that at which the single-plane correction was calculated and applied. Note that the result in this case is almost identical to the straightforward single-plane approach. This perhaps needs a little clarification: trial masses were added equally to the two mass and the corresponding correction masses were calculated using the single-plane procedure. This relies on a single measurement point normally, but not necessarily, a bearing pedestal.

The two-plane procedure utilizes measurements from two distinct locations and is consequently able to take due recognition of the modal contributions to the vibration signal. Since a rigid rotor has only two modes of vibration (in each direction), the full complexity of the motion can be dissected and this is why a rigid rotor balanced in two planes is balanced for all speeds.

So far, the discussion has focused on rigid rotors, yet much of it is directly applicable to flexible rotors also. It has been observed that a rigid rotor can be balanced for all speeds by two-plane balancing, but the single-plane approach can at best be effective only over a restricted speed range.

This is crucial to understanding the distinction between rigid and flexible rotors. A rigid rotor can operate over any speed range after two-plane balancing because both (i.e. all) of its modes of vibration have been balanced. But a flexible rotor has more than two modes: many machines have only two modes within the running range but, very often, the third mode is only a little higher in frequency and consequently has significant influence on the operation. The implication of this is that two-plane balancing will only yield satisfactory results at a specific rotational speed or at best, a narrow range of speeds. With this caveat, however, two-plane balancing can be an effective procedure for flexible rotors and the approach presented before is called the influence coefficient method.

The influence coefficient approach, as before, involves two measurement points, two trial masses (involving two separate locations) and two trial runs at the same speed as the original condition. While the implementation of this approach is exactly the same as in the rigid rotor case, it is important to appreciate the difference in behavior of a flexible rotor. For instance, while ideal balance may be attainable at the running speed, the configuration yielding this condition may give an unacceptable vibration level as the machine goes through critical speeds during transient operation. Hence, it may be necessary to adopt some compromise between the requirements over the speed range as a whole.

4.3.2.3 Modal Balancing

A rather different method of balancing of a flexible rotor is the modal approach in which balancing is undertaken at several speeds, normally as close as possible to the critical speeds of the system. In principle, the balancing speeds chosen should be the critical speeds but sometimes this is not possible owing to unacceptably high vibration levels. A single measurement point is required but the number of balance planes used equals the number of modes to be balanced, normally two or three. Unlike the influence coefficient method, however, modal balancing requires prior knowledge of the machine in the form of the mode shapes and corresponding critical speeds. This knowledge may be derived from a theoretical model or previous operational data.

The procedure amounts to a multiple application of the single-plane approach, except that at each speed, trial masses must be added to each of the balance planes in proportion to the mode shape corresponding to the speed being studied. The premise of the approach is that each mode is independent of all others (a point discussed next) and hence the response at one critical speed is dominated by the corresponding mode and consequently each mode can be balanced independently. This is now illustrated with a Example 4.4.

Example 4.4: Modal Balancing

It is known that critical speeds occur at 1,200 and 2,000 rpm, respectively, the two balance planes are on two discs which are symmetric about the center line as shown in Figure 4.8 but in this case, the rotor is flexible. The two mode shapes take the form $\begin{Bmatrix} 1 \\ 1 \end{Bmatrix}$ and $\begin{Bmatrix} 1 \\ -1 \end{Bmatrix}$. After a trial run at 1,200 rpm (or as close to it as the machine can safely run), the calculated correction is 40 g at 30° ahead of the phase reference and at a radius of 100 mm. Equal correction masses are then attached to the machine and the speed is then increased to 2,000 rpm, and after noting the initial vibration reading, equal trial masses are added in antiphase to the two balance planes (at the same radius of 100 mm). From this, the required correction is computed to be 20 g at 45° after the phase reference. Table 4.3 shows the added mass and positions on each of the two balance planes. The four masses will balance the imbalance excitation of the first two modes and in many circumstances this will yield satisfactory operation. Only rarely do machines in current service require balancing for higher modes and in such cases more balance planes are required. Of course, on each of the two planes in this example, the two masses may be replaced by a single one representing their (vector) sum, but in many cases, this may not be convenient as it would require an additional rundown of the machine.

Modal balancing is more satisfactory for machines which operate over a range of rotor speeds. The concepts were discussed by Parkinson *et al.* (1980) and Parkinson (1991) following the pioneering work of Bishop and Gladwell (1959), yet the influence coefficient approach is still the more common method in industry. Both approaches rely on the concept of linearity and both methods can be easily applied in a routine approach, yet there are some pitfalls of which plant operators must be aware. One of the most common reasons for inconsistencies in balancing procedures is the presence of a bend. The behavior of bent rotors gives rise to a synchronous vibration, but it does not behave in the same way as

TABLE 4.3

Modal Balancing Example

	Plane 1		Plane 2	
	Mag	Phase	Mag	Phase
Mode 1	40	30	40	30
Mode 2	20	−45	20	135
Resultant	49	7	40	59

imbalance because the variation of force with rotor speed is quite different and the implications of this are quite profound. It is discussed in detail in Section 4.4. In essence, the forcing due to a bend is constant as opposed to the dependence on speed squared for imbalance.

The problem of nonlinearity has already been discussed and can be effectively overcome by balancing in an iterative manner, but often this is not necessary. A rather more profound problem is that the bearings provide an overwhelming proportion of the system damping which means that the damping is not "classical" which in turn means that the mode shapes are complex and that they are not orthogonal. The important point here is that the lack of orthogonality implies that some of the fundamental requirements of modal balancing are not strictly satisfied. An enhanced procedure, called bimodal balancing, has been suggested by Kellenburger and Rihak (1988), but to the best of the author's knowledge this has yet to be applied in an industrial setting. In fact, despite the theoretical doubts, the evidence suggests that acceptable modal balancing can be obtained on machine mounted on oil journal bearings.

It is here that the importance of the distinction between rigid and flexible rotors becomes clear since a rigid rotor, balanced in two planes, will be balanced for all speeds. This is clear in modal terms since all the modes (i.e. two rigid body modes) have been balanced. With a flexible rotor however, this will not be the case as other modes assume importance with increasing rotor speed. Of course, there is no such thing as a truly rigid rotor, but the criterion is based on the closeness of the first flexural free–free natural frequency to the running range. The state of balance of a flexible rotor must be considered at its speed of operation.

4.3.3 A Comparison of Approaches

At this point it is appropriate to illustrate the difference between some common approaches to balancing. In Figures 4.11–4.13, the results are shown of three ways in which a single-plane balance procedure has been applied. It is clear that these simple procedures can be appropriate in some circumstances. However, for higher speeds, some form of multiplane balancing is usually necessary and in this subsection a comparison is given of the effects of two-plane and modal balancing. The same sample rotor is used for these studies.

Figure 4.9 shows the response on both bearings in the "as found" condition. The system was then (two plane) balanced at two different speeds 2,340 and 6,200 rev/min, respectively, and both results (for bearing 1) are shown in Figure 4.14. If the rotor was rigid, a two-plane balance would remain valid at all speeds, but this is not so in this case. It should be noted that in each case, the vibration becomes zero at the balancing speed but becomes quite significant at other speeds.

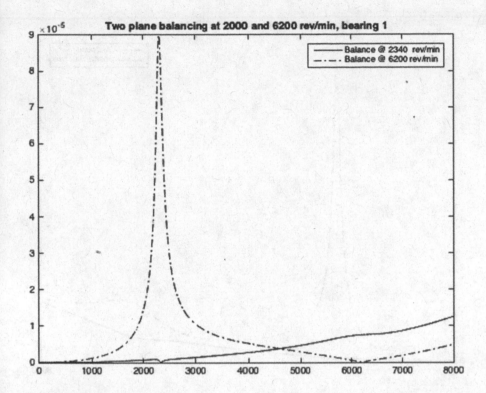

FIGURE 4.14
Typical Effects of Two-Plane Balancing.

The alternative procedure is modal balancing and the result of applying this approach is shown in Figure 4.15. It is clear that with modal balancing, while there is no speed at which the vibration is zero, there is a lower response over a range of speeds. Therefore, it is generally better to use modal balancing for a machine which operates at a range of speeds, but conventional two-plane balancing may be superior for a fixed speed rotor.

4.3.4 Nonlinear Effects

On many machines, in particular those mounted on oil film journal bearings, there are some nonlinear effects. To explore this, the system described in Table 4.2 is considered with the imbalance increase in a range of factors. A series of calculations were carried using a range of imbalance levels and Table 4.4 gives the imbalance response at three different speeds when the reference imbalance is multiplied by 1, 2, 4, 8, and 16, respectively. The calculations have been carried out using short bearing theory: while this

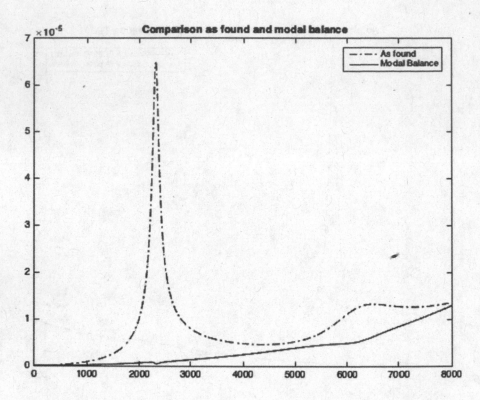

FIGURE 4.15
Balancing Using Two Modes.

has a number of assumptions and is not precise, it is representative of many of the phenomena observed in oil film bearings. The feature of note here is the inherent nonlinearity. At 900 rev/min, the nonlinear behavior only becomes apparent as the imbalance increases. More precisely, we note that the nonlinear behavior becomes important when the displacement at the bearing is commensurate with the clearance which would seem to be physically reasonable. Note from the table that at higher speeds this characteristic appears at lower values of imbalance.

At this point consider the effect this has on the balancing procedure: it is immediately clear that the response in all cases shown in Table 4.4 increases less rapidly than the applied imbalance. This behavior may be characterized by the bearings which effectively "stiffen" with increasing amplitude. Of course, this has direct relevance to balancing since by the application of trial masses, the sensitivity of the system is established prior to the application of the appropriate correction masses. A consequence of this apparent stiffening of the bearings is that the first estimate of the required correction will be, in general, subject to error but it will yield

TABLE 4.4

Amplitude Range Relative to Clearance

Imbalance Scale	900 rev/min	1,800 rev/min	2,700 rev/min
1	0.054	0.155	0.51
2	0.106	0.299	0.826
4	0.209	0.539	1.15
8	0.394	0.84	1.40
16	0.672	1.12	1.58

reduced vibration levels enabling a second balancing to be undertaken. This may not be necessary in all cases but is dependent on the extent of the "as found" amplitudes and the required tolerances. Where two-stage balancing is undertaken it yields valuable practical data on the nonlinear bearing behavior.

4.3.5 Recent Developments

Over recent years a number of authors have proposed methods for balancing in a single step (i.e. without any "trial runs"). In an industrial setting, such a technique could prove very valuable. For a particular machine, given extensive and detailed records, an engineer may well have good insight into the particular machine's sensitivity and such prior information will help reduce the need for test runs.

Novel approaches are based on system identification techniques which are discussed at some length in Chapter 7. The estimation of imbalance in this way is a particular case of the fitting of a number of parameters of a numerical model to measured data. In this case, a vector of imbalances at a predetermined set of balance plane is included as parameters to be identified. The approach is given by Lees and Friswell (1997). The technique is still relatively new, but laboratory trials and limited site application have given very promising results.

4.4 Rotor Bends

Section 4.3 gives an account of mass imbalance and this is undoubtedly the most prevalent "fault" condition in rotating machinery. It is rather doubtful whether imbalance is properly termed as a machine fault as all machines have it to some degree. What is a machine fault is an excessive level of imbalance when judged against some suitable criterion. As discussed in

Section 4.3, imbalance always gives rise to synchronous vibration but the converse is not true, that is, a once per revolution signal is not necessarily the result of imbalance. It may be that the shaft has a bend. A bend, like imbalance gives rise to synchronous vibration, but there are some significant differences in the characteristics of an unbalanced rotor and one which is bent: of course, a real rotor may be unbalanced and bent at the same time, but it is convenient to separate the two effects to illustrate the principles.

First consider the rotor shown in Figure 4.16.

The figure depicts a simple rotor with a bend near the centre span. It is important, however, to clarify what is meant by the term bend. What is being discussed here is quite different from the catenary of a flexible rotor which is discussed in Chapter 5 in the study of misalignment. In the case of a bend, the distorted shape of the rotor rotates about the center line of the shaft, whereas the catenary is fixed in space and time, at least for the case of an axisymmetric rotor.

The equation of motion of a rotor with a bend can be derived by means of Lagrange's equation. If the bend vector is $\mathbf{b}$, then the potential energy of the rotor, with position vector $\{\mathbf{x}\}$ is given by

$$U = \frac{1}{2}\{\mathbf{x} - \mathbf{b}\}^{T}[\mathbf{K}][\mathbf{x} - \mathbf{b}] \qquad (4.36)$$

The physical validity of this expression may be verified by observing that, in some coordinate system, a straight shaft could be arranged so that $\mathbf{x} = \mathbf{0}$ but if the rotor has a bend, that would not be a configuration with zero potential energy. This zero energy state can only be achieved when $\{\mathbf{x} - \mathbf{b}\} = \mathbf{0}$. The bend $\mathbf{b}$ is just the shape the rotor would take up when resting on a frictionless table. It should be added that in many cases such an imaginary table would be very large!

The kinetic energy is straightforward and is given by

$$T = \frac{1}{2}\{\dot{\mathbf{x}}\}^{T}[\mathbf{M}]\{\dot{\mathbf{x}}\} \qquad (4.37)$$

A straightforward application of Lagrange's equation yields

FIGURE 4.16
A Bent Rotor.

$$[\mathbf{M}]\{\ddot{\mathbf{x}}\} + [\mathbf{K}]\{\mathbf{x} - \mathbf{b}\} = 0 \tag{4.38}$$

as the equation of motion for a bent shaft with no damping and no unbalance. It is instructive to examine the behavior of this simplified case before proceeding to more complicated situations. Considering a rotor now with a steady angular speed of Ω, the vibration level (with the inclusion of a damping matrix $\mathbf{C}$) may be obtained directly giving

$$\mathbf{x} = \left[-\mathbf{M}\Omega^2 + j\Omega\mathbf{C} + \mathbf{K}\right]^{-1} \mathbf{Kb} e^{j\Omega t} \tag{4.39}$$

Hence, it is clear that a bend in a rotor gives rise to a once per revolution excitation, but the amplitude variation of this with rotor speed is fundamentally different from that caused by rotor unbalance. This is clear from the absence of a term proportional to Ω^2 in the numerator of Equation 4.39 (as compared to Equation 4.14). To illustrate this point, consider a rotor with an unbalance distribution Me, where $\mathbf{e}$ is the mass eccentricity, vibration level is

$$\mathbf{x}_{bal.} = \left[-\mathbf{M}\Omega^2 + j\Omega\mathbf{C} + \mathbf{K}\right]^{-1} \mathbf{Me}\Omega^2 e^{j\Omega t} \tag{4.40}$$

It is obvious that we can express this in terms of Fourier components such that

$$\mathbf{x}_{bal}(\Omega) = \left[-\mathbf{M}\Omega^2 + j\Omega\mathbf{C} + \mathbf{K}\right]^{-1} \mathbf{Me}\Omega^2 \tag{4.41}$$

For the sake of simplicity, we assume that the damping is small and consequently the peak response occurs at $\Omega = \Omega_c$ given by the condition $\left|\mathbf{K} - \Omega_c^2\mathbf{M}\right| = 0$. At the critical speed, the response due to bend and unbalance will be equal if

$$\mathbf{Kb} = \mathbf{Me}\Omega_c^2 \tag{4.42}$$

Assuming this to be the case, let us now compare the displacements due to unbalance and bends at zero speed and high (infinite) speed. Directly from Equations 4.41 and 4.42 it is observed that

$$X_{bal}(0) = 0 \qquad \lim_{\Omega \to \infty}(X_{bal}(\Omega)) = -\varepsilon$$

$$X_{bend}(0) = b \qquad \lim_{\Omega \to \infty}(X_{bend}(\Omega)) = 0 \tag{4.43}$$

Figure 4.17 shows a rotor model with unbalance distributed on the two discs. The response of this system with a unit unbalance on the disk is shown together in Figure 4.18 with a simple bend having the same

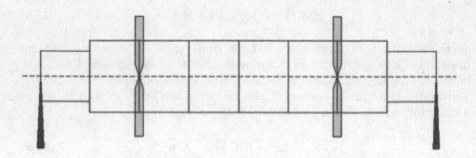

FIGURE 4.17
Basic Layout of Test Model.

response at resonance. The comparison of the two curves illustrates the very important point that while bends and unbalance both give rise to synchronous signals (once per revolution), the variation as rotor speed varies is quite different. A consequence of this is that, although the effects of a bend can be compensated by balancing, this can only be done at a single rotor speed. In Figure 4.18, a bend has been introduced, the force of which exactly matches that from the unbalance distribution. Notice that the unbalance and bend components match at this single frequency (21 Hz), but nowhere else. Note that below 21 Hz the unbalance response is below that of the bend while the opposite is the case at higher frequencies. While this argument has been explained implicitly using a single degree of freedom system, it is equally valid for a more complex system. In this subsection, the only assumption (apart from linearity) is classical damping so that modes are real. It is now assumed that both unbalance and a bend are present

$$[\mathbf{M}]\{\ddot{\mathbf{x}} + \ddot{\mathbf{e}}\} + [\mathbf{K}]\{\mathbf{x} - \mathbf{b}\} = 0 \tag{4.44}$$

so that at constant speed Ω, the equation of motion becomes

$$[-M\Omega^2 + \mathbf{K}]\{\mathbf{x}\} = M\Omega^2\mathbf{e} + \mathbf{Kb} \tag{4.45}$$

This equation can be either solved directly or in modal terms.

As is clear from Equation 4.45, a bent rotor can only be "balanced out" at a single speed.

In fact, a bend in a rotor can be easily diagnosed by the presence of significant vibration at very low speed (i.e. far below the first natural frequency). This means that a bend can only be compensated by a balance mass for a single speed and consequently, provided a machine is to operate only at one speed, then it may be balanced

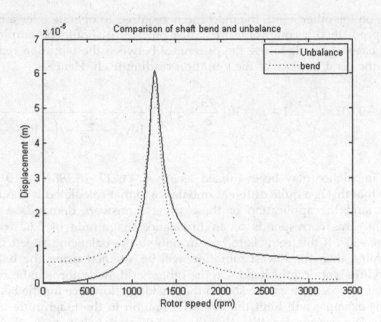

FIGURE 4.18
Imbalance and Bend Responses.

without regard to the bend. In such cases, however, due care must be taken to ensure that vibration levels are acceptable during run-up and rundown.

An alternative procedure is available for machines which are required to operate over a range of rotor speeds. In this case, the bend must be first "measured" by observing the response at low speed. These levels are then subtracted (vectorially) from the measure response in balancing with correction masses calculated on this basis. This approach is illustrated in Example 4.5.

Example 4.5: Balancing with a bend

To examine the effect of a possible bend, let us return to Example 4.2, but with the added provision that at very low speed, the response was measured as 10 μm, 30° ahead of the phase reference. The question now is how to best balance the rotor. In the earlier example, the required correction balance was calculated as 9.03×10^{-4} kg m at 134°. If the rotor is to be run only at 1,500 rev/min, then the presence of the bend may be ignored but at any other speed the rotor will not be correctly balanced. However, there is a danger in this approach as the machine may go through very high vibration during rundown.

If, on the other hand, the machine is required to operate over a range of speed, then a compromise balance procedure should be employed. The basis of this is to use the difference between the vibration reading and the bend in place of the vibration reading itself. Hence,

$$me = -0.001 * \left(\frac{25}{\sqrt{2}}(1+j) - 10\left(\frac{\sqrt{3}}{2} + \frac{j}{2}\right) \right) * \frac{1}{10\sqrt{3} - \frac{25}{\sqrt{2}} + j\left(-10 - \frac{25}{\sqrt{2}}\right)}$$

This imbalance may be evaluated as $me = (4.6217 - 3.1984j) \times 10^{-4}$ kg. Note that this is a quite different imbalance to that calculated without the bend, and the application of these diverse answers demands a little insight. This corresponds to an imbalance magnitude of 5.62×10^{-4} kg m@$-35°$. If this rotor is to be run only at the balancing speed (1,500 rev/min), then the lowest vibration will be attained using the balance procedure ignoring the bend. On the other hand, the rotor is to be run at a range of speeds, it can be shown that making allowance for the bend as in this example will limit the overall vibration to the magnitude of the bend. The reason for this slightly complicated behavior is that the response of bends and unbalance is rather different even though they both give rise to synchronous (once per revolution) vibration.

The underlying concept of this approach is clear: by subtracting the bend, the operator is isolating that part of the vibration which is caused by imbalance and then proceeding by basing calculations on that proportion of the excitation. The nature of oil journal bearings also leads to some problems with regards to imbalance response and the balancing process.

Before leaving this topic, consider a system in which the second mode is within the running range. Figure 4.19 shows a four-bearing model used for this study.

In this case, the model has been arranged to have equal response to unbalance and the bend at the first resonance (in this case around 10 Hz), at the second peak, the response of the bend is negligible. Of course, one could have a bend to give a response at the second peak – but one cannot have both. Figure 4.20 shows the effects of different balancing strategies. Taken over the full speed range, the "corrected" approach appears more satisfactory.

4.5 Concluding Remarks

This chapter has discussed the topics of imbalance and rotor bends. The two are closely associated together as they both give rise to synchronous

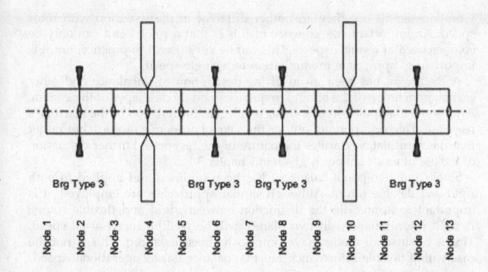

FIGURE 4.19
A Four-Bearing Model.

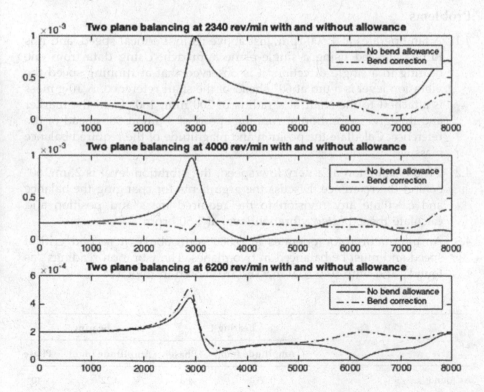

FIGURE 4.20
Different Balancing Options with a Bend.

vibration signals but they are rather different in the variation with rotor speed. An important consequence of this is that a rotor bend can only be compensated at a single speed. This can be very useful in practice, but it is important to appreciate the limitation to a single speed.

A discussion has been given of the description of imbalance and other excitation in terms of the natural frequencies and mode shapes of the system. After some manipulation this yields an elegant formulation of the machine's response. This is important, not for the sake of academic niceties, but rather that this formulation clarifies the nature of the response. Further discussion of the use of mode shapes is given in Chapter 3.

Single and two-plane balancing has been outlined and applied to both rigid and flexible rotors. Although similar approaches are employed, it is important to appreciate the distinction between rigid and flexible rotors. A rigid rotor, balanced in two planes, will be well balanced at all speed. This is because all modes have (implicitly) been balanced. This is not the case with a flexible rotor which must be balanced at its operational speed.

Problems

4.1 A fan operates at 1,500 rpm, just above its first critical speed, and it is to be balanced using a single-plane approach. Using data from one bearing in a single direction, it is observed that at running speed the vibration level is 5 μm at 60° ahead of the shaft reference. A 10 g mass is attached to the shaft at a radius of 200 mm, 180° from the marker and this gives vibration at running speed of 3 μm at 30° ahead of the reference. Calculate the position and magnitude of the required balance mass.

4.2 On the same fan at a very low speed, the vibration level is 2 μm, 60° behind the reference. Discuss the arguments for changing the balance and calculate any revision to the required mass and position and calculate the expected vibration level at 1,500 rpm.

4.3 An alternator rotor operates at 3,000 rpm, above it second critical speed and must be balanced in two planes. The vibration readings "as found" and with two trial masses are shown in the table

	Bearing 1		Bearing 2	
	Amplitude (mm)	Phase	Amplitude (mm)	Phase
As found	0.35	80°	0.22	−40°
100 gm@0°, 200 mm radius Disk 1	0.20	45°	0.30	30°
100 gm@0°, 200 mm radius Disk 2	0.30 mm	30°	0.10	0°

Calculate the magnitude and position of the required balance masses.

4.4   The rotor in Question 4.3 is found to have a bend. When rotating at 50 rpm, it is observed to vibrate with amplitudes of 5 and 10 μm, respectively, at the two bearings. Should the balance be modified? If so, what will be the vibration at 3,000 rpm with the new balance configuration?

4.5   A machine operating over a broad speed range is to be balanced modally. Explain why this is the most appropriate approach. It is known that, referred to the two balance planes, the two modes are $\{1\ 1\}^T$ and $\{1\ -1\}^T$. Measurements are made on the cap of bearing 1. The vibration "as found" on bearing 1 is 50 μm@30° ahead of phase reference. Close to the first critical speed, with 50 g masses attached to both discs at the phase reference mark and 100 mm radius, the vibration is 35 μm@ −45°. The mass on the second disc is then moved by 180° (at the same radius) and the machine is then run-up to its second critical speed. The vibration is now measured as 20 μm@60°. Determine the magnitude and position of appropriate balance masses.

References

Bishop, R.E.D. and Gladwell, G.M.L., 1959, The vibration and balancing of an unbalanced flexible rotor, *Journal of Mechanical Engineering Science*, 1, pp. 66–77.

Foiles, W.C., Allaire, P.E. and Gunter, E.J., 1998, Review: Rotor balancing, *Shock and Vibration*, 5, pp. 325–336.

Friswell, M.I., Penny, J.E.T., Garvey, S.D. and Lees, A.W., 2010, *Dynamics of Rotating Machines*, Cambridge University Press, New York.

Kellenburger, W. and Rihak, P., 1988, Bimodal (complex) balancing of large turbo-generator rotors having large or small unbalance, *Institution of Mechanical Engineers Conference on Vibrations in Rotating Machinery*, Edinburgh, paper 292/88.

Lees, A.W. and Friswell, M.I., 1997, The evaluation of rotor imbalance in flexibly mounted machines, *Journal of Sound and Vibration*, 208(5), pp. 671–683.

Parkinson, A.G., 1991, Balancing of rotating machines, *Proceedings of the Institution of Mechanical Engineers, Part C, Journal of Mechanical Engineering Science*, 205, pp. 53–66.

Parkinson, A.G., Darlow, M.S. and Smalley, A.J., 1980, A theoretical introduction to the development of a unified approach to flexible rotor balancing, *Journal of Sound & Vibration*, 68, pp. 489–506.

Reiger, N.F., 1986, *Balancing of rigid and flexible rotors*, The Shock and Vibration Information Center, U.S. Department of Defense, Arvonia, VA.

5

Faults in Machines (2)

5.1 Introduction

Chapter 4 discusses the important issues arising with mass imbalance and the ways in which this can be alleviated by balancing. Rotor bends are also analyzed: this is a fault which is often confused with imbalance, and in some instances, it may be appropriate to correct a bend by balancing, but in general it is important to recognize the essential differences. In this chapter some further "common" faults in rotating machine are discussed. Whilst these are generally less common than imbalance (the most prevalent single fault), they tend to be somewhat more complex. Furthermore, faults interact and it sometimes becomes difficult to identify the root cause with any degree of certainty.

The chapter begins with a discussion of misalignment, in itself a complicated issue that remains an active area for researchers as there is not yet a satisfactory overarching theoretical foundation which supports the variety of observed behavior. After describing the basic features, some recent thoughts on the topic are outlined. Another important fault is cracked rotors. While (thankfully) much less common than other defects outlined, the consequences can be disastrous both in terms of plant and a threat to the life of operators. The topics of torsion and rotor rub are then described. A discussion of the main sources of instability is given.

The last section of the chapter gives a discussion of the various ways in which the faults interact. It is these interactions that sometimes make it difficult to ascribe a prime cause to a machine defect. So many of these effects become interrelated and this is, perhaps, the main reason why automated detection systems have made only limited progress despite numerous attempts.

5.2 Misalignment

Rotor misalignment is second in importance only to mass imbalance in the list of significant faults on rotating machinery. Despite its major industrial significance, it is as yet incompletely understood, and later in this section some current ideas are discussed. The incomplete understanding to date is reflected in the lack of research publications in this area: for mass imbalance several thousand papers are readily available whereas for misalignment one has difficulty in finding more than a few hundred, although there has been an increased study in recent years. It is interesting to reflect on why this should be so and it becomes clear that this field of study is indeed complex and fraught with difficulty.

The simplest case to consider is a rigid shaft mounted on two bearings. Consideration of moments on the system will dictate that appropriate loadings fall on the two bearings, but if they are not correctly aligned the loads will not be taken evenly across the face of the bearing and rapid wear may be the result. Depending on the type of bearing used, substantial deviations in bearing performance can be observed, but the true situation is significantly more complex insofar as the operation of the complete machine may be significantly impaired. Depending on the type of machine in question, there will be fine clearances within it through which working fluid flows and hence any change in the shaft position within these clearances will have a profound effect on machine operation. So even in the simplest case, it is important to ensure that a) the shaft and the bearings are coaxial and b) the shaft passes axially through internal clearances and neck rings. This latter influence is touched upon in Chapter 6. Where machines of this type have to be coupled together (e.g. an electric motor and a pump) this is commonly achieved with a flexible coupling.

With large machines, however, the coupling between rotors is usually rigid. For instance, a large steam turbo-generator may comprise several rotors (e.g. many turbo-generators have seven rotors) each mounted usually on two bearings (although recently there has been a trend toward single bearings on an individual rotor). The seven in a typical large turbo-generator would be high-pressure (HP) turbine, intermediate pressure (IP) turbine, three low-pressure (LP) turbines, the alternator, and the exciter. Large units are typically 50 m in length, with a total rotor mass of about 250 tonnes. This is just an example and the combination of rotors will vary with the design of unit. These rotors will operate at widely different temperatures, and the need to maintain the complete unit in the correct configuration presents an engineering challenge.

Figure 5.1 illustrates the two types of misalignment in which the abutting rotor faces can meet with lateral or angular relative displacement. In reality, most cases will be a combination of these two as shown in the third diagram. It is, however, convenient to view parallel and angular misalignment separately as some distinct effects arise.

5.2.1 Key Phenomena

It is very common for a system to display the geometries illustrated in Figure 5.1. The mismatch can be overcome in three ways

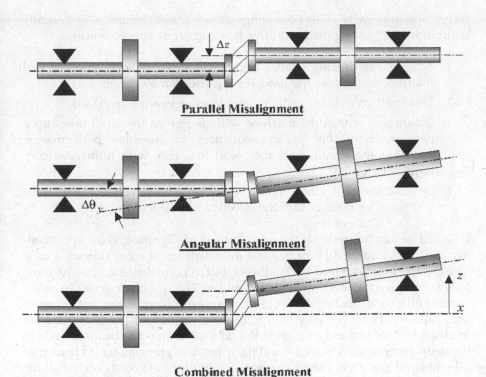

FIGURE 5.1
Types of Misalignment.

(Reproduced with permission from Sinha *et al.*, 2004)

a) The use of a flexible coupling which allows for this mismatch but transmits the torque. These devices come in a range of types but most involve some nonlinear behavior. Flexible couplings are used predominantly on smaller machines but a full discussion of their (interesting) properties is outside the scope of the present text.

b) The location of adjacent bearings is changed in such a way as to substantially reduce the mismatch, both laterally and angularly.

c) Under some circumstances it may be necessary to apply forces and moments to the coupling to force a rigid connection. This should be a last resort as it is always necessary to minimize the forces and moments applied. Some of the possible consequences are discussed below.

The ways of correcting misalignment in flexible rotors are addressed in Section 5.2.3, but the current discussion focuses on the effects observed due to misalignment. The primary cause of difficulty is that the anomalous forces and moments at the couplings cause a redistribution of (usually inappropriate) bearing loads and this has numerous consequences.

1) Because the bearing (static) loads change the bearing stiffness and damping and hence the natural frequencies may change.
2) This may give rise to stability problems or bearing overload.
3) Clearances within the machine will change as the shaft takes up a new position: this has consequences on machine performance. Together these influences may lead to a rub, with numerous consequential issues.
4) Because the shafts now rotate about an axis which is not its true geometric or mass center, the effective imbalance may alter.

It should be emphasized that the crucial aspect of alignment is the apportioning of correct load on all bearings and minimization of rotor stresses. Excessive bearing loads can lead to rapid wear, but in many instances, unduly low loading can lead to even greater difficulties. This problem arises in many types of oil journal bearings, often used in large machines owing to their good load capacity. However, they tend to become unstable if the (static) load is inadequate. Under these conditions the oil flow within the bearing tends to break up and rotate en block around the rotor. The phenomenon is known as oil-whirl and this gives rise to appreciable vibrations at speeds up to half the shaft rotation speed (see e.g. Muszynska, 2005); this is discussed in Section 5.6. The bearing stiffness and damping become effectively zero, and consequently very high vibration levels arise which can be extremely dangerous in terms of machine integrity and safety of operational staff.

There are essentially two ways to resolve this difficulty: the first is to connect the mating shafts with a flexible coupling which is a device that transmits torque whilst having a low flexural stiffness. Flexible couplings have some associated problems but can provide a very effective means of connecting machine components. The basics of this approach are discussed in Section 5.2.3.

The alternative approach is a rigid coupling of the two rotors and this is employed on most large machines. Whilst avoiding some of the problems associated with flexible couplings, it does lead to other difficulties as outlined in Section 5.2.3.

5.2.2 Flexible Couplings

Machines may be coupled using a variety of mechanisms but we will illustrate some of the modeling problems by reference to a single membrane coupling, a type which is in common use. The principle is illustrated in Figure 5.2. Each of the two shafts is connected to different parts of a flexible membrane which becomes the only component through which the rotors are connected. Hence torque is transmitted, but bending moments are

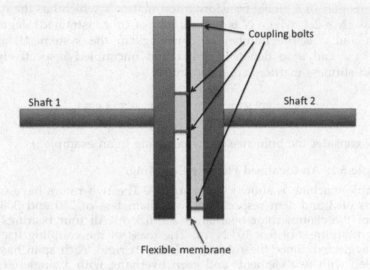

FIGURE 5.2
A Membrane Flexible Coupling.

not because the membrane can bend. This amounts effectively to the insertion of a hinge in the rotor. In practice, pairs of couplings are often used, separated by a spacer shaft to obtain better vibration isolation. Here, however, the modeling difficulties are discussed by reference to a single coupling.

Let us now consider the model of a turbine and alternator discussed earlier, but with a single flexible coupling between them. Note that although flexible couplings have been used predominantly on smaller machines, their application is now widening to encompass large plant items. We begin by modeling the two rotors separately; the lowest natural frequencies of both rotors are shown in Table 5.1. The nodal coordinates at the end of rotor 1 are $\{u_{1n} \quad v_{1n} \quad \theta_{1n} \quad \psi_{1n}\}^T$ while those for the connected end of rotor 2 are given by $\{u_{21} \quad v_{21} \quad \theta_{21} \quad \psi_{21}\}^T$. The coupling is modeled by applying the constraints $u_{1n} = u_{21}$, $v_{1n} = v_{21}$. In this way the two adjoining nodes which before connection had eight associated degrees of freedom now have six. This is achieved by means of a transformation matrix

$$\begin{Bmatrix} u_{1n} \\ v_{1n} \\ \theta_{1n} \\ \psi_{1n} \\ u_{21} \\ v_{21} \\ \theta_{21} \\ \psi_{21} \end{Bmatrix} = \begin{bmatrix} 1 & 0 & 0 & 0 & 0 & 0 \\ 0 & 1 & 0 & 0 & 0 & 0 \\ 0 & 0 & 1 & 0 & 0 & 0 \\ 0 & 0 & 0 & 1 & 0 & 0 \\ 1 & 0 & 0 & 0 & 0 & 0 \\ 0 & 1 & 0 & 0 & 0 & 0 \\ 0 & 0 & 0 & 0 & 1 & 0 \\ 0 & 0 & 0 & 0 & 0 & 1 \end{bmatrix} \begin{Bmatrix} u \\ v \\ \theta_{1n} \\ \psi_{1n} \\ \theta_{21} \\ \psi_{21} \end{Bmatrix} \qquad (5.1)$$

This forms part of a global transformation matrix $\mathbf{T}$ which has the dimensions $N \times (N - 2c)$, where N is the number of (unconstrained) degrees of freedom and c is the number of couplings in the system. Using the subscripts c and u to denote coupled and uncoupled respectively, the mass and stiffness matrices are given by

$$\mathbf{K}_c = \mathbf{T}^T \mathbf{K}_u \mathbf{T} \qquad\qquad \mathbf{M}_c = \mathbf{T}^T \mathbf{M}_u \mathbf{T} \qquad\qquad (5.2)$$

We now consider the influence of the coupling in an example.

Example 5.1: An Idealized Flexible Coupling.

A simple machine is shown in Figure 5.3. The two rotors have overall lengths of 3 and 4 m respectively and diameters of 250 and 300 mm. Both of the central rotor overhangs are 0.5 m. All four bearings have constant stiffness of 5×10^6 N/m. The mass of the coupling itself has been neglected. Since the rotors are (almost) rigid, each span has been modeled with two elements and each overhang with a single element. The natural frequencies of the system have been calculated under three conditions, namely uncoupled, coupling of only displacement, and finally full coupling (of displacement and slope). The results for the first two modes in each case are shown in Table 5.1.

Note that the effects of the coupling are different on the two modes depending on the particular mode shape.

Couplings play a crucial role in the dynamic behavior of rotating machinery and the brief description given here is just an overview of methods used to adequately model the situations. In reality, couplings have mass and inertia properties which must also be modeled: furthermore, because the masses are

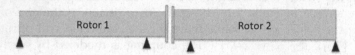

FIGURE 5.3
Sample System.

TABLE 5.1

Natural Frequencies (Hz) under Different Coupling

	Uncoupled			
	Rotor 1	Rotor 2	Hinge Coupling	Solid Coupling
Mode 1 (Hz)	13.47	10.06	10.86	13.47
Mode 2 (Hz)	21.62	16.31	14.83	15.10

sometimes considerable and they load the shaft some distance from the bearings, appreciable effects on dynamic behavior can result.

There is a variety of flexible coupling types and it is not the aim here to give a comprehensive survey, but rather a representative sample of their operation together with some indication of some of the practical difficulties which arise. The example of the Hooke's joint is not only important in itself but also provides an excellent introduction to important concepts of flexible couplings. In essence, the requirement of a coupling is to connect some, but not all, degrees of freedom: usually the requirement is to transmit torque but not lateral forces or bending moments. In its simplest form, the Hooke's joint, or Carden shaft, connects the two rotors simply by two pivots, one on each of the two rotors, set at right angles. Basic kinematics shows that if the angular velocity of the first rotor is held constant at Ω_1, and β is the angular mismatch between the rotors, then the velocity of the second rotor becomes

$$\Omega_2 = \frac{\Omega_1 \cos\beta}{1 - \sin^2\beta \cos^2\Omega_1 t} \tag{5.3}$$

And the resulting angular acceleration is given by the derivative of this expression, namely

$$\dot{\Omega}_2 = \frac{\Omega_1^2 \cos\beta \sin^2\beta \sin 2\Omega_1 t}{\left(1 - \sin^2\beta \cos^2\Omega_1 t\right)^2} \tag{5.4}$$

Taking as an example an input speed of 50 Hz (3,000 rev/min) and an angular mismatch of 5°, the variation in output speed is shown in Figure 5.4.

However, it is clear that the speed variations are modest, amounting in this case to about $\pm\,^1/_2\%$. The angular accelerations are appreciable and this will impose high stresses on components such as coupling bolts and membranes. This figure of course is for a coupling in perfect condition. A defective bolt or failed membrane will change the situation completely.

There are various types of coupling in operation but they fall into three main groups:

a) Membrane couplings
b) Gear couplings
c) Rubber block coupling

There is also a fourth type, based on a fluid clutch principle, but these tend to be more specialized applications. Each type of coupling has its own virtues and faults. The gear type, for instance, has the benefit of being lighter than other types but it is essential to maintain a good lubrication system. Balancing may also present some difficulties.

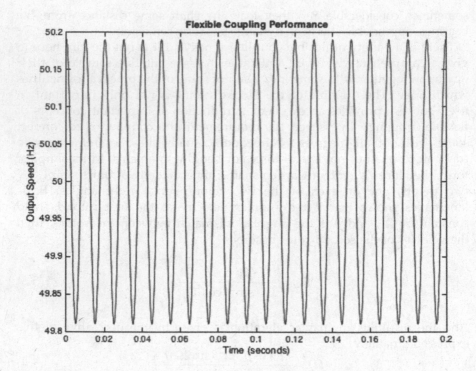

FIGURE 5.4
Speed Variations.

5.2.3 Solid Couplings

In large, high-power machines, such as turbo-generators, individual rotors are often rigidly connected. This is usually achieved by bolting flanges. To some extent, this is a simple situation to model since the assembly of rotor can, in this case, be modeled as a single rotor with, of course, appropriate provision for the mass and inertia of the flange coupling. As discussed in Section 5.2.4 accurate modeling of these coupling is a little more complicated than this simple formulation suggests.

There are also some problems to examine concerning the alignment of the overall machine. In selecting the position for the bearings, the choice is made on two criteria:

a) Elimination of cyclic stresses in the shaft
b) Good distribution of bearing loads

We illustrate the problem with a single stage turbine that is connected to a large alternator (in practice, of course, there will be several turbine stages, but this has been simplified to a four-bearing arrangement for ease of illustration) as shown in Figure 5.5.

5.2.3.1 The Catenary

Both rotors in this example are heavy and flexible, implying that both will "sag" due to gravity. The extent of this static defection is readily calculated using the mass and stiffness matrices for the system. Hence, we may write

$$\delta = \mathbf{K}^{-1}\mathbf{M}\mathbf{S}g \tag{5.5}$$

where $S = \{0 \ 1 \ 0 \ 0 \ 0 \ 1 \ 0 \ 0 \ \cdots \ \cdots \ \cdots \ \cdots \ 0 \ 1 \ 0 \ 0\}^T$ and g is the acceleration due to gravity. In the case considered, the turbine rotor is much stiffer than the long, relatively slender, alternator. Assume that the bearings have been arranged to be in a straight line. This would appear to be a reasonable starting point in the absence of more information, but it soon becomes clear that such a simple arrangement is not satisfactory. To appreciate this, consider the catenary of the two rotors if they are not coupled. As shown in Figure 5.6 the two shafts have a mismatch in terms of both the vertical position and the angle at which the mating faces meet. The two shafts fail to meet and to couple the two in this configuration would induce high shear forces and bending moments in the shafts. Furthermore, the forces and moments change the distribution of bearing loads and hence alter the dynamic characteristics.

If the rotors are coupled in this configuration, the resulting shaft catenary is shown in Figure 5.7, where it will be noticed that although the maximum deflection has reduced slightly, there is very high curvature of the shaft near the coupling, indicating high bending moments (and, consequently, high stresses). Note also that these forces and moments would be stationary in space and therefore cyclic on the rotor,

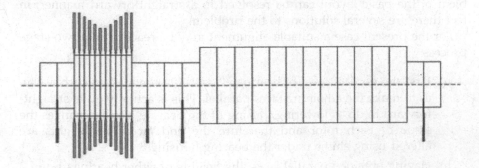

FIGURE 5.5
Idealized Turbo-Generator.

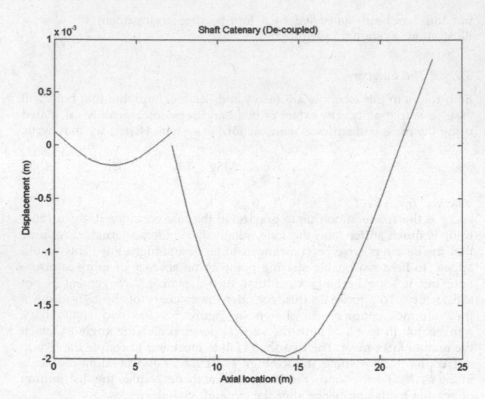

FIGURE 5.6
Uncoupled Rotors.

a fact that could (and occasionally has) lead to major problems, including possible shaft failure.

Note, however, that in these figures, the vertical scale has been greatly magnified for the purpose of illustration. Setting aside for a moment the problems introduced by thermal differentials during operation, the problem of the basic layout can be resolved in a straightforward manner; in fact there are several solutions to the problem.

For the present case a suitable alignment may be reached in a two-stage process:

a) First the vertical level of bearings 2 and 3 is changed in such a way as to make the adjoining faces parallel. This is actually quite straightforward to do as lowering/raising of the bearing simply changes the slope of each rotor and therefore the end face. The heights are altered using shims under the bearing housings.

b) Having obtained parallel faces, the heights of either bearings 1 and 2 or 3 and 4 are changed in such a way as to ensure that at the junction the faces are both parallel and concentric.

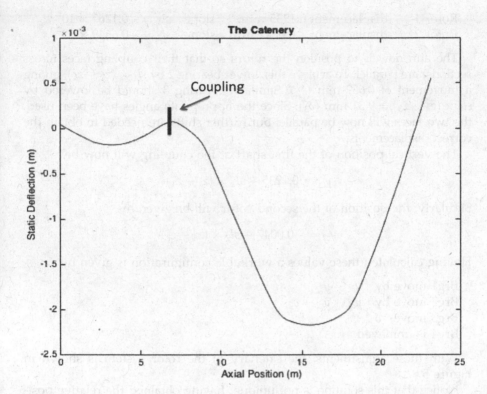

FIGURE 5.7
Rotors Forced into Coupling.

Both the steps are a matter of geometry and, given the dimensions of the machine, the calculation is easily completed.

In the case of a solidly coupled system, common in large machines, the basic parameters required are the slopes of the neighboring rotors in the uncoupled state. In the current, admittedly rather simplified, example, the axial positions of the bearings (as measured from the end of the rotor) is

Bearing 1	$z_1 = 1$ m
Bearing 2	$z_2 = 6.2$ m
Bearing 3	$z_3 = 7.2$ m
Bearing 4	$z_4 = 22.2$ m

And the coupling is midway between bearings 2 and 3 at $z_c = 6.7$ m.

The bearings are all initially in line and the slopes and heights of the two faces are as calculated in Figure 5.6. On real plant they can be readily measured with the angular discrepancies being quantified using feeler gauges. In this case the displacement and slope at the couplings were calculated as

Rotor 1   displacement 0.1235 mm   slope $sl_1 = -0.1267 \times 10^{-3}$
Rotor 2   displacement -0.0047 mm   slope $sl_2 = +0.6649 \times 10^{-3}$

The aim now is to position the rotors so that their coupling faces (cross sections) are parallel. To achieve this, lower bearing 2 by $sl_1 \times (z_2 - z_1)$, giving a movement of 0.659 mm (δ_2). Similarly, bearing 3 should be lowered by $sl_2 \times (z_4 - z_3) = 9.31$ mm (δ_3). Since the appropriate angles have been used, the two faces will now be parallel but further shifts are needed to obtain the correct displacements.

The vertical position of the first shaft at the coupling will now be

$$y_{c1} = 0.1235 - sl_1 \times (z_c - z_1)$$

Similarly, the position of the second rotor will be given by

$$y_{c2} = -0.0047 - sl_2 \times (z_4 - z_c)$$

Having calculated these values a workable configuration is given by

Brg1 move by $y_{c1} - y_{c2}$
Brg2 move by $y_{c1} - y_{c2} + \delta_2$
Brg3 move δ_3
Brg4 is unmoved

With these adjustments, the catenary of the rotor system is shown in Figure 5.8.

Notice that this solution is not unique; having obtained the relative positions of the two rotors, the complete assembly may be rotated. For example, it may be more convenient to move some bearings rather than others. An alternative solution may be reached by raising bearing 4 rather than lowering bearing 3, together with appropriate movements of the other position. In effect the engineer is free to "rotate" the arrangement to suit operational needs provided the relative positions and orientations of the rotors are maintained. While the numbers change the effect is just the same – rotating the orientation of the rotor so as to make the mating faces parallel and concentric. In this example some negligibly small movements have been considered, but on a large unit very appreciable offsets arise, often of several centimeters' difference in height between the HP turbine and alternator. However as illustrated by comparing Figures 5.7 and 5.8, the alignment adjustment eliminates, or at least significantly reduces, the bending moments in the rotor train.

5.2.4 Misalignment Excitation – A New Model

Misalignment difficulties are most significant on large multibearing machines with two bearings per rotor. Some more recent turbo-generators have only a single bearing per rotor and these machines are less vulnerable to alignment

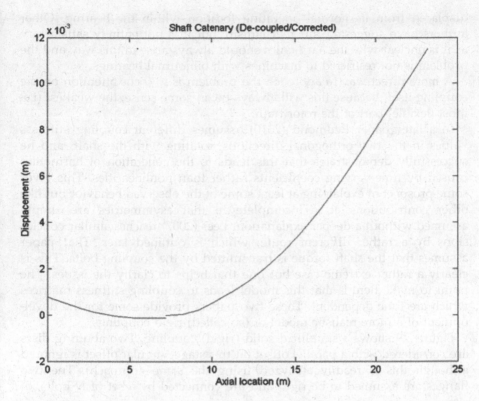

FIGURE 5.8
Coupled Rotors with Appropriate Correction.

issues, although this type of design does introduce other problems. To a lesser
or greater extent, alignment is an issue for all rotating machinery and in view
of this it is surprising that the research findings are rather limited. In marked
contrast to the imbalance case, there is not even complete agreement on the
symptoms and even more diversity on the explanation. This is probably
because there are a number of variables influencing the way in which a rotor
will respond to misalignment and the overall analysis of the situation in
somewhat complex.

Most authors discuss the generation of harmonics of shaft speed, predominantly the 2× component, yet Patel and Darpe (2009) found in rig
trials that 3× was the dominant harmonic. Where flexible couplings are
involved, it is easy to appreciate how nonlinearities arise from relatively
straightforward kinematics, but with rigidly coupled rotors, the situation is
much less clear. Al-Hussein and Redmond (2002) developed the equations
of motion of two rotors with parallel misalignment, but could not establish
any mechanism for directly generating harmonics. The paper suggests that
harmonics arise from the nonlinearity of the bearings as the rotor is

displaced from its normal operating location within the bearing. Other authors have suggested this mechanism, but it is not entirely satisfactory as it is unclear why the harmonics should always arise in this way and the problem is not restricted to machines with oil journal bearings.

A more direct way to approach the problem is to focus attention on the coupling itself because this will always be, in some sense, the weakest (i.e. most flexible) part of the rotor train.

In a later paper Redmond (2010) assumes different coupling stiffness values in the two orthogonal directions, rotating with the shaft, and he successfully demonstrates that this leads to the generation of harmonics caused by time-varying coefficients rather than nonlinearities. This gives some prospect of explaining at least some of the observed behavior but like other contributions, it is incomplete in that asymmetries are simply assumed without a deeper explanation. Lees (2007) reaches similar conclusions by a rather different route, which is outlined later. That paper assumes that the shaft torque is transmitted by the coupling bolts. This is clearly a rather extreme case but one that helps to clarify the issues. The point to make here is that this model leads to coupling stiffness matrices which are time dependent. These two papers provide some for the development of a more realistic model of (so-called) rigid couplings.

Figure 5.9 shows a simplified solid (rigid) coupling. Two abutting discs are considered with a parallel offset (at this stage, angular offset is ignored although this is readily analyzed using the same approach). The two flanges are assumed to be rigid and are connected by a set of N bolts, of

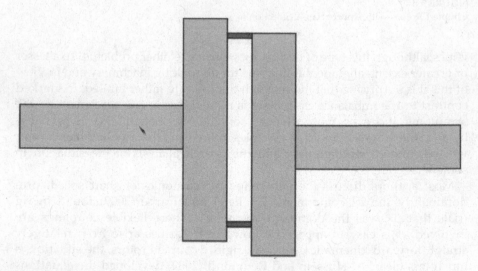

FIGURE 5.9
Simplified Rigid Coupling.

(Reproduced with permission from Lees, 2007)

which only two are shown in the figure for the sake of clarity. Assume that on the flange of the first rotor, the bolt holes are on a circle of radius r, concentric with the shaft. On the second rotor, however, since the shaft is displaced yet the bolts connect without stress, the center of the circle of holes must be displaced from the rotor center by δ say. This is the model developed by Lees (2007), but here a rather more complete picture of the motion is needed.

Under normal operation, the N bolts will be tightened to a specified degree, monitored in terms of either torque or bolt extension. Let us assume a bolt torque set at τ and this will ensure a thrust P between the faces. The torque (T) will be transmitted partly by friction and partly via shear of the bolts. The case considered by Lees (2007) represents a limiting case in which the frictional contribution tends to be zero. But to take the analysis further, the total stresses arising in the bolts must be considered and these will include terms arising from the bending moments at the coupling due to imbalance excitation. However, there is clearly no direct means of assessing this effect and a model is needed to estimate the bending moments from measured vibration data. Given that at least part of the torque is transmitted by the friction between the abutting flanges and the frictional forces will be substantially influenced by moments acting at the coupling, these are now considered.

5.2.4.1 Moments due to Imbalance

Rotor imbalance will give rise to a bow which may exert forces and bending moments at the couplings. The imbalance response at speed Ω is given by

$$\mathbf{x} = \left[\mathbf{K} + j\Omega\mathbf{C} - \Omega^2\mathbf{M}\right]^{-1}\mathbf{M}\Omega^2\varepsilon \tag{5.6}$$

where ε is the mass eccentricity vector. Usually there will only be measurements of the rotor motion available at or close to the bearings, but the model may be used to "recover" estimates of the motion at all the other nodal points, subject of course to all the usual concerns on model accuracy and other uncertainties. Let us say the coupling being studied is at node m of the model, then the four degrees of freedom at the coupling will be $[4m - 3, 4m - 2, 4m - 1, 4m]$ and to calculate the bending moment the motion of nodes $m-1$ and $m+1$ must be considered. To derive the bending moment, recall the derivation of the basic beam finite element set out in Chapter 3. For the sake of clarity, the Euler beam formulation is used here. The displacements within the chosen elements are given by

$$\begin{aligned} u(z_e) &= a_0 + a_1 z_e + a_2 z_e^2 + a_3 z_e^3 \\ v(z_e) &= b_0 + b_1 z_e + b_2 z_e^2 + b_3 z_e^3 \end{aligned} \tag{5.7}$$

z_e is the axial position within the element. Now the bending moments in the two orthogonal directions are given by

$$M_x = EI\left(\frac{\partial^2 v}{\partial z_e^2}\right) = 2b_2 + 6b_3 z_e \qquad\qquad M_y = EI\left(\frac{\partial^2 u}{\partial z_e^2}\right) = 2a_2 + 6a_3 z_e \quad (5.8)$$

So, the problem now is to determine the parameters a and b and to do this it is convenient to consider the two directions separately. Taking the first direction, knowing the displacements at the two nodes of the element, the motion can be written as

$$\begin{Bmatrix} u_{4m-7} \\ u_{4m-4} \\ u_{4m-3} \\ u_{4m} \end{Bmatrix} = \begin{bmatrix} 1 & 0 & 0 & 0 \\ 0 & 1 & 0 & 0 \\ 1 & z_e & z_e^2 & z_e^3 \\ 0 & 1 & 2z_e & 3z_e^2 \end{bmatrix} \begin{Bmatrix} a_0 \\ a_1 \\ a_2 \\ a_3 \end{Bmatrix} \qquad (5.9)$$

Hence the solution may be written as

$$\begin{Bmatrix} a_0 \\ a_1 \\ a_2 \\ a_3 \end{Bmatrix} = \begin{bmatrix} 1 & 0 & 0 & 0 \\ 0 & 1 & 0 & 0 \\ 1 & z_e & z_e^2 & z_e^3 \\ 0 & 1 & 2z_e & 3z_e^2 \end{bmatrix}^{-1} \begin{Bmatrix} u_{4m-7} \\ u_{4m-4} \\ u_{4m-3} \\ u_{4m} \end{Bmatrix} \qquad (5.10)$$

Similarly, motion in the v directions yields

$$\begin{Bmatrix} b_0 \\ b_1 \\ b_2 \\ b_3 \end{Bmatrix} = \begin{bmatrix} 1 & 0 & 0 & 0 \\ 0 & 1 & 0 & 0 \\ 1 & z_e & z_e^2 & z_e^3 \\ 0 & 1 & 2z_e & 3z_e^2 \end{bmatrix}^{-1} \begin{Bmatrix} u_{4m-6} \\ u_{4m-5} \\ u_{4m-2} \\ u_{4m-1} \end{Bmatrix} \qquad (5.11)$$

With these results the bending moments in the element immediately before the coupling is determined and using the same approach those in the element after the coupling can also be evaluated. The shear force can also be calculated from these parameters as will be mentioned later in the discussion. Note that since only displacement and slope are continuous across element boundaries, different estimates for bending moments and forces are obtained in the two elements. The optimum estimate is simply the mean of the corresponding values. Hence, given an imbalance on the rotor, there will be a rotating bending moment vector at the coupling, or, to be precise, at the flanges of the coupling.

5.2.4.2 The Effect of Bending Moments

Clearly the couplings play a crucial role in determining the dynamic behavior of a misaligned machine. Ideally, they will transmit only torque but no forces or bending moment – but often this will not be the case. A simplified view of a flange coupling is shown in Figure 5.9: In this sketch only two bolts are shown but on a real machine there will be many, perhaps 16 or 32, but their essential role is to maintain an axial load between the mating surfaces of the coupling.

Clearly the performance of the coupling is dependent on the interfacial pressure, P, between the mating surfaces, and this is determined by the tension in the bolts and the coupling geometry. Given this uniform pressure distribution across the face, it is straightforward to calculate the maximum torque that can be transmitted. This is given by

$$T_c = \frac{2}{3}\mu P \pi r^2 \tag{5.12}$$

where μ is the coefficient of friction; P is the interfacial pressure (as set by the bolt tightness); r is the face radius.

In any real rotor there will be bending moments orthogonal to the rotor which will distort the pressure profile from uniformity and reduce the torque which can be transmitted by friction. Generally, there will be a bending moment acting, one arising from imbalance and this will rotate with the shaft, the other (if present) arises from misalignment and will be fixed in space. These two bending moments are designated Γ_u and Γ_s respectively and using a finite element model of the rotor, they can be readily calculated as outlined above. Note that the misalignment moment will be subject only to a minor variation with rotor speed (arising from bearing properties) whereas the imbalance term varies substantially and in a complex manner. Hence at any time instant the rotor is at orientation $\theta = \omega t$; the resultant bending moment is

$$\Gamma = \left\{ \begin{array}{l} \Gamma_{sx} + \Gamma_{ux}\cos(\theta - \varphi) \\ \Gamma_{sy} + \Gamma_{uy}\sin(\theta - \varphi) \end{array} \right\} \tag{5.13}$$

where the phase shift φ is determined by the dynamics of the system.

Because of the combination of static and rotating bending moments both the magnitude and direction of the resultant vary over a cycle of the shaft and this modifies the way in which torque is transmitted. Furthermore, the mechanism by which this occurs depends on the torque: at modest torque levels transmission will be principally by friction whereas at high levels the coupling bolts play a role.

The bending moment imposes a gradient on the interfacial pressure distribution on the flanges and this has the effect of displacing the center of rotation by some distance δ, whose magnitude and direction ϕ depend on both the degree of misalignment and the imbalance excitation. The equation of motion will then become

$$Kx + Cx + M\ddot{x} = M\dot{\Omega}^2 e + F_c(\delta, \phi) \qquad (5.14)$$

δ and ϕ may, of course, be strongly time dependent and since it has arisen from the interaction of imbalance and misalignment it will often have a number of harmonics of the rotation speed Ω and so the response will contain harmonics of shaft speed. F_c is the lateral force generated at the coupling.

The friction per unit area at any point of the mating surface is taken to be less than or equal to the coefficient of friction multiplied by the interfacial pressure. Hence, using a local frame of reference in which the y-axis is aligned with the maximum bending moment, it may be assumed that the profile of the friction is as shown in Figure 5.10, that is constant for

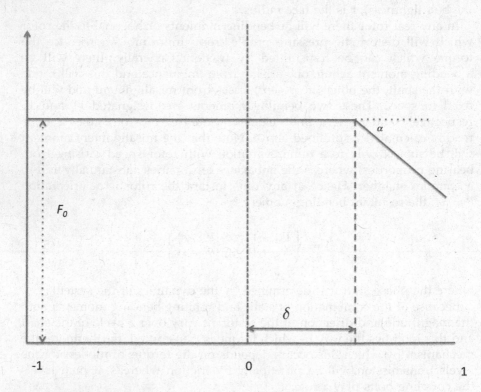

FIGURE 5.10
Idealized Friction Profile.

$r' < 1 - \delta$ and reducing linearly for $r' > 1 - \delta$ (so long as it stays positive). This distribution of the friction would appear reasonable but it is a point that should at some stage be checked experimentally; however, this may require some careful design of the facility.

At this stage, the transition level, $r' = \delta$, between constant friction and linear variation is unknown but it is clear that at this point, the friction force per unit area is

$$F(\delta) = \mu P(\delta) \tag{5.15}$$

Given a linear pressure variation due to the bending moments, it is convenient to express this as

$$P(x) = P_0 \left(1 - \frac{|\Gamma|}{\Gamma_0} x \right) \tag{5.16}$$

where the moment scaling factor Γ_0 is the bending moment which reduces the pressure to zero at the outer edge of the coupling face. This can be calculated as

$$\Gamma_0 = \int\limits_0^\pi \int\limits_0^r P_0(1 - r'\sin\phi)r' \sin\phi \cos\phi \, dr' d\phi = \frac{4}{9} P_0 r^3 \tag{5.17}$$

x is measured along the (local) direction orthogonal to the bending moment. Letting $\alpha = \frac{|\Gamma|}{\Gamma_0}$ this is rewritten as

$$P(x) = P_0(1 - \alpha x) \tag{5.18}$$

Note that δ is measured from the center of the disc and hence $\delta = 1$ means there is no transition.

What does vary in the two regions (i.e. $x < \delta$ and $x > \delta$) is the friction which transmits the torque, provided this is below a critical value. The force per unit area takes the form

$$
\begin{aligned}
F(x) &= F_0 & x < \delta \\
F(x) &= F_0(1 - \alpha[x - \delta]) & x > \delta
\end{aligned}
\tag{5.19}
$$

The factor $F_0 \leq \mu P_0$ takes on either the values required to transmit the torque or its limiting value. In this expression the coordinate system is local and instantaneous, so that the overall bending moment is about the x-axis. This seemingly complex procedure reduces the problem to the determination of the two parameters F_0, δ, since α is determined by the imposed bending moment. Note, however, that this nonuniform pressure profile will reduce the torque capacity of the coupling, in which case load may be transferred to the bolts.

For any position the transition point, δ, and the normalized bending moment, α, the torque capacity of the coupling is readily calculated by integrating the frictional forces across the face. This capacity reduction factor is denoted R_c. For any instant in the motion, the bending moment is calculated and using this, the appropriate values of α, δ are evaluated using the approach described below. Once these two parameters are established, however, several features follow as a consequence. If $\delta < 1$ then the center of torque (i.e. the point in the cross section at which there is no net lateral force) will deviate from the (local) center of gravity. As indicated above, the forces of friction will not be symmetrical about the center line and there will be a net force in the direction of the resultant bending moment, which varies, but not in a uniform manner. Because the effective pressure varies across the face, the torque capacity of the coupling will be reduced and the reduction factor is given by

$$R_c(\alpha, \delta) = \int\limits_{-1}^{1} \int\limits_{-\sqrt{1-y^2}}^{\sqrt{1-y^2}} \frac{F(x)}{F_0} \frac{y^2}{\sqrt{x^2 + y^2}} \, dx \, dy \qquad (5.20)$$

$R_c(\alpha, \delta)$ is shown in Figure 5.11 and this function can be tabulated prior to the main calculations. For clarity, this figure shows only a reduced resolution of the actual tabulation. Since this (or any other credible) pattern of friction is not symmetrical about the (local) x-axis, the frictional forces will exert a resultant force on the geometric center of the coupling, meaning that a force will act on the shaft. The local force (at right angles to the direction of the pressure gradient) may be tabulated as

$$f_{loc}(\alpha, \delta) = \int\limits_{\delta}^{1} \int\limits_{-\sqrt{1-y^2}}^{\sqrt{1-y^2}} F_0 \alpha \frac{y^2}{\sqrt{x^2 + y^2}} \, dx \, dy \qquad (5.21)$$

The forces acting at the coupling are completely determined by the three parameters F_0, α, δ. Of these, the bending moment applied, α, is known but the other two parameters must now be established.

The force level F_0 is readily calculated as

$$F_0 = \frac{T}{T_c R_c(\alpha, \delta)} \mu P \qquad (5.22)$$

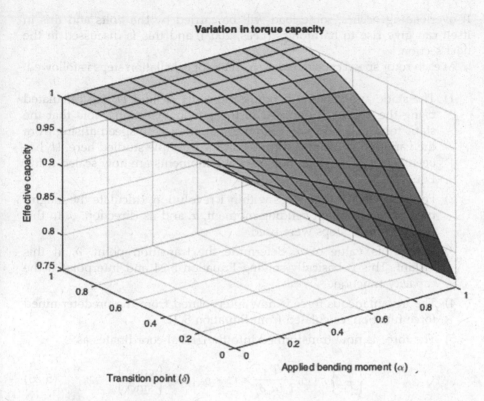

FIGURE 5.11
Reduction in Torque Capacity.

But at the transition point, the interfacial pressure will be $P(1 - \alpha\delta)$ and as friction here will be limiting

$$F_0 = \frac{T}{T_c R_c(\alpha, \delta)} \mu P = \mu P(1 - \alpha\delta) \qquad (5.23)$$

Hence

$$\left(1 - \frac{T}{T_c R_c(\alpha, \delta)}\right) = \alpha\delta \qquad (5.24)$$

Using interpolation, this nonlinear equation is readily solved to yield the transition point δ. Once this is established, the resulting forces are completely determined. Note that solutions for δ exist only for cases in which the coupling is not overloaded, that is,

$$1 \geq \frac{T}{T_c R_c(\alpha, \delta)} \qquad (5.25)$$

If overloading occurs, some load will be carried by the bolts and this in itself can give rise to harmonics (Lees, 2007), and this is discussed in the next section.

At each rotor speed considered, a sequence of calculation steps is followed.

1) The static and rotating bending moments U_s and U_r are calculated using the approach outlined in the previous section. Note that the static term may only show a minor variation with speed arising from the variation in bearing stiffness. In the example studied here, U_s has been taken as constant. Both bending moments are now scaled using Equation 5.17.

2) For a number of time steps within 1 revolution, calculate the magnitude of the resultant bending moment, α, and its direction, φ. In this study, 10 time steps were used.

3) Using this value of α, determine the transition point, δ, at this instant. This is basically solving Equation 5.24 and interpolated the capacity function.

4) The instantaneous force is now interpolated from the predetermined force function, calculated from Equation 5.21.

5) The force is now transformed into the global coordinates as

$$\begin{Bmatrix} F_x \\ F_y \end{Bmatrix} = P_0 \frac{T}{T_c(\alpha,\delta)} \times r^2 \times f_{loc}(\alpha,\delta) \begin{Bmatrix} \cos\phi \\ \sin\phi \end{Bmatrix} \tag{5.26}$$

6) Having completed these calculations at each step in the cycle, a number (say 10) of identical records are concatenated in order to improve the resolution of the ensuing Fourier transform. The assembled signal then goes through FFT to yield the harmonics of the particular rotor speed, denoted as $\tilde{F}_n$.

Example 5.2: Application to a "Rigid" Coupling

The machine shown in Figure 5.12 is considered to examine the effects of the process outlined above. For the sake of convenience, attention is restricted to a four-bearing machine comprising a single stage turbine driving an alternator. The overall length of the machine is 17 m, and the bearings are plain journals with a clearance of 60 μm. The mass of the two rotors is 24 and 50 tonnes, respectively. The power transmitted at 3,000 rpm was 125 MW and the transmitted torque for the example was held constant, but this condition may be easily varied. The principal dimensions of the model are shown in Table 5.2. In practice, of course, a turbine generating this power would have more than a single stage, but this model is taken for the sake of simplicity. Imbalance is imposed at nodes 3, 5, and 18.

Clearly, the parameters of the coupling are of crucial importance here and the parameters chosen were

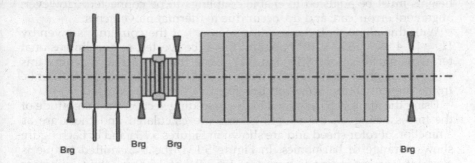

FIGURE 5.12
Rotor Model.

TABLE 5.2

Main Parameters of Model

Bearing Locations	1	1 m
	2	4 m
	3	5.5 m
	4	15.5 m
Rotor 1 details		
Main body diameter		0.8 m
Main body length		2 m
Stub diameter		0.6 m
Rotor 2 details		
Main body diameter		1.2 m
Main body length		6 m
Stub diameter		0.6 m
Bearing clearance		0.5 mm
Oil viscosity		0.04 Nsec/m^2

Number of bolts 16
Pitch diameter 0.8 m
Bolt diameter 40 mm
Bolt tightness 90% yield
Coefficient of friction 0.6

The bearings in this example are set in a straight line. Clearly this is inappropriate as self-weight bending imposes moments at the coupling, which is at nodes 9 and 10 in the model. This model is rather stiffer than a typical machine and consequently the catenary of the alternator section is reduced. This is suitable to demonstrate the principles as the couplings can be connected in a simple straight-line datum. In a real machine, bearing heights must be adjusted to enable couplings to be connected. However, alignment errors can, and do, occur due to thermal movements.

With this alignment, the bending moment at the coupling is given by $U_s = \{ -4.87 \quad 1.53 \} \times 10^4$ Nm, and this is converted to nondimensional terms using Equations 5.16 and 5.17. Note that on a real machine this will vary slightly with speed owing to the varying bearing coefficients. In the present study, however, this fixed value has been used.

Using the approach described in the preceding section the magnitude of the forces acting on the coupling faces are calculated and these are as a function of rotor speed and are shown in Figures 5.13a and b. Each figure shows a range of harmonics. In Figure 5.13a, the transmitted torque is constant (corresponding to power of 125 MW at 3,000 rev/min), whereas in Figure 5.13b, power is constant up to the limit set by the capacity T_c, the two cases coinciding at 3,000 rev/min. The comparison of these two figures illustrates the importance of the transmitted torque.

Equal and opposite forces are applied to the adjoining rotors but, because of the coupling flexibility, these do not cancel. In the modeling here, forces are transmitted through the bolts represented as a composite beam element.

It is shown here that flange couplings (and, no doubt, other types) show nonlinear behavior that can give rise to harmonics of shaft speed. To analyze this situation fully would require a complete three-dimensional finite element model of the coupling which may be of limited applicability in practice. To overcome this, a simplified analysis is presented here aimed at clarifying the physics of the situation. Further work is needed in two areas firstly to confirm or modify the friction distribution across the face of the coupling (Figure 5.10) and secondly to combine this work with nonlinear bearings. A third activity may also be added aimed at incorporating the analysis of the high torque case discussed by Lees (2007). An interesting question arises as to the possibility of incorporating the complete range without the need for a fully three-dimensional analysis.

The friction profile shown in Figure 5.10 is an assumption. A significant problem is that on real plant there is little detailed knowledge of how components will behave. This indicates the pressing need for some rig testing on this topic.

Whilst any direct diagnostic technique for machines appears to be some way off (if at all possible), the ideas presented here form a useful basis. For example, the variation of the second harmonic with transmitted torque may provide important information to a machine operator. The study also provides some insights into some design requirements of the couplings used.

5.2.4.3 Influence of the Bolts

The situation described so far supposes that torque is transmitted entirely by way of the interfacial friction. In all cases the bolt tightness will set the limits of bending moment that can be sustained, but in general torque will

(a)

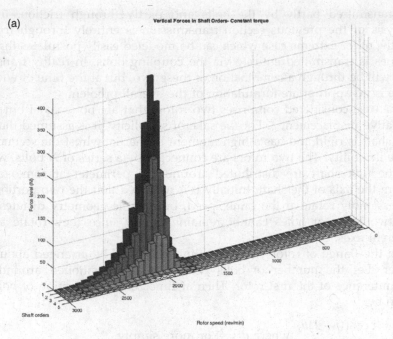

(b)

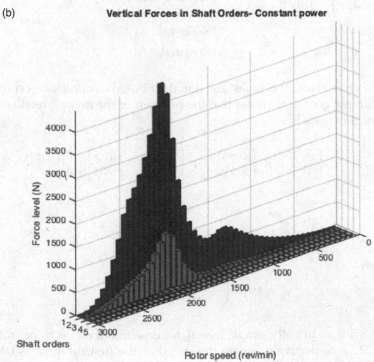

FIGURE 5.13
Vertical Forces – Constant and Varying Torque.

be transmitted partly by the bolts and partly through friction. In the analysis of the previous section transmission is entirely through friction but the other extreme case which can be modeled easily postulates that the torques is transmitted entirely via the coupling bolts. In reality transmission will be through a combination of these two, but at the time of writing, there is no appropriate formulation of the overall problem.

The case considered comprises two rotors that are not coaxial but have a relative displacement δ. For the sake of simplicity, it is assumed that the first shaft is rigid and has a high moment of inertia, whilst the second has some flexibility. The two rotors are connected by a series of N bolts, which on the first shaft are distributed around the perimeter at some radius r from the axis of the shaft. Initially it is assumed that the two portions of the assembly rotate at the same speed, but simple geometry dictates that the two pairs of holes cannot remain aligned since they rotate about different axes.

On the flange of rotor 1, the coupling bolts will be arranged about the center. Let the number of bolts be N, equally positioned around the circumference of the first rotor. Then at time zero, the position of bolt j is given by

$$x_j = r\cos((j-1)\theta)$$
$$y_j = r\sin((j-1)\theta)$$

where $\theta = \frac{2\pi}{N}$ or more simply,

$$x_j = r\cos\theta_j$$
$$y_j = r\sin\theta_j \qquad (5.27)$$

On the other rotor, the holes are not distributed around the center, but offset. Simple geometry shows that the position of the jth bolt relative to the center of this rotor is

$$\begin{Bmatrix} X_j \\ Y_j \end{Bmatrix} = R_j \begin{Bmatrix} \cos\varphi_j \\ \sin\varphi_j \end{Bmatrix} + \begin{bmatrix} \cos\phi & \sin\phi \\ -\sin\phi & \cos\phi \end{bmatrix} \begin{Bmatrix} 0 \\ \delta \end{Bmatrix} \qquad (5.28)$$

where

$$R_j = \sqrt{1 + \delta^2 + 2 \times \delta \times \cos(j-1)\theta}$$
$$\varphi_j = \tan^{-1}\left(\frac{\delta + r\sin((j-1)\theta)}{r\cos((j-1)\theta)}\right) \qquad (5.29)$$

To start with an initially simple model, assume that the two rotors rotate at speeds Ω, $\dot{\phi}$ respectively, where the speed of the first rotor, Ω, is taken to be constant. Then the kinetic energy is

$$T = \frac{1}{2}J_1\Omega^2 + \frac{1}{2}J_2\dot{\phi}^2 \qquad (5.30)$$

where J_1 and J_2 are the respective polar moments of inertia of the two rotors (assuming them to be rigid – this is easily generalized once the formulation is completed). The potential energy is

$$U = \frac{K_b}{2}\sum_{j=1}^{N}\left[(r\cos(\theta_j + \Omega t) - R_j\cos(\varphi_j + \phi) - \delta\sin\phi)^2\right] +$$
$$\frac{K_b}{2}\sum_{j=1}^{N}\left[(r\sin(\theta_j + \Omega t) - R_j\sin(\varphi_j + \phi) - \delta\cos\phi)^2\right] \qquad (5.31)$$

Taking nominal values for the stiffnesses of the coupling bolts as K_b the stored energy varies as the rotor system rotates. The motion of the two flanges is illustrated in Figure 5.14, showing the positions of the bolt locations (only three bolts are shown for the sake of clarity). This variation is not sinusoidal and has some harmonic content. Note the locus of the coupling bolts: those on rotor 1 simply follow a single circle, whereas those on rotor 2 each follow a different circle.

The analysis so far has implicitly considered only torsional motion, but in general we wish to examine a system with lateral motion as well. The full development of the equations of motion can be found in Lees (2007).

We now introduce a simplified model to illustrate the nature of the parametric excitation arising in a misaligned system. The two shafts are connected by a series of N bolts. The first shaft is rigid and has a very large torsional inertia. At some point in time, each coupling bolt is perfectly positioned at radius r on rotor 1, but on the coupling face of the second rotor, the bolt holes are offset. The second rotor has torsional inertia J_2 and stiffness values k_x, k_y and this rotor has displacements u, v, φ. The arrangement is shown in Figure 5.9.

The analysis of the dynamics is somewhat involved, but after some algebra Lees (2007) showed that the governing equation is

$$\mathbf{K}\mathbf{x} + \Delta\mathbf{K}(t)\mathbf{x} + \mathbf{C}\dot{\mathbf{x}} + \mathbf{M}\ddot{\mathbf{x}} = \mathbf{F}(t) \qquad (5.32)$$

where

$$\mathbf{F} = \frac{NK_b\delta}{2}\begin{Bmatrix} \cos\Omega t \\ \sin\Omega t \\ \delta \\ -\cos\Omega t \\ -\sin\Omega t \\ -\delta \end{Bmatrix} \qquad (5.33)$$

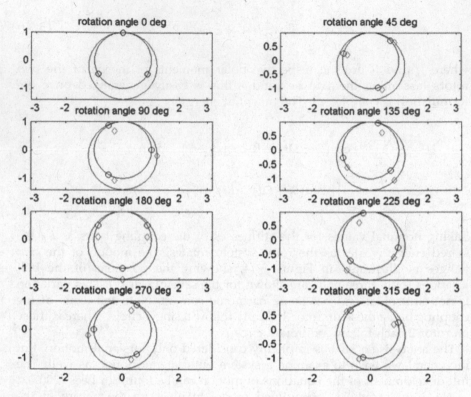

FIGURE 5.14
Motion of Adjoining Flanges.

(Reproduced with permission from Lees, 2007)

and

$$
\Delta K(t) = \frac{NK_b\delta}{2}
\begin{bmatrix}
0 & 0 & \sin\Omega t & 0 & 0 & -\sin\Omega t \\
0 & 0 & \cos\Omega t & 0 & 0 & -\cos\Omega t \\
\sin\Omega t & \cos\Omega t & 0 & -\sin\Omega t & -\cos\Omega t & 0 \\
0 & 0 & -\sin\Omega t & 0 & 0 & \sin\Omega t \\
0 & 0 & -\cos\Omega t & 0 & 0 & \cos\Omega t \\
\sin\Omega t & -\cos\Omega t & 0 & \sin\Omega t & \cos\Omega t & 0
\end{bmatrix}
$$

$$(5.34)$$

N is the number of connecting bolts, bolt stiffness K_b, and δ is the offset. Similar expressions can be derived for angular offsets.

The key point to observe in this is that the equation, although linear, has coefficients which vary in time and this will give rise to harmonics of shaft speed in the vibration measurements. Physically this comes from the offset between adjoining shafts, or sections, which means that the transmitted torque on of shaft is modified and has a resultant force on its neighbor. On some systems this might simply impose requirements, albeit demanding ones, on machine set-up, but in many cases the situation is a little more complicated.

5.2.5 Toward an Overall Model

Amid this fairly extensive mathematical analysis, it is worth reflecting on the physical basis of the phenomena. Fundamentally, because the center of rotation and geometric center do not coincide, an input torque is resolved into a couple and a force. Furthermore, the way in which this resolution is achieved is a function of time. Nevertheless, the discussion is valid in principle for any system in which the transmission has some angular variability and it is difficult to imagine a practical machine in which this is not the case. Of course, the mechanism of coupling between torsional and lateral motion discussed above is not the sole source of the second harmonic excitation.

Other factors contributing to harmonic excitation in misaligned systems include static change of loading position changes the restoring loads, torque vector rotation, nonlinearity in the bolts, bearing nonlinearity, and changes of internal clearance leading to variations of pressure or electric field depending on the type of machine. Indeed, this is an area in which there remain a number of unresolved issues and it is perhaps the multiplicity of sources which has stifled progress on a thorough understanding of misalignment. The confused situation is more pertinent since misalignment remains one of the most significant practical problems in rotating machinery. In summary, both of the cases for torque transmission have been shown to give rise to vibration components at harmonics of the rotation frequency. There is a need for a comprehensive formulation in order to give quantitative predictions of behavior in real machines and laboratory studies to verify the models.

5.3 Cracked Rotors

The dynamics of cracked rotors has been an active area of study for a considerable time and this activity is driven by the severe problem of assuring rotor integrity. Over the years there have been an appreciable number of cracked rotors, most of which have been diagnosed in time to avoid catastrophic consequences. Such failures incur very high costs in terms of capital damage, loss of production, and not least danger to personnel, and consequently it is of the utmost importance to recognize incipient faults from operational data. A prerequisite to such a facility is an understanding of the dynamics of a cracked rotor. This is a topic on which there have been many papers over the last few decades, but real progress in understanding has been somewhat limited in the last 20 years – it would appear that the field is in a phase of diminishing returns.

Rotating machines are often subject to high stresses and on occasions this has become manifest via cracking of the main rotor. Whilst rotor cracking is a relatively rare fault, it is potentially very dangerous presenting a severe physical and economic danger. Because of this, it is important to understand the characteristics of the vibrational behavior of a cracked rotor. Research carried out in this area over the last 50 years has analyzed the effect of a transverse crack, the most common category, although torsional loadings give high-stress components at an angle to the axis of the rotor and corresponding crack geometries have been observed.

Let us now consider the dynamics of a rotor with a transverse crack. This is illustrated in Figure 5.15. We assume in this section that the axis of the rotor is horizontal. (In the case of a vertical shaft, the problem of detection is somewhat more complicated.) As the shaft rotates, the stiffness rotor will vary. This is because when the orientation is such that there is a compressive stress across the crack faces (due to the self-weight bending), the crack will be closed and the shaft will have its full stiffness, that is, the crack is effectively absent. However, half a revolution later, the crack will be open under the action of tensile stresses. This causes a reduction in the effective stiffness of the rotor. The portion of the crack

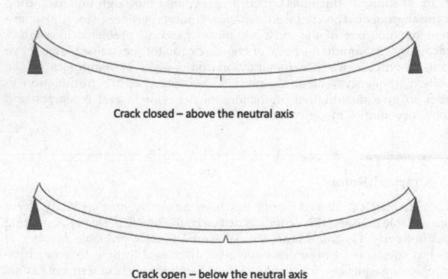

Crack closed – above the neutral axis

Crack open – below the neutral axis

FIGURE 5.15
Alternate Opening and Closing of a Crack.

below the rotor's neutral axis will be open and this degree of opening will determine the dynamic characteristics. The calculation of the dynamics can be significantly simplified by observing that in many instances, the vibration amplitude will be significantly less than the shaft deflection due to gravity (i.e. the so-called self-weight bending). With this assumption the nonlinear equation of motion is converted to a linear equation with time-varying coefficients, making the analysis very much more tractable.

The behavior of a rotating cracked shaft received extensive discussion from 1975 onward with notable early contributions from Mayes and Davies (1976, 1984), Davies and Mayes (1984), Gasch (1993), and Henry and Okah-Avae (1976). The topic was reviewed by Dimaragonas (1996). It was noted that, in the case of a horizontal rotor, the self-weight bending plays a very important role in the dynamics. At any stage in the cycle, the crack may be open or closed – indeed if operating very close to a critical speed, the crack may remain open throughout a full shaft revolution if the unbalance is very high. However, for modest levels of unbalance and away from critical speeds, the equations governing the shaft motion may be linearized by observing that the bending moment due to self-weight bending will dominate over inertial terms. It is assumed that only the open portions of the crack change the rotor's stiffness and hence this stiffness will be a function of the orientation of the rotor.

This is also true for the case where the rotor has been taken out of the machine for examination and hung (horizontally) from a cradle. There are two key steps to be established for the understanding of cracked rotors: First the reduction in stiffness due to a given crack is established, and second, the size of effective crack for each orientation is discussed.

Mayes and Davies (1976) were among the first authors to address the issue of cracks in real machines and in a series of elegant papers introduced significant progress in the analysis of the topic. Two very significant advances were in the treatment of natural frequencies and the analysis of the forced response.

5.3.1 Change in Natural Frequencies

The study of the change in natural frequency due to a chordal crack stems from basic fracture mechanics. The connection between crack depth and change in the rotor's dynamic properties was derived by considering the change in energy as a static force was applied to a rotor. The situation is shown in Figure 5.15.

This figure shows the shaft in two different orientations. The key point is the role played by gravity, causing the shaft to sag to some extent between the bearings. As the shaft rotates, for the part of the cycle when the crack is in the upper portion of the rotor, bending moments from the weight will tend to hold the crack closed and at this instant the full stiffness will be retained.

Half a cycle later, the crack will be below the neutral axis of the rotor and the weight-induced bending moments will tend to open the crack, perhaps inducing further growth. In this case the shaft will be weakened to some extent and it is clear that as the shaft rotates, its stiffness will vary with some cyclic pattern, although at this stage it is not clear what the precise amplitude or pattern of this variation takes. Although there have been a very large number of research papers on this topic, in truth, the detailed pattern of the variation is not of great importance as shown by Penny and Friswell (2002): every crack in a real machine will have a different size and profile and will give rise to effects which differ in detail. What is crucially important for the practicing engineer is the recognition of the salient effects characteristic of a rotor crack.

In their paper, Mayes and Davies (1976) relate the change in natural frequency to the crack depth and location. The analysis considered the energy required to increase the deflection of a crack beam (rotational effects were neglected in this analysis). The energy (equal to the force applied multiplied by the increase in deflection) is accounted for partly by the potential energy stored in the beam and party by the energy required to grow the crack. A key concept here is the strain energy release rate, G, given by

$$G = -\frac{\partial U}{\partial A} \tag{5.35}$$

where U is the stored energy and A is the area of the crack. This can be related to the stress intensity factor by the relationship (for plane stress)

$$G = \frac{K_I^2}{E} \tag{5.36}$$

K_1 is the stress intensity factor. The mathematics of the stress intensity factor is somewhat involved but for the present discussion it is not necessary to delve into this. A range of expressions are available for different specimen geometries and crack profiles but the important point to note is that this factor essentially describes the stress distribution around the crack. It is not surprising that this is needed to obtain the energy stored in a cracked rotor and hence the change in natural frequencies.

It was shown that (Mayes and Davies, 1976)

$$\Delta\left(\omega_n^2\right) = -4\left(\frac{EI^2}{\pi r^3}\right)\left(1 - \nu^2\right)F(\mu)\left(y''_N(s_c)\right)^2 \tag{5.37}$$

where $y''_N(s_c) = \left[\frac{d^2 y_N}{dz^2}\right]_{z=s_c}$, that is, the second derivative of the mode shape at the crack location s_c, and F is a function of nondimensional crack depth.

y_N is the Nth mode shape and μ is the nondimensional crack depth, that is, crack depth divided by shaft radius. The argument was developed by considering the energy required to extend the crack by some small extent and comparing this to the energy stored in the deformed beam.

The other parameters in this equation are

E   Young's modulus
I   second moment of area
r   shaft radius
v   Poisson's ratio

In principle, the function $F(\mu)$ can be derived from the appropriate stress concentration factor, if this is known. A different approach was taken in Mayes and Davies (1984) where the authors inferred the values of F from a series of experiments with chordal cracks. It was shown that to a very good approximation this function is just equal to the fractional change in the second moment of area as measured at the crack face. $F(\mu)$, the *compliance function*, has two components corresponding to two orthogonal directions in line with and at right angles to the crack front. The compliance function F is essentially the second moment of area of the remaining (i.e. uncracked) portion of the cross section. What gives this function a highly nonlinear dependency on nondimensional crack depth, μ, is the fact that the second moment of area is taken about the new mean plane, not that of the original centerline of the shaft. Indeed, it is the form of this (which is a function of the circular geometry) which determines the sensitivity. For a circular cross-section rotor, a chordal crack will need to have an appreciable depth before it can be detected by means of vibration alone. There have been numerous attempts in the literature to develop methods with increased resolution, with some modest successes, but there is a fundamental limitation that a microscopic crack will have little influence on the rotor-dynamic behavior. This is because F is a very nonlinear function of μ as illustrated in Figure 5.16.

Having established the change in stiffness and natural frequency arising from a crack, it is straightforward to represent these factors in a finite element model. In Mayes and Davies (1984) it is shown that a crack may be represented by reducing the second moment of area of a single element by ΔI. Using a Rayleigh–Ritz type of approach it can be shown that if the second moment of area of the element concerned is I_0, then

$$\frac{\Delta I / I_0}{1 - \Delta I / I_0} = \frac{r}{L_e}(1 - v^2)F(\mu) \tag{5.38}$$

where L_e is the length of the section with reduced properties and r is the radius of the shaft at that cross section. Note that this parameter is at the

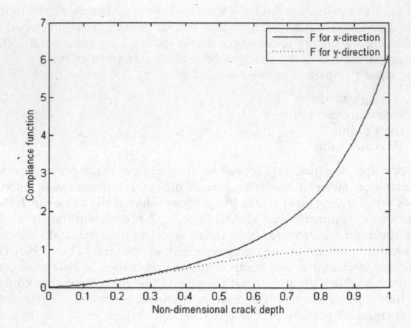

FIGURE 5.16
The Compliance Function.

discretion of the modeler within a reasonable range but the choice determines the value of second moment of area. For any given chordal crack, there are two values of *F*, and two orthogonal directions in the plane of the crack. Hence the parameters of a representative model are now fully specified.

In the prediction of the dynamic behavior of a cracked rotor for each orientation of the rotor, a suitable crack model must be derived and there are several steps to achieving this. For the rotor under nonrotating conditions it is reasonable to assume that the portion of a chordal crack lying below the neutral axis will be opened under the influence of self-weight bending. The actual portion of the crack open under these circumstances will be a rather complicated shape having its own stress intensity factor. In general, these factors are not known, but the details are not of great importance. For most purposes it suffices to base calculations of the effects on total opening with a sinusoidal variation. The details of the opening will influence only the relative amplitude of the higher harmonics (see Penny and Friswell, 2002).

At each orientation of the rotor, for a given size of chordal crack, the area of open crack must be calculated. The direction of the weaker axis bisects the angle of the open crack portion. The compliance functions in the two ortho-gonal directions are then evaluated by assuming equivalence with the chordal crack of equal effective area. It is argued that this relatively simple procedure is sufficient to reflect the main physics of the situation. Whilst it could be assumed that the equivalence should be based on second moments rather than area, it must be remembered that we do not have the appropriate stress concentration factors to use and this being the case, it is desirable to keep the calculation as simple as possible. The two extremes of crack open and closed are modeled correctly, and, to a large extent, precise evaluation of the intermediate values is not necessary. Figure 5.17 shows graphically the open and closed portions for a particular crack depth and orientation.

Example 5.3: A Cracked Rotor

As an example, consider the effects of a crack on the natural frequencies of the rotor shown in Figure 5.18. This is a large rotor, similar to a generator rotor with an overall length of 16 m and a mass of 52,000 kg. The bearings are taken as rigid Example 5.3. The diameter of the rotor 1.2 m over most of the length but note that on a real machine this would not all be solid steel. To account for this the density in the model has been reduced.

Chordal cracks are considered with depths of 0.25, 0.5, 0.75, and 1 times the shaft radius, and therefore, even the smaller cases are very significant defects. Each of these cracks is considered at three different axial locations, namely 4, 6, and 8 m from the end. The results were calculated by reducing the stiffness of relevant sections of the rotor in accord with Equation 5.38 while holding the mass constant.

In this study, 16 Timoshenko beam elements were used and in each case the element of either side of the nominal crack location was reduced.

Table 5.3 shows the results of the calculation and a number of important features are apparent. An important first point is that although all the cracks considered are significant, in most cases the change in natural frequency is modest. For instance, a crack of radius depth (i.e. half the cross section) at the 4 m location (or quarter span) reduces the natural frequency by less than ½%. A small change such as this would not be detected on real plant. For

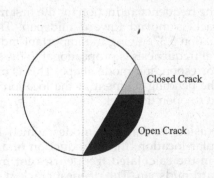

FIGURE 5.17
The Equivalent Crack.

FIGURE 5.18
Test Rotor L=16 m, D_{max}=1.2 m, Mass 52,000 kg.

TABLE 5.3

Effects of a Crack on Natural Frequencies

Crack Depth (μ)	Mode 1 (Uncracked 14.61 Hz)		
	$L/4$ 4 m	$3L/8$ 6 m	$L/2$ 8 m
0.25	14.56	14.52	14.49
0.5	14.46	14.33	14.26
0.75	14.25	13.97	13.82
1	13.61	12.95	12.65
	Mode 2 (Uncracked 46.79 Hz)		
0.25	46.51	46.61	46.77
0.5	45.94	46.24	46.72
0.75	44.97	45.63	46.63
1	42.71	44.27	46.42

each crack depth, the frequency reduction for the first mode increases as the position of the crack is moved toward midspan. This pattern is to be expected from Equation 5.37, giving the important indicator that the effect of a crack on natural frequencies is proportional to the square of the second derivative of the corresponding mode shape. This is physically reasonable as this relates to the bending moment at the location of the defect which will, of course, tend to open the crack.

The logical conclusion is that, for modes which have zero bending moment at a particular location, the introduction of a crack will have no effect. This is clear in the calculated frequencies for mode 2, as the crack location moves toward midspan. The deepest crack at the midspan shows a frequency change of less than 0.8%. Given the mode shapes shown in Figure 5.19, it is perhaps surprising that there is any change in natural frequency for this mode. The reason for this lies in the modeling: the crack's influence is spread over two elements, or 2 m in this model and so covers regions of the rotor on which the crack has an effect. This highlights the

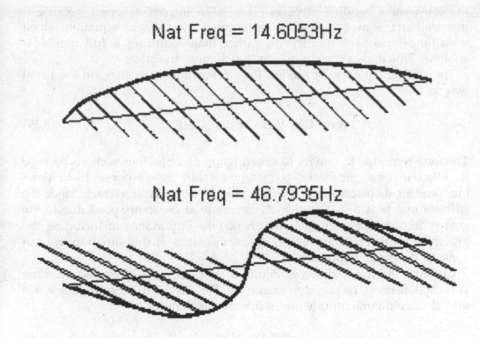

Nat Freq = 14.6053Hz

Nat Freq = 46.7935Hz

FIGURE 5.19
Modes Shape of First Two Modes (in One Direction).

importance of having regard to the potential and limitations of any mathematical idealization.

Even with very large cracks, the changes in natural frequencies do not provide an adequately sensitive indicator of machine health for use on operational machines, but as discussed in Chapter 9, these small changes can provide a useful tool in crack localization under laboratory/workshop testing. For initial fault detection, however, the forced response provides a better guide. This is discussed in the next subsection and attention is focused on machines with horizontal rotors. Those with vertical shafts are a little more complicated to study because without gravity dominating the crack dynamics, nonlinear effects assume greater significance.

In short, forced response is used to detect a crack, and natural frequency changes help to locate it. It should be emphasized that the problem of crack location is not trivial. Only rarely will a crack be obvious to visual inspection without the removal of surrounding components.

5.3.2 Forced Response

Mayes and Davies (1984) observed that for large machines with modest degrees of imbalance, the stress imposed by imbalance is much less than the self-weight bending stresses. This is an important point because it means that the motion can be represented by a linear equation, albeit with time-dependent coefficients, rather than requiring a full nonlinear analysis. This makes the study very much more tractable.

The general equation of motion for a rotor may be written, in the usual way as

$$\mathbf{K}x + \mathbf{C}\dot{x} + \mathbf{M}\ddot{x} = \mathbf{F}_{imbalance} + \mathbf{F}_{gravity} \tag{5.39}$$

The force term due to gravity has been ignored in previous sections because its effect in other problems is merely to shift the response by a time-independent displacement. However, this is not so with a crack: since the stiffness matrix is time varying, the response at constant speed due to the gravity force is itself time varying, hence the importance of including the gravity term. Before delving into the mathematics of this situation, we can consider the problem physically.

Consider the rotor shown in Figure 5.18. There will, of course, be some vertical deflection, or sag, due to gravity. For a rotor that has no crack, we may deduce the magnitude and shape of this deflection

$$\mathbf{K}\delta = \mathbf{M}\mathbf{S}g \tag{5.40}$$

where g is the acceleration due to gravity and $\mathbf{S}$ is a selection matrix picking only the vertical components, as used in Section 5.2.3. Inverting the stiffness matrix gives

$$\delta = \mathbf{K}^{-1}\mathbf{M}\mathbf{S}g \tag{5.41}$$

Provided that the stiffness is constant, the deflection in the shaft will also be constant and in fact the shaft rotates about this "distorted" axis rather than about the straight line connecting the bearings.

Assuming the crack is at the top of the shaft at an instant in time, there will be compressive stresses in the region of the crack and hence the crack will have no effect on the catenary. If we now consider the situation after half a revolution of the rotor, then the crack will be at the lower side of the rotor and will be subject to a tensile stress distribution. Hence the stiffness of the rotor will be reduced by the presence of the rotor. The situation now is as shown in Figure 5.15.

This will result in a greater gravity deflection or "sag" and hence the deflection of the rotor will be seen to have a once per revolution deflection, even in the absence of any unbalance forces. Now consider this from a more mathematical viewpoint. Let us first assume that there is no unbalance on the shaft and therefore the deflection is only due to gravity. The stiffness will now take the form

$$K = K_0 - K_c(t, u, v) \tag{5.42}$$

where u, v are the deflections due to imbalance. However, in machines the deflection due to gravity is much larger than the imbalance deflection and therefore the self-weight bending stresses dictate the dynamics of the crack and consequently, the stiffness can be assumed to be a function only of shaft orientation or equivalently, of time. Following Mayes and Davies (1984) the time dependence is taken (for an ax-symmetric rotor) as

$$K = K_0 - K_c \times (1 + \cos \Omega t)\big/_2 \tag{5.43}$$

This form assumes that the crack is fully open at time zero, but this is easily generalized by the inclusion of a phase parameter. For the present discussion, a phase of zero is assumed. At this stage, the motion can be analyzed by considering separately the deflection due to imbalance x, and that due to gravity, δ. With this notation, the equation of motion may be written as (neglecting gyroscopic terms for the sake of simplicity – but their inclusion is straightforward).

Then we may write

$$\left[K_0 - K_c \frac{(1 + \cos \Omega t)}{2} \right] (x + \delta) + C(\dot{x} + \dot{\delta}) + M(\ddot{x} + \ddot{\delta}) = MSg + M\Omega^2 \varepsilon \tag{5.44}$$

The stationary part of the solution is given by Equation 5.41.

In Equation 5.44 the stiffness (at constant speed Ω) is a function time only and, although the coefficients are time varying, the equation remains linear. This is a valid assumption for modest levels of vibration in which the bending moment due to self-weight is much greater than that due to imbalance. This is important because it means that the principle of super-position applies (i.e. solutions may be added together). Hence a solution is possible by summing the terms at different frequencies. Noting that the cosine term will give rise to higher-order excitation, consider the forcing at angular speed Ω. Having determined the static response in Equation 5.41, the solution at rotor speed is given by

$$\left[K_0 - \frac{K_c}{2} \right] (x_\Omega + \delta_\Omega) + C(\dot{x}_\Omega + \dot{\delta}_\Omega) + M(\ddot{x}_\Omega + \ddot{\delta}_\Omega) = M\Omega^2 \varepsilon + \frac{K_c}{2} \delta_\Omega \tag{5.45}$$

Separating out the part of the solution depending on gravity leads to

$$\left[K_0 - \frac{K_c}{2} \right] \delta_\Omega + C\dot{\delta}_\Omega + M\ddot{\delta}_\Omega = \frac{K_c}{2} \delta_0 = \frac{K_c}{2} K_0^{-1} MSg \tag{5.46}$$

This now the gravity deflection at shaft speed so the solution is given by

$$\delta_\Omega = \left[\mathbf{K}_0 - \frac{\mathbf{K}_c}{2} - \mathbf{M}\Omega^2 + j\mathbf{C}\Omega \right]^{-1} \frac{\mathbf{K}_c}{2} \mathbf{K}_0^{-1} \mathbf{M} \mathbf{S} g \qquad (5.47)$$

Thus, there will be a once per revolution variation in the rotor position due to purely the effect of gravity and the variation in shaft stiffness due to the crack. This will be of the form $\delta_\Omega e^{j\Omega t}$. This will usually be very small. There will also be higher-frequency terms but these are of less practical importance.

The important issue is the response due to imbalance. From Equation 5.45 the synchronous response to imbalance is given by $\mathrm{Re}(\mathbf{x}_0 e^{j\Omega t})$ where

$$\mathbf{x}_0 = \left[\mathbf{K}_0 - \frac{\mathbf{K}_c}{2} - \mathbf{M}\Omega^2 + j\mathbf{C}\Omega \right]^{-1} \mathbf{M}\Omega^2 \varepsilon \qquad (5.48)$$

This is just the standard expression for imbalance response. Of greater significance with regard to characteristics of crack rotors is the response at twice shaft speed. Referring back to Equation 5.45 and extracting second-order terms, the vibration at twice rotation speed is given by $\mathrm{Re}(\mathbf{x}_{2\Omega} e^{2j\Omega t})$ where

$$\mathbf{K}_0 \mathbf{x}_{2\Omega} + 2j\Omega \mathbf{C} \mathbf{x}_{2\Omega} - 4\mathbf{M}\Omega^2 \mathbf{x}_{2\Omega} = \frac{\mathbf{K}_c}{2}(\mathbf{x}_\Omega + \delta_\Omega) \qquad (5.49)$$

This is easily solved to give

$$\mathbf{x}_{2\Omega} = \left[\mathbf{K}_0 - 4\mathbf{M}\Omega^2 + 2j\Omega \mathbf{C} \right]^{-1} \frac{\mathbf{K}_c}{2}(\mathbf{x}_\Omega + \delta_\Omega) \qquad (5.50)$$

almost invariably the second term is negligible, that is, $|\mathbf{x}_\Omega| \gg |\delta_\Omega|$

An analysis has been presented by Sinou and Lees (2005) where a more rigorous discussion is given based on the harmonic balance approach. In that calculation the nonlinear influence of the imbalance deformation was included and the proportion of the crack open was recalculated at each step. The procedure outlined above is considered sufficient for most studies.

Equation 5.46 is, of course, central to any study of the dynamics of a cracked rotor, and there have been an enormous number of papers addressing the topic. The second harmonic, evaluated in Equation 5.50, provides an excellent indicator of crack damage. Although there are other, benign, sources of excitation at twice rotor speed (particularly of electrical machines) the corresponding levels of vibration tend to be very stable. Therefore, any unexplained change in the twice per revolution vibration component (increase or decrease) warrants an urgent investigation. Chapter 9 has a detailed account of the diagnosis of a crack on a turbo-alternator.

5.4 Torsional Excitation

The lateral vibrations of a rotating shaft are usually the most demanding in terms of potential damaging influence on a machine's operation and for this reason many machines carry the appropriate instrumentation. In any case, these features are readily quantifiable using standard techniques. The same is not true of torsional vibrations which are easily overlooked yet can lead to high stresses and major problems in machines. They can be overlooked as these vibrations occur strictly within the rotor and usually do not transmit to the surrounding structure or components: the exception to this is the case of machines incorporating gearboxes and a case involving this is discussed in Chapter 9.

The fact that these oscillations are not transmitted to the surroundings implies that they are not readily measured, but given appropriate instrumentation this difficulty can be overcome. A further consequence of considerable concern is that damping of torsional vibration is often restricted to the material damping of the rotor itself and will often be very light. This can be overcome in some systems by the introduction of special flexible couplings which provide damping by either the compression of rubber blocks or the movement of fluids as in the case of hydraulic clutches.

Torsional excitation is an area where further developments are needed to enhance monitoring capabilities. Usually, at least in the initial stages of an investigation, no quantitative information will be available of the torque variation, but it should not be overlooked as a potential source of problems. So, what can cause variations in transmitted torque? There are a number of possibilities:

a) In fluid handling machines (including pumps and turbines) the torque generated or absorbed will not be absolutely uniform but will inevitably be composed of a series of pulsations which occur each time an impeller or rotor blade passes a passage in the stator. Hence a detailed study of the torque profile in time shows a set of patterns, the precise form of which is determined by the detailed hydrodynamic behavior of the machine. In an ideal system these pulsations will occur many times per revolution (depending on the number of blades, vanes, and stages) but clearly any departure from uniformity in the components can give rise to excitation at significantly lower frequencies but with increased amplitude.

b) Similarly, in electrical machines there will always be a small high-frequency variation in torque related to the passage of rotor and stator windings. Once again, any irregularity of the spacings will reduce the frequency but increase the amplitude of variation.

c) Variations can arise from build imperfections or the effects of wear and increase clearances. They can also arise from changes in alignment – one of the many ways in which apparently distinct error mechanisms become related. Indeed, it is the interrelationships of most of the fault mechanisms which make the ultimate resolution of machine problems such a challenging area of activity.

d) Another source of torsional vibration arises in any system which includes a gearbox. Gears, like any mechanical system, inevitably are imperfect. Gear errors are subject to a series of national and international standards, but the effect of even modest errors can be very significant and the sensitivity of a system to gear errors depends on the characteristics of the complete machine, particularly the torsional critical speeds and their corresponding mode shapes.

There are two main ways in which torsional vibrations can be measured, but both are rather awkward for differing reasons. The first, and perhaps most obvious, of these is the use of a single toothed wheel on the rotor coupled to a proximity transducer forming a detailed tachometer signal. Because of the high inertia of many machines, the actual variation in angular velocity will be small even with high torque fluctuations and consequently considerable signal processing may be required to produce information of usable accuracy. However, the effort may lead to invaluable insights into machine operation. An extension to this concept is the use of two (or more) toothed wheels at different axial positions on the shaft. With careful calibration, such a system can directly measure the twist of a rotor. To put this in context, however, consider the order of magnitude of twist in, say, a 500 MW alternator rotor with a length of 12 m. Assuming that the torque acts across half the length, a straightforward calculation shows that the twist is of order 0.025 radians, corresponding to a displacement at the surface of the rotor of 6.4 mm or a time difference of 82 μsec at 3,000 rev/min. Note that the fluctuations are likely to be much smaller than this figure and consequently a combination of rapid data acquisition and effective signal processing is likely to be needed. Chapter 8 shows some appropriate techniques.

A more direct approach can be used to directly measure the transmitted torque on a rotor either by incorporating a torque meter into the machine or by fitting a coupling shaft with an appropriate strain gauge bridge arrangement. Such an approach has been used in the investigation of a long-standing problem on a boiler feed pump and was reported by Lees and Haines (1978). Because torque is measured directly, this approach is much less demanding in terms of resolution, but one has the practical problem of retrieving the information. At the time of the investigation (1976), this was transmitted from the shaft by a telemetry system, but a modern alternative would be a shaft mounted logger for subsequent recovery. In any event, the requirements go well beyond what might be considered standard instrumentation, but the information (and insight) gained is often very valuable. The study is discussed in Chapter 9, but here the underlying principles and methods of analysis are examined.

5.4.1 Dynamics of Gearboxes

The generation of torque fluctuations in gears is basically a displacement-controlled process rather than the more usual situation of force control. This means that the gears must move in such a manner as to accommodate the errors in their profile and twisting moments are generated which ensure that this motion can take place.

The system shown in Figure 5.20 comprises a driver unit (designated by inertia I_1) which drives a single stage gearbox (inertias I_2, I_3) through a flexible coupling, to a driven machine (inertia I_4) again coupled via a flexible coupling. This is a simple idealization of a system yet it retains sufficient detail to be representative of many practical arrangements. It should be noted, however, that this model only represents the torsional motion. In reality there will be coupling to the lateral motion which will be

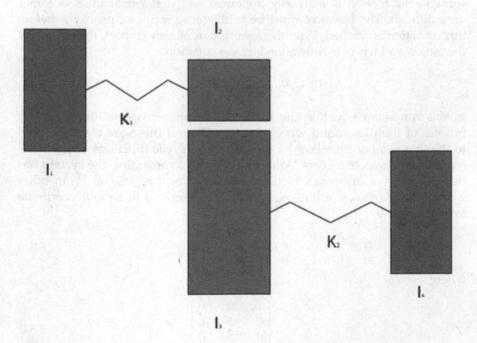

FIGURE 5.20
Idealized Torsional Model.

reflected by vibration monitored on the gearbox bearings. This measurement often provides a good indication of the machine condition, although
accurate modeling is frequently difficult owing to the problem of obtaining
an accurate assessment of stiffness properties. Concentrating on the torsional motion, the equation of motion may be written as

$$
\begin{bmatrix}
k_1 & -k_1 & 0 & 0 \\
-k_1 & k_1 & 0 & 0 \\
0 & 0 & k_2 & -k_2 \\
0 & 0 & -k_2 & k_2
\end{bmatrix}
\begin{Bmatrix}
\theta_1 \\ \theta_2 \\ \theta_3 \\ \theta_4
\end{Bmatrix}
+
\begin{bmatrix}
I_1 & 0 & 0 & 0 \\
0 & I_2 & 0 & 0 \\
0 & 0 & I_3 & 0 \\
0 & 0 & 0 & I_4
\end{bmatrix}
\frac{d^2}{dt^2}
\begin{Bmatrix}
\theta_1 \\ \theta_2 \\ \theta_3 \\ \theta_4
\end{Bmatrix}
=
\begin{Bmatrix}
T \\ P \\ \gamma P \\ D
\end{Bmatrix}
\tag{5.51}
$$

For the sake of the present discussion, T and D, representing the relevant
frequency components of input and output torques, may be taken as
steady values as here we wish to focus on the fluctuations introduced by
errors in the gear profile. Hence set $T = D = 0$. P is the transmitted torque
arising from errors in the gearbox and γ is the gear ratio.

Although Equation 5.51 shows four angular positions varying, there are in
fact only three degrees of freedom because the two inertias representing the
gear pinion and wheel are constrained to remain in contact. This is a reasonable assumption because if mating gears loose contact during operation, high-
impact stresses will develop on recontact, resulting in very rapid wear. In this
scenario the motion is markedly nonlinear and detailed analysis becomes
very difficult. The objective must be to diagnose incipient problems before
this condition is reached. With the assumption of gear contact, the motion of
the pinion and wheel is constrained by the equation

$$
\theta_2 + \gamma\theta_3 = \varepsilon_2(\Theta_2) + \gamma\varepsilon_3(\Theta_3) \tag{5.52}
$$

In this constraint equation the two error components $\varepsilon_2, \varepsilon_3$ are the error
profiles of the pinion and wheel respectively and these are clearly locked
to the two components. Both of these must be cyclic functions of positions
around the respective gears. After imposing this constraint, the system has
three degrees of freedom which are chosen to be $\{\theta_1 \quad \theta_2 \quad \theta_4\}^T$. In other
words, the dynamics of the system is to be described in a new coordinate
system given by

$$
\theta = \mathbf{S}\theta' + \mathbf{E} \tag{5.53}
$$

where

$$
\mathbf{S} =
\begin{bmatrix}
1 & 0 & 0 \\
0 & 1 & 0 \\
0 & -1/\gamma & 0 \\
0 & 0 & 1
\end{bmatrix}
\tag{5.54}
$$

and the error term is given by $\mathbf{E} = \{0 \quad 0 \quad \varepsilon_2(\Theta_2) + \varepsilon_3(\Theta_3) \quad 0\}^T$.

Now substituting Equation 5.53 into the equation of motion (5.51) yields (neglecting damping)

$$\mathbf{KS}\theta' + \mathbf{KE} + \mathbf{MS}\ddot{\theta}' + \mathbf{M\ddot{E}} = \mathbf{F} \qquad (5.55)$$

This equation is now premultiplied by $\mathbf{S}^T$ to give

$$\mathbf{S}^T\mathbf{KS}\theta' + \mathbf{S}^T\mathbf{KE} + \mathbf{S}^T\mathbf{MS}\ddot{\theta}' + \mathbf{S}^T\mathbf{M\ddot{E}} = \mathbf{S}^T\mathbf{F} = 0 \qquad (5.56)$$

Note that to a first approximation the error function $\mathbf{E}$ is a function of time. In cases of substantial errors, it may become nonlinear since it is dependent on the actual position of the gear teeth hence including the oscillation. On occasions this relatively minor effect may become important giving rise to frequency components in the measured signal which are not related to the rotational speed of the rotor. Of course, even greater nonlinear effects are observed if there is a loss of contact in the gears. In either scenario the basic conclusion would be that torsional oscillations are excessive and some palliative actions would be required.

Flexible couplings are discussed in Section 5.2 in connection with misalignment but at this point some general comments on the role of flexible couplings in machinery with regard to torsion are appropriate. There are various types of coupling, some based on the Hooke's joint principles while other types rely on transmission through rubber inserts. Their most common purpose is the reduction of transmitted forces and moments allowing misalignment between driver and driven units which may arise from large differential temperatures or external loads. Ideally, they possess zero lateral and axial stiffness, although this ideal is never fully realized. Some do, however, incorporate considerable torsional stiffness, and these may dominate the dynamics of a small machine such as the model presented in this section. On systems incorporating (torsionally) flexible rotors, the calculation of the stiffness matrix is slightly more involved but easily accomplished using standard FE methods. Flexible couplings of all types inevitably involve some degree of nonlinearity in their behavior.

The dynamics of the geared system is quite complicated and this is reflected in Equation 5.55. It is seen that the motion is driven by $\mathbf{E}$ and its second time derivative. In the model presented, only E_3 is nonzero and this takes the form

$$E_3 = \sum_{k=1}^{n_p} a_k \cos k(\Omega_p t + \theta_2) + \sum_{m=1}^{n_w} b_m \cos m(\Omega_w t + \theta_3) \qquad (5.57)$$

n_p, n_w are the numbers of teeth on the pinion and wheel respectively while the two rotational speeds are Ω_p and Ω_w in radians/sec.

For small oscillations, the nonlinear components may be neglected and this just yields sinusoidal variations at the two shaft speeds and harmonics, the amplitude of which depends on the error profiles on the two gears. However, if wear has become appreciable, and the machine is operating close to a torsional critical, the nonlinear effects of Equation 5.57 may become significant. The solution of Equation 5.56 under these circumstances is somewhat involved (not to say tedious!) but the physical form of the result can be readily appreciated. Some analysts in this area represent the forcing term as white noise, but although that way forward is reasonably effective, it is not very illuminating. Observe that because both θ terms have sinusoidal forms E_3 involves a series of products of sine waves. The net result of this is that the overall excitation involves many frequency components ranging up to twice the mesh frequency, that is, $2n_p\Omega_2/(2\pi)$. The spacing of the components is the reciprocal of the remeshing time, that is, the time for a given tooth on the pinion to meet a specified tooth on the wheel. Typically, this time will be given by $T_{rm} = {2\pi n_p}/{\Omega_2}$ assuming n_p, n_w do not have a common factor and this time may be many seconds. Hence the excitation is a wide band spectrum whose shape is dependent on the error profiles, expresses in the parameters a_k, b_m.

It is not surprising therefore that the system will select those components of excitation close to a natural frequency. Consequently, in addition to harmonics of shaft speed and harmonics thereof, the system may also generate components of vibration at the natural frequencies, irrespective of shaft speed.

The appearance of harmonics is characteristic of either nonlinearity or sometimes time-varying stiffness or mass terms. Both of these are highly pertinent to the understanding of rotating machinery and these issues are discussed in the next section. A rigorous mathematical analysis is not presented, but this brief discussion conveys some physical insight into the phenomena. The problem is discussed further by Lees et al. (2011).

5.4.1.1 Tooth Stiffness Effects

In the analysis presented in the preceding subsection, the gear teeth were assumed to be rigid. This approach is suitable for problems which arise at low to moderate frequencies, and the precise demarcation is dependent on the plant involved. At higher frequencies, for example in studying noise emission, the gear mesh flexibility should be incorporated into the study. This is achieved by modifying the equation of constraint (Equation 5.52) to be

$$\theta_2 + \gamma\theta_3 = \varepsilon_2(\Theta_2) + \gamma\varepsilon_3(\Theta_3) - {P}/{(k_m \times r_2)} \tag{5.58}$$

where r_2 is the radius of the appropriate gear pinion. Note that the analysis changes a little in that there will be, in this case five variables, but with five equations. k_m is the gear mesh stiffness and there are published methods for determining appropriate values, although it remains a topic of considerable discussion. The logic of the analysis remains the same. It is often difficult, if not impossible, to obtain adequate information to perform a really accurate calculation but the real value of such calculations is to give insight into sensitivities.

However, the issue of mesh flexibility is rather involved and can vary as teeth mesh and remesh. This is a specialized topic and can give rise to instability (see e.g. Lin and Parker, 2002)

5.4.2 Mixing of Flexural, Axial, and Torsional

The analysis given in the preceding section is clearly very simplified, but this is often sufficient to gain a real understanding of plant behavior. In particular, the study involves only torsional motion, but very often other modes of vibration will also be involved. Often the torsional fluctuations will be reflected in vibration at the gearbox bearings but with single-helical and bevel gears, axial motion is also excited.

The analysis of 5.4.1 is readily extended to include lateral and axial degrees of freedom but clearly the study becomes unsuitable for an analytical discussion. Numerical studies, however, provide a convenient means of gaining understanding of key parameters.

5.5 Nonlinearity

To begin this discussion, consider a single degree of freedom system described by the equation

$$Kx + C\dot{x} + M\ddot{x} + K_1 x^2 = Pe^{j\omega t} \tag{5.59}$$

If the nonlinear term $K_1 = 0$ then the solution for the steady state is easily achieved by assuming a form $x = x_0 e^{j\omega t}$. Substituting this into Equation 5.59 gives

$$(K + j\omega C - \omega^2 M)x_0 e^{j\omega t} = Pe^{j\omega t} \tag{5.60}$$

The simplification is straightforward because the first and second derivatives of the function $e^{j\omega t}$ are just constants multiplied by the original function: this basically is the reason why Fourier methods are so useful in the study of vibration problems.

This assumes of course that K, M, C are all constant, varying with neither x nor time. The sinusoidal variations cancel and the expression is easily inverted to give

$$x_0 = \frac{P}{K - \omega^2 M + j\omega C} \tag{5.61}$$

From this the full solution is easily recovered. Note that a transient solution of Equation 5.59 (with $K_1 = 0$) is only marginally more arduous. The real difficulty arises when there is nonlinearity present (i.e. $K_1 \neq 0$) because in that case the equation analogous to 5.60 becomes

$$(K - \omega^2 M + j\omega C)x_0 e^{j\omega t} + K_1 x_0^2 e^{j2\omega t} = P e^{j\omega t} \tag{5.62}$$

Unlike the linear case, it is not possible to cancel the sinusoidal term to simplify the solution procedure. This dilemma is solved by appreciating that the substitution made for x is not adequate for this case as there cannot be a single frequency component involved in the solution. Indeed, this is the important point of this section.

Therefore, a new form of solution is taken as

$$x = \sum_{n=-\infty}^{\infty} a_n e^{jn\omega t} \tag{5.63}$$

where, of course, the coefficients a_n are not restricted to being real (complexity of the coefficients a_n just corresponds to a shift of phase). The technique for finding a solution comprises substitution of Equation 5.63 into 5.62 then seeking to match each harmonic term. This procedure is known as the harmonic balance method and is discussed in many text books. It is, however, illuminating to consider the form of the equation with this form of solution substituted. Equation 5.62 becomes

$$\sum_{n=-\infty}^{\infty} (K - n^2\omega^2 M + jn\omega C)a_n e^{jn\omega t} + K_1 \left(\sum_{n=-\infty}^{\infty} a_n e^{jn\omega t} \right) \left(\sum_{n=-\infty}^{\infty} a_n e^{jn\omega t} \right) = P e^{j\omega t} \tag{5.64}$$

Owing to the double summation, the balancing of the different frequency components is a little involved. The essential point is that the different values a_n are interdependent and consequently nonlinear systems will generate harmonics of the forcing frequency and this indeed is normally the factor, which reveals the presence of nonlinear features in a system. It should be noted that although the system described by Equation 5.59 represents the simplest possible form of nonlinearity, the conclusion of the interdependency of various harmonic terms is completely general.

This is similar to the behavior of gears discussed in the previous section and more fully in Chapter 9. In that case, the nonlinearity means that the effective (displacement controlled) forcing involves a product of two series of Fourier terms (in this case the gear error profiles). This gives rise to excitation at all sum and difference frequencies. The detection of this type of spectrum interaction presents one technique of detecting nonlinear behavior via the so-called bispectrum (see Sinha, 2006). This is discussed in Chapter 8.

5.6 Instability

In Chapter 2 it is shown how an Argand diagram can effectively be used to identify unstable modes. It is now appropriate to discuss some of the main causes of instability in rotating machines. These instabilities are one of the crucial distinctions between the dynamics of machines and that of structures. Fundamentally this is because energy is being supplied to the rotor and under some circumstances some energy can be channeled into vibration. A mode is unstable if a perturbation grows in time rather than decays. This will occur if the damping becomes negative. In the following three subsections a brief discussion is given of the predominant sources of instability.

5.6.1 Oil-Whirl and Oil-Whip

These two closely associated phenomena stem from the dynamics of oil film bearings. They both arise from the nonlinear behavior of oil-bearings, although their onset can be discussed in linear terms. Oil-whirl is often termed "half-speed whirl" but this is a rather misleading terminology. The phenomenon of oil-whirl does give rise to nonsynchronous vibration, typically at a frequency 25–40% of shaft speed. The often quoted "half speed" is just the upper limit of frequency at which it can, but very rarely does, occur.

Simple linear models of the operation of oil film bearings are given in Chapter 6, where it can be seen that both stiffness and damping are strongly dependent on the load imposed on the bearing. Hence if a bearing carries little or no load, due to alignment errors, then the damping will be very much reduced. In this instance accurate prediction can be gained only from a fully nonlinear study – a demanding requirement. Physically this whirl is caused by a rotation of the full oil film and hence the normal mode of operation is suspended.

A closely associated fault is the very dangerous condition oil-whip. Fundamentally this has the same origin as oil-whirl but rather than giving rise to a vibration at some fraction of shaft speed, it locks into a free–free mode of the rotor. Because of this only the internal damping of the rotor itself is effective and this will usually be very small and consequently very high levels of vibration can be excited. This can be very damaging to plant but fortunately it is quite a rare condition.

Both oil-whirl and oil-whip can be avoided by ensuring appropriate alignment. Where necessary different bearing geometry can be used or, sometimes, simply changing the oil viscosity.

5.6.2 Rotor Asymmetry

Many rotors have some degree of asymmetry in their construction: this is particularly true of electrical rotors. To examine the effects of this, it is appropriate to express the motion in a coordinate system rotating with the shaft. To transfer between stationary and rotating systems, the transformation matrix is given by

$$\mathbf{T} = \begin{bmatrix} \cos\Omega t & -\sin\Omega t \\ \sin\Omega t & \cos\Omega t \end{bmatrix} \tag{5.65}$$

If the displacements in the stationary frame are u, v, in the rotating frame these become u_r, v_r and differentiating

$$\begin{Bmatrix} \ddot{u} \\ \ddot{v} \end{Bmatrix} = \mathbf{T}\left(\begin{Bmatrix} \ddot{u}_r \\ \ddot{v}_r \end{Bmatrix} - \Omega^2 \begin{Bmatrix} u_r \\ v_r \end{Bmatrix} \right) + 2\dot{\mathbf{T}} \begin{Bmatrix} \dot{u}_r \\ \dot{v}_r \end{Bmatrix} \tag{5.66}$$

Similarly, transforming in the other direction

$$\begin{Bmatrix} \ddot{u}_r \\ \ddot{v}_r \end{Bmatrix} = \mathbf{T}^T\left(\begin{Bmatrix} \ddot{u} \\ \ddot{v} \end{Bmatrix} - \Omega^2 \begin{Bmatrix} u \\ v \end{Bmatrix} \right) + 2\dot{\mathbf{T}}^T \begin{Bmatrix} \dot{u} \\ \dot{v} \end{Bmatrix} \tag{5.67}$$

At this point the equation of motion of the shaft may be written in the stationary coordinates as

$$m \begin{Bmatrix} \ddot{u} \\ \ddot{v} \end{Bmatrix} = \begin{Bmatrix} f_x \\ f_y \end{Bmatrix} \tag{5.68}$$

which is readily translated into the rotating system as

$$m\mathbf{T}^T\mathbf{T}\left(\begin{Bmatrix} \ddot{u}_r \\ \ddot{v}_r \end{Bmatrix} - \Omega^2 \begin{Bmatrix} u_r \\ v_r \end{Bmatrix} \right) + 2m\mathbf{T}^T \dot{\mathbf{T}} \begin{Bmatrix} \dot{u}_r \\ \dot{v}_r \end{Bmatrix} = \mathbf{T}^T \begin{Bmatrix} f_x \\ f_y \end{Bmatrix} \tag{5.69}$$

But in this rotating frame, the elastic forces become

$$T^T \begin{Bmatrix} f_x \\ f_y \end{Bmatrix} = \begin{bmatrix} k_{xr} & 0 \\ 0 & k_{yr} \end{bmatrix} \begin{Bmatrix} u_r \\ v_r \end{Bmatrix} \tag{5.70}$$

Dividing this equation by m, and multiplying out the matrices gives the equation of motion as

$$\left\{ \begin{array}{c} \ddot{u}_r \\ \ddot{v}_r \end{array} \right\} + 2\Omega \left[\begin{array}{cc} 0 & -1 \\ 1 & 0 \end{array} \right] \left\{ \begin{array}{c} \dot{u}_r \\ \dot{v}_r \end{array} \right\} + \left[\begin{array}{cc} \omega_x^2 - \Omega^2 & 0 \\ 0 & \omega_y^2 - \Omega^2 \end{array} \right] \left\{ \begin{array}{c} u_r \\ v_r \end{array} \right\} = \left\{ \begin{array}{c} 0 \\ 0 \end{array} \right\} \quad (5.71)$$

where $\omega_x^2 = k_{xr}/m$, $\omega_y^2 = k_{yr}/m$.

For free response, searching for a solution of the form $\left\{ \begin{array}{c} u_r \\ v_r \end{array} \right\} = \left\{ \begin{array}{c} u_0 \\ v_0 \end{array} \right\} e^{st}$, the eigenvalues of the system are given by the solution of

$$det \left[\begin{array}{cc} s^2 + \omega_x^2 - \Omega^2 & -2\Omega s \\ 2\Omega s & s^2 + \omega_y^2 - \Omega^2 \end{array} \right] = 0 \quad (5.72)$$

This leads to the quartic equation

$$s^4 + \left(\omega_x^2 + \omega_y^2 + 2\Omega^2 \right) s^2 + \left(\omega_x^2 - \Omega^2 \right) \left(\omega_y^2 - \Omega^2 \right) = 0 \quad (5.73)$$

Clearly there are four solutions to the equation which are either real or in conjugate pairs and their product is $\left(\omega_x^2 - \Omega^2 \right) \left(\omega_y^2 - \Omega^2 \right)$.

If this is quantity is less than zero then there must be a root with a positive real part, implying system instability. Of course, stability is aided by the introduction of external damping and the argument may be reworked and it is easily shown that with damping c, the criterion becomes

$$\left(\omega_x^2 - \Omega^2 \right) \left(\omega_y^2 - \Omega^2 \right) + c_0^2 \Omega^2 < 0 \quad (5.74)$$

where $c_0 = c/m$. Friswell *et al.* (2010) give a full discussion of this with some example plots. This rather complicated analysis gives the simple result that, in the absence of external damping, a rotor will be unstable in the frequency range between the natural frequencies in its two orthogonal directions, that is, if $\omega_x < \Omega < \omega_y$.

5.6.3 Rotor Damping

Perhaps the most counterintuitive source of instability is damping on the rotor itself. While external damping sources always aid stability, this is not true of damping on the rotating structure and damping of this type can arise from various sources. The difficulty arises basically because of the changing phase of the forces. To understand this point, consider the damping force in the rotating frame, denoted as f_{dr}, which are given as

$$f_{dr} = -c_i \left\{ \begin{matrix} \dot{u}_r \\ \dot{v}_r \end{matrix} \right\} \tag{5.75}$$

Using Equation 5.67, the damping force in the stationary frame may by written as

$$f_d = -c_i \mathbf{T} \left(\mathbf{T}^T \left\{ \begin{matrix} \dot{u} \\ \dot{v} \end{matrix} \right\} + \dot{T}^T \left\{ \begin{matrix} u \\ v \end{matrix} \right\} \right) \tag{5.76}$$

In the fixed frame of reference, the equation of motion becomes

$$\left\{ \begin{matrix} \ddot{u} \\ \ddot{v} \end{matrix} \right\} + 2\omega_n(c_e + c_i) \left\{ \begin{matrix} \dot{u} \\ \dot{v} \end{matrix} \right\} + \begin{bmatrix} k & c_i\Omega \\ -c_i\Omega & k \end{bmatrix} \left\{ \begin{matrix} u \\ v \end{matrix} \right\} = 0 \tag{5.77}$$

where c_e is the external damping. Dividing this equation by m gives

$$m \left\{ \begin{matrix} \ddot{u} \\ \ddot{v} \end{matrix} \right\} + (\xi_e + \xi_i) \left\{ \begin{matrix} \dot{u} \\ \dot{v} \end{matrix} \right\} + \begin{bmatrix} \omega_n^2 & 2\omega_n\xi_i\Omega \\ -2\omega_n\xi_i\Omega_i & \omega_n^2 \end{bmatrix} \left\{ \begin{matrix} u \\ v \end{matrix} \right\} = 0 \tag{5.78}$$

In this equation the substitutions have been made $\omega_n = \sqrt{k/m}$, $\xi_e = \frac{c_e}{2m\omega_n}$ and $\xi_i = \frac{c_i}{2m\omega_n}$.

Seeking a solution of the form $\left\{ \begin{matrix} u(t) \\ v(t) \end{matrix} \right\} = \left\{ \begin{matrix} u_0 \\ v_0 \end{matrix} \right\} e^{st}$ leads to the expression

$$\begin{bmatrix} s^2 + 2\omega_n(\xi_e + \xi_i)s + \omega_n^2 & 2\xi_i\omega\omega_n \\ -2\xi_i\omega\omega_n & s^2 + 2\omega_n(\xi_e + \xi_i)s + \omega_n^2 \end{bmatrix} \left\{ \begin{matrix} u_0 \\ v_0 \end{matrix} \right\} = 0 \tag{5.79}$$

Taking the determinant of the matrix and then taking the square root gives an equation for the eigenvalues as

$$s^2 + 2(\xi_e + \xi_i)\omega_n js + (\omega_n^2 \pm 2j\xi_i\omega\omega_n) = 0 \tag{5.80}$$

Observing that at the threshold of instability, the real part of the solution must be zero, then taking the negative sign in Equation 5.80,

$$2(\xi_e + \xi_i)\omega_n\omega - 2j\xi_i\omega\omega_n = 0 \tag{5.81}$$

Since it is clear that the rotor is stable at low speed, following this fairly complicated analysis, the straightforward condition is reached that the rotor will be stable provided that

$$\Omega < \omega_n \left(1 + \frac{\xi_i}{\xi_e} \right) \tag{5.82}$$

This is simply saying that internal damping is destabilizing, but external damping may be sufficient to overcome the effect. This plays an important part in many machines.

5.7 Interactions and Diagnostics

Inevitably the various faults discussed in these chapters are discussed individually but it should be appreciated that many, if not all, of them can be interrelated. This has important practical consequences as it means that it can be difficult to determine the sequence of events giving rise to observed behavior.

5.7.1 Synchronous Excitation

If the symptom is synchronous once per revolution vibration, the root cause could be imbalance or a rotor bend and Chapter 4 discusses the distinction between the two. A bend, however, may present a complicated problem: is it a permanent bend or one brought on by thermal effects? If thermal, then is the case a rotor rub (discussed in Chapter 6), or is it caused by irregularities in the fluid flow in the machine. If the cause is rub, then there are a number of possible causes

a) An initial bend
b) An imbalance
c) Misalignment of bearings
d) Incorrect clearance
e) Thermal bend

It is difficult to give precise rules by which the engineer can distinguish between these possibilities. Often additional data such as bearing oil temperatures will come into consideration, as this can give some insight into bearing loads. Performance data can help give indication of any variation in clearances.

But the possible causes of the problem are not independent. For instance, a rotor may have a small initial imbalance but be in a machine with tight clearance which also suffers some misalignment. It is often these sorts of problems where a number of issues interact that give rise to a fault condition. It may well be that none of the contributory factors, in themselves, would result in unacceptable behavior.

5.7.2 Twice per Revolution Excitation

This is a little less common but still may originate from a combination of root causes. Apart from the possibility of external forcing, the primary reasons for harmonic generation are either some form of nonlinear beha-vior (discussed in Section 5.5) or some periodic variation in the system parameters. In many respects the effects of these two is similar, but from an analysis point of view there is an important distinction: if parameters vary cyclically but the equation remains linear, the concept of superposi-tion applies and the contribution of different modes may be added. When the equation is truly nonlinear this concept is no longer valid.

The conditions that show harmonic excitation include:

a) Rotor crack
b) Misalignment
c) Other types of nonlinearity
d) Electric influence
e) Rotor asymmetry

The difficulty is that many of these faults are interrelated both within this group and with the factors giving rise to synchronous vibration. Mis-alignment and imbalance can both give rise to rotor rub which can in turn lead to a thermal bend and/or the Newkirk effect (discussed in Chapter 6) of a slow cyclic behavior and it is often difficult to clarify the root cause. While simple rules can help, the only real solution is the development of as complete a picture of the machine's operation as possible. This will include design, build, and maintenance practices as well as vibration records.

5.7.3 Asynchronous Vibration

In addition to the components of vibration occurring at shaft speed and its multiples, it is important to give due attention to parts of the vibration signal without a simple relationship to shaft speed. Particular attention is needed to subharmonic components below half shaft speed. The principal causes for these problems are:

a) External influences
b) Fluid film, or other instabilities
c) Component looseness
d) Rubs

Here again, the categorization is by no means clear-cut. Rubs can generate asynchronous excitation; indeed, under some circumstances it can be chao-tic, but as discussed, it can also give synchronous and harmonic behavior. Fluid film instabilities may arise from misalignment or, of course, they may be errors in build or some component failure. Of the four categories here, the easiest to identify is normally the external influences, simply by applying some limited measurement to neighboring plant.

With the rather complicated set of interrelationships between the vibration signals of varying faults, insight into the physical operation of the machine provides the best prospect for problem resolution.

5.8 Closing Remarks

Chapters 4 and 5 together have described some of the main faults observed in rotating machinery, but so far one important category of fault has not been discussed in any depth. This is rub, or, to put it more generally, the interaction of a rotor with its stator. This can give rise to a variety of effects and needs some detailed explanation: this is why a complete chapter – Chapter 6 – is assigned to this topic.

Problems

5.1 A machine comprises two identical rotors connected by a rigid coupling. The diameter of the rotor is 200 mm and at the midspan of each rotor is a disc of mass 250 kg. The rotors are rigidly coupled and the influence of gyroscopic effects is negligible. Each of the bearings has a constant stiffness of 10^6 N/m and damping of 1,000 Nsec/m. If the imbalance on disc 1 is 0.004 kgm, and proportional damping can be assumed giving 1% in the first mode, calculate the response at the bearings over the running range of 0–3,000 rev/min.

5.2 Explain the difficulties of aligning a large multibearing machine with flexible rotors.

   a) Describe the circumstances under which a machine mounted on oil film bearings can show vibration whose frequency is below half the rotor speed.

   b) With the aid of an appropriate diagram, illustrate a method to obtain correct alignment

   c) Two identical rotors are to be connected together. Under the influence of their own weight, the ends of the shaft make an angle of 0.17 min of arc with the vertical. If each end has an overhang of 0.5 m and the spacing between bearings on each shaft is 4 m, what should the alignment be? *[Keep the first rotor fixed]*

   d) If it is now required to keep the two end bearings at the same level as each other, what should be the positions of the other two bearings?

5.3   Assuming that the machine in Question 5.1 has been arranged with all the bearings in line, determine the bending moment at the coupling when the machine is stationary. With the rotors decoupled, determine the catenaries of the two separate rotors.

5.4   A motor whose rotor has a moment of inertia of 200 kgm^2 is connected through a flexible coupling of stiffness 3×10^6 Nm/rad to a gearbox which is then connected via a coupling of stiffness 8×10^5 Nm/rad to a pump with a rotor with a moment of inertia 150 kgm^2. The inertia of the pinion and wheel are 20 and 300 kgm^2 respectively and the gear ratio is 2. Determine the torsional natural frequencies. If there is a cyclic error on the wheel of 10 μm, what is the magnitude of torque oscillations when the input shaft is rotating at 3,000 rev/min.

5.5   A uniform rotor 8 m in length is supported on rigid short bearings at either end. A 400 kg disc is mounted at midspan. Determine the change of the first natural frequency due to a chordal crack, 2 m from end 1 with depth of 0.25, 0.5, 0.75, and 1 times the radius.

5.6   A generator is known to have critical speeds at 600 and 2,200 rpm, and this shows on the once per revolution run-down curve. However, over a period of 24 h, the signal at twice running speed gradually increases and a second run-down shows that the twice per rev component has a small peak at 1,100 rpm, together with a very much larger increase at 300 rpm. What can you conclude about the nature and location of the fault? Give a reasoned argument for your conclusions. (Assume the rotor can be modeled as a uniform beam on rigid supports.)

References

Al-Hussein, K.M. and Redmond, I., 2002, Dynamic response of two rotors connected by rigid mechanical coupling with parallel misalignment, *Journal of Sound and Vibration*, 249, pp. 483–498.

Davies, W.G.R. and Mayes, I.W., 1984, The vibrational behaviour of a multi-shaft, multi-bearing system in the presence of a propagating transverse crack, *Journal of Vibration, Acoustics, Stress and Reliability in Design*, 106, pp. 146–153.

Dimaragonas, A.D., 1996, Vibration of cracked structures: A state of the art review, *Engineering Fracture Mechanics*, 55(5), pp. 831–857.

Gasch, R., 1993, A survey of the dynamic behaviour of a simple rotating shaft with a transverse crack, *Journal of Sound and Vibration*, 160, pp. 313–332.

Henry, T.A. and Okah-Avae, B.E., 1976, Vibrations in cracked shafts, *Institution of Mechanical Engineers Conference on Vibrations in Rotating Machinery*, Cambridge, UK, pp. 15–19.

Lees, A.W., 2007, Misalignment in rigidly coupled rotors, *Journal of Sound and Vibration*, 305, pp. 261–371.

Lees, A.W. and Friswell, M.I., 2001, The vibration signature of chordal cracks in asymmetric rotors, *19th International Modal Analysis Conference*, Kissimmee, FL.

Lees, A.W., Friswell, M.I. and Litak, G., 2011, Torsional vibration of machines with gear errors, *9th International Conference on Damage Assessment of Structures (DAMAS) Location*, University of Oxford, St Anne's College, Oxford, UK.

Lees, A.W. and Haines, K.A., 1978, Torsional vibrations of a boiler feed pump, *Transactions of American Society of Mechanical Engineers*, 77-DET–28, 100(4), pp. 643–677.

Lin, J. and Parker, R.G., 2002, Planetary gear parametric instability caused by mesh stiffness variation, *Journal of Sound & Vibration*, 249(1), pp. 129–145.

Mayes, I.W. and Davies, W.G.R., 1976, The behaviour of a rotating shaft system containing a transverse crack, *Institution of Mechanical Engineers Conference on Vibrations in Rotating Machinery*, Cambridge, UK, pp. 53–64.

Mayes, I.W. and Davies, W.G.R., 1984, 'Analysis of the response of a multirotor-bearing system containing a transverse crack, *Journal of Vibration, Acoustics, Stress and Reliability in Design*, 106, pp. 139–145.

Muszynska, A., 2005, *Rotordynamics*, CRC Press, Taylor & Francis, Boca Raton, FL.

Patel, T.J. and Darpe, A.K., 2009, Vibration response of misaligned rotors, *Journal of Sound & Vibration*, 325, pp. 609–628.

Penny, J.E.T. and Friswell, M.I., 2002, Simplified model of rotor cracks, *ISMA Conference on Noise & Vibration 27*, Leuven, Belgium, pp. 607–615.

Redmond, I., 2010, Study of a misaligned flexibly coupled shaft system having nonlinear bearings and cyclic coupling stiffness – Theoretical model and analysis, *Journal of Sound and Vibration*, 329, pp. 700–720.

Sinha, J.K., 2006, Bi-spectrum for identifying crack and misalignment in shaft of a rotating machine, *Smart Structures and Systems*, 2, pp. 47–60.

Sinha, J.K., Lees, A.W. and Friswell, M.I., 2004, Estimating unbalance and misalignment of a flexible rotating machine from a single run-down, *Journal of Sound and Vibration*, 272(3–5), pp. 967–989.

Sinou, J.J. and Lees, A.W., 2005, The influence of cracks in rotating shafts, *Journal of Sound and Vibration*, 285, pp. 1015–1037.

6

Rotor–Stator Interaction

6.1 Introduction

Up to this point, rotating shafts have been considered in isolation, somehow remote from their surroundings. While this is a necessary starting point in consideration of such systems, it neglects important facets of machine behavior. In many instances, the rotor will be surrounded by a working fluid and it is the conditions within and around these fluid regions that determine the operational effectiveness of the machine. In this sense, the study of rotor-dynamics is an enabling technology that paves the way for effective machine operation: it is not an end in itself.

Whilst the rotor is at the heart of any rotating machine, useful work is provided only through the interaction of the rotating component with its stator, and hence the interaction between the two is of crucial significance. Chapters 4 and 5 discuss a range of common faults related to the rotor but in this chapter, the scope is widened to consider the rotor's relationship with the stator. There are several ways in which the interaction between stator and rotor can be categorized, but a useful division is:

a) Interaction through the bearings – to some extent this has been discussed in Chapter 4, but here further details are outlined as to the performance of oil film bearings in particular.

b) The interaction via the working fluid –through both internal channels that exert force and moments on the rotor and fluid-filled bushes and seals which act as subsidiary bearings.

c) Direct interaction with stator through contact – this may take on one of two forms.

   i) Collision and recoil, involving impulsive forces.

   ii) Slow rubbing leading to heating.

A study of these factors is essential in understanding the operation of machines and it is crucial in recognizing a number of fault conditions. Overall, this is an area of considerable complexity and only an overview can be given here.

6.2 Interaction through Bearings

6.2.1 Oil Journal Bearings

While many smaller machines are supported on rolling element bearings, oil journal bearings (in a range of subtle variations) predominate on large machines. Their predominance is primarily due to their high load bearing capacity. The analysis of these bearings is somewhat involved and in general requires a numerical solution of Reynold's equation. An added difficulty is that often knowledge of the system parameters, such as clearance and surface condition, is limited. For a given bearing geometry, three parameters (or, rather, sets of parameters) are essential to characterize the influence on a machine's behavior, namely the stiffness, damping, and load capacity. Of these, the stiffness and damping terms are matrices which are functions of shaft rotation speed.

Nevertheless, closed-form solution can be established for short bearings (i.e. $L \ll D$, where L is the axial length of the bearing and D its diameter) as in this case the assumption can be made that the variations in pressure are much greater in the axial direction as compared to the circumferential. Hamrock *et al.* (2004) give a full analysis of this and other cases and a more specific discussion is given by Friswell *et al.* (2010). These analytic descriptions do make some important assumptions, namely that the process is isothermal, that the flow is laminar, and that there is no cavitation within the bearing. A number of books give the analysis of short bearings but it is worth quoting the results here for the sake of illustration. Childs (1993) shows that the radial and tangential forces are given by

$$f_r = -\frac{D\Omega\eta L^3 \varepsilon^2}{2c^2(1-\varepsilon^2)^2} \qquad f_t = -\frac{\pi D\Omega\eta L^3 \varepsilon}{2c^2(1-\varepsilon^2)^{3/2}} \qquad (6.1)$$

where D is the journal (shaft) diameter, Ω is the rotational speed (in radians per second), η is the oil viscosity, L is the (axial) width of the bearing, c the radial clearance, and ε is the shaft eccentricity as a proportion of c.

From the relations in 6.1, it is clear that the modulus of f is given by

$$|f| = \sqrt{f_r^2 + f_t^2} = -\frac{\pi D\Omega\eta L^3 \varepsilon}{8c^2(1-\varepsilon^2)^2}\left(\left(\frac{16}{\pi}-1\right)\varepsilon^2 + 1\right)^{1/2} \qquad (6.2)$$

It is convenient to plot the normalized force f_N such that $|f| = \frac{\pi D\Omega\eta L^3}{8c^2}f_N$

Figure 6.1 shows the variation of force with eccentricity, clearly a nonlinear relationship.

The derivative of the force (i.e. the stiffness) is also nonlinear, and this has major importance. Therefore, given a load on the bearing it is important to establish the equilibrium running position of the shaft within the bearing. After a little manipulation it can be shown that if a vertical force f is applied the bearing, the shaft will have an eccentricity ε which satisfies the equation

$$\varepsilon^8 - 4\varepsilon^6 + \left(6 - S_s^2(16 - \pi^2)\right)\varepsilon^4 - \left(4 + \pi^2 S_s^2\right)\varepsilon^2 + 1 = 0 \qquad (6.3)$$

where S_s is the *modified Sommerfeld number* (or the *Ocvirk number*) and is given by $S_s = \frac{D\Omega\eta L^3}{8fc^2}$. Although there are eight solutions to this equation, only one of these will be between zero and one and this solution gives the shaft eccentricity relative to the clearance.

Figure 6.2 shows the variation of the equilibrium position of the shaft within the bearing assuming that the load is vertical. (Note that this is usually, but not always, the case.)

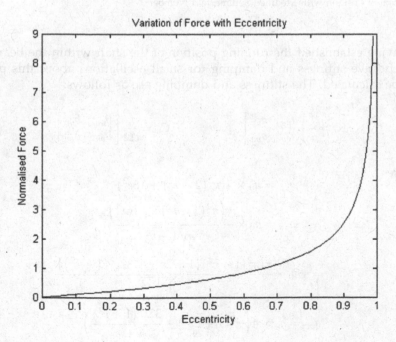

FIGURE 6.1
Normalized Force Variation.

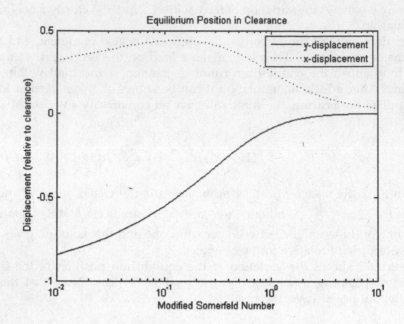

FIGURE 6.2
Variation of Position with Modified Sommerfeld Number.

Having established the running position of the shaft within the bearing, the effective stiffness and damping for small oscillations about this point can be calculated. The stiffness and damping are as follows:

$$\mathbf{K}_e = \frac{f}{c}\begin{bmatrix} a_{uu} & a_{uv} \\ a_{vu} & a_{vv} \end{bmatrix} \qquad\qquad \mathbf{C}_e = \frac{f}{c\Omega}\begin{bmatrix} b_{uu} & b_{uv} \\ b_{vu} & b_{vv} \end{bmatrix} \qquad (6.4)$$

where

$$a_{uu} = h_0 \times 4\left(\pi^2\left(2 - \varepsilon^2\right) + 16\varepsilon^2\right)$$

$$a_{uv} = h_0 \times \frac{\pi\left(\pi^2\left(1 - \varepsilon^2\right)^2 - 16\varepsilon^4\right)}{\varepsilon\sqrt{1 - \varepsilon^2}}$$

$$a_{vu} = -h_0 \times \frac{\pi\left(\pi^2\left(1 - \varepsilon^2\right)\left(1 + 2\varepsilon^2\right) + 32\varepsilon^2\left(1 + \varepsilon^2\right)\right)}{\varepsilon\sqrt{1 - \varepsilon^2}}$$

$$a_{vv} = h_0 \times 4\left(\pi^2\left(1 + 2\varepsilon^2\right) + \frac{32\varepsilon^2\left(1 + \varepsilon^2\right)}{1 - \varepsilon^2}\right)$$

$$b_{uu} = h_0 \times \frac{2\pi\sqrt{1-\varepsilon^2}\left(\pi^2\left(1+2\varepsilon^2\right)-16\varepsilon^2\right)}{\varepsilon}$$

$$b_{uv} = b_{vu} = -h_0 \times 8\left(\pi^2\left(1+2\varepsilon^2\right)-16\varepsilon^2\right)$$

$$b_{vv} = h_0 \times \frac{2\pi\left(\pi^2\left(1-\varepsilon^2\right)^2+48\varepsilon^2\right)}{\varepsilon\sqrt{1-\varepsilon^2}}$$

and

$$h_0 = \frac{1}{\left(\pi^2(1-\varepsilon^2)+16\varepsilon^2\right)^{3/2}} \tag{6.5}$$

These rather complicated expressions suffice to give insight in the behavior of hydrodynamic bearings. At zero speed, the shaft will be in contact with the housing and gradually move away as pressure is developed in the oil film. It will be clear that these expressions for stiffness and damping are highly nonlinear and hence any calculated stiffness and damping terms are valid only for small vibrations about the equilibrium position. Any motion in which the amplitude with the journal is significant relative to the bearing clearance will become nonlinear and it is important that such motion is recognized. It is interesting to note the variation in stiffness for different positions within the clearance and Figure 6.3 shows the variation of bearing stiffness with vertical position (all displacements are considered relative to the radial clearance).

The stiffness is expressed in relative terms, but it is apparent how the stiffness increases dramatically at high eccentricity values.

The study presented here considers plain journal bearings but there is a range of closely related types of bearings for which similar methods may be applied but clearly with some different results. Grooving of the bearings produces some differences in pressure profiles and such cases are best analyzed numerically with suitable software. Similarly, a fairly common variation is the inclusion of tilting pads to support the oil film. These pads tend to reduce the bearing asymmetry and help with stability considerations.

6.2.2 Rolling Element Bearings

While hydrodynamic journal bearings predominate in large machines, other types of bearings are important. Many small- and medium-sized machines are supported on rolling element bearings and these are generally much stiffer than hydrodynamic bearings. The deflection of the balls or rollers can be calculated from Hertzian contact principles. From Kramer (1993), the effective stiffness of ball bearing is

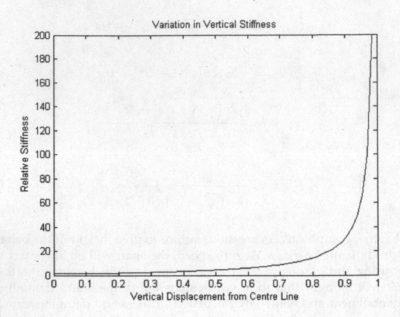

FIGURE 6.3
Bearing Stiffness in the Vertical Direction.

$$k_{vv} = k_b n_r^{2/3} d^{1/3} f_s^{1/3} \cos^{5/3} \alpha \qquad (6.6)$$

where n_b is the number of balls of diameter d, α is the contact angle within the race, f_s is the vertical force, and the constant is $k_b = 13 \times 10^6 \text{N}^{2/3} \text{m}^{-4/3}$.

For a roller bearing the expression is

$$k_{vv} = k_r n_r^{0.9} l^{0.8} f_s^{0.1} \cos^{1.9} \alpha \qquad (6.7)$$

Here there are n_r rollers of length l, and the constant $k_r = 1 \times 10^9 \text{N}^{0.9} \text{m}^{-1.8}$.

Equations 6.6 and 6.7 give the stiffness in the vertical direction and Kramer (1993) gives the ratio of horizontal to vertical stiffness (assuming the load is vertical) as

Ball Bearing: k_{uu}/k_{vv}=0.46, 0.64, 0.73 for 8, 12, 16 balls respectively.
Roller Bearing: k_{uu}/k_{vv}=0.49, 0.66, 0.74 for 8, 12, 16 rollers respectively.

In practice, these bearings are often much stiffer than either the rotors or the supporting structures and may be taken as rigid for the purposes of modeling. Indeed, it is often desirable to take this approach to avoid numerical instabilities arising from the very high stiffness values.

The essential point with rolling element bearings is that they give rise to different frequencies in the vibration spectrum. Two frequencies of significance correspond to the rotation of the cage (or train) and that of the balls (or rollers). The angular velocity of the cage is given by

$$\Omega_c = \frac{\Omega}{2}\left\{1 - \frac{d}{D}\cos\alpha\right\} \tag{6.8}$$

and that of the balls is

$$\Omega_r = \frac{\Omega}{2}\left(\frac{D}{d}\right)\left\{1 - \left(\frac{d}{D}\right)^2\cos^2\alpha\right\} \tag{6.9}$$

where D is the pitch diameter of the rotor (or inner raceway) and d is the diameter of the balls (or rollers), while α is the angle of contact. Ω_c is called the *fundamental train frequency* (FTF) and Ω_r is the *ball spin frequency* (BSF). Two further frequencies can be generated if there is a defect in the inner or outer raceways; these are

$$\Omega_i = n_b\frac{\Omega}{2}\left\{1 + \frac{d}{D}\cos\alpha\right\} \qquad \Omega_o = n_b\frac{\Omega}{2}\left\{1 - \frac{d}{D}\cos\alpha\right\} \tag{6.10}$$

These two frequencies are the *ball passing frequencies* in the inner and outer raceways (BPFI/BPFO) respectively. The relative amplitudes of these four frequency components provide a good indication as to the condition of rolling element bearings. These form the major bearing types on a wide range of light and low-speed machinery.

6.2.3 Other Types of Bearing

6.2.3.1 Active Magnetic Bearings

While the bearings described in the previous two subsections predominate, there are other types in operation. Of these, perhaps the most important are magnetic bearings, of which there are two main types: passive and active. In both cases, the big advantage is that by using a magnetic field to levitate and locate the rotor, all contact with the rotor is avoided and frictional losses are extremely low. In the passive type, permanent magnets are used to impose a fixed field on the rotor.

Of somewhat greater interest are active magnetic bearings. In these bearings the magnetic field is governed by a control system using active feedback operating on shaft position data. They provide negligible drag and are suitable therefore for high-speed machines. Costs are high, but for a growing number of applications, they provide an excellent solution to a number of practical problems. The use of active magnetic bearings has been reviewed by Schwietzer (2005)

6.2.3.2 *Externally Pressurized Bearings*

The externally pressurized or hydrostatic bearing is another type of oil journal bearing. The rotor is supported by an oil film but the pressure field within the oil is generated by an external pump rather than the rotation of the shaft itself and this has both advantages and disadvantages. The benefits are that load can be taken at low speed and that operational parameters can be varied externally through the external pump. Any means of applying external control to a rotor system is interesting in the context of developing autonomous machines and this topic is discussed in Chapter 10. The penalty paid for this facility is additional complication and cost in the oil supply pump.

6.2.3.3 *Foil Bearings*

Foil bearings have been known for a considerable time but interest has increased in recent years as part of general attraction toward oil free systems. Foil bearings are basically air bearings but with the addition of a flexible foil or membrane which takes the load at low speeds. As speed increases the film of air, the pressure field of which this develops viscous and inertia forces arising from the rotation supports load. The main areas of application are light, high-speed rotors. There has been increased interest in these bearings in recent years and Bonello (2019) has discussed the difficulties in evaluating their effective stiffness.

6.3 Interaction via Working Fluid

6.3.1 Pump Bushes and Seals

As an introduction to this section, let us consider the specific example of a large centrifugal pump. This is a good example to consider because the fluid has a high density and therefore has a considerable influence on the dynamics of the rotor.

The purpose of a boiler feed pump in a power plant is to raise the pressure of a large quantity of water. This is achieved by passing the water through a series of impellers that give the fluid kinetic energy and at the outer edge of the impeller, the fluid passes into the diffuser section of the stator. There may be several impellers within a pump but the trend is toward reducing this number with the most recent feed pumps having only two stages. Each impeller has a number of channels divide by vanes (often six) through which the fluid passes.

The diffuser of each stage comprises of a set of divergent channels and hence as the fluid passes through, its velocity reduces and consequently the kinetic energy is converted to potential energy which is this case is manifested as pressure. This high-pressure fluid is then directed to the entry of the next stage. The number of diffuser vanes is normally one greater than the number of impeller vanes in order to even out the wearing process. In addition to the impellers there are two neck rings at each stage: these are narrow clearances through which there is a small leakage path for the fluid separating neighboring pump stages. Some clearance is essential for the smooth operation of the unit but if it becomes excessive, the efficiency of the pump is impaired: large clearances can also markedly increase the axial thrust on the pump rotor, but a discussion of this is beyond the scope of this text. Other components on the pump rotor are the thrust balance device, the seals (preventing leakage between pump and casing), and the bearings. All of these items exert forces on the rotor and thereby modify the dynamic properties of the machine, often quite dramatically.

Figure 6.4 shows the internals of a modern boiler feed pump. This is a two-stage machine (i.e. there are two impellers) and it operates at a high speed of about 7,000 rpm. Although there has been a marked trend toward higher speeds and fewer pump stages, many four-stage units remain in service.

Before proceeding, it is worth examining the pressure differential acting across the seals as this is in itself quite a complicated issue. In essence, within a pump there are two fine clearances that isolate one stage from its neighbor and each of these act as "mini-bearings." As discussed above their stiffness, damping, and inertial characteristics are strongly dependent on the pressure differential across the seal, but this, in turn, is a rather complicated function of

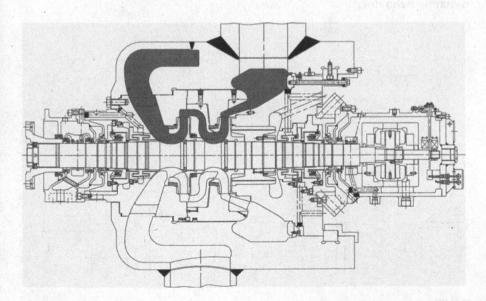

FIGURE 6.4
Internal Components of a Boiler Feed Pump.

(Reproduced with permission of Trillium Flow Services (UK) Ltd.)

the running conditions and it is worth examining the fundamental issues underlying this situation. In Figure 6.4 the area shaded in grey shows the main flow path within a single stage of the centrifugal pump.

The objective here is to obtain some understanding of the forces around the impeller of a centrifugal pump. There are two aspects to this task: The first objective is to establish the pressure differentials across the two seals (denoted A, B, and C in Figure 6.5) and then we consider the cyclic forces which arise. To be a little more specific, the objective is to consider the parameters influencing behavior rather than a precise evaluation since, as we shall shortly observe, a detailed calculation is difficult and of limited value since any wear leading to clearance changes will be observed in marked performance changes. Nevertheless, it is important to understand the trends.

Water from the entry (E) flows through the first impeller and most of it will enter the diffuser section (D). This forms part of the stator and comprises divergent channels through which the water slows and consequently increases pressure. From D, however, a small proportion of the fluid leaks back to E via the neck-ring (A). The flow then proceeds through the diffuser channels to be delivered to the inlet of the second impeller at F. At this point there is another leakage route via the seal at (B) through which a small flow will return to point D. Similarly, some of the flow at the outlet of the second impeller can return to F via the neck-ring at C. The balance drum, on the right-hand side of the figure, also has a significant influence, but this is not considered here.

The task now is to consider how these neck-rings and seals influence the dynamic behavior.

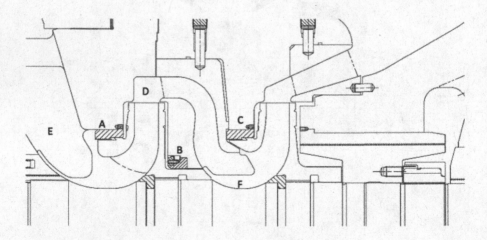

FIGURE 6.5
A Single Pump Stage.

(Reproduced with permission of Trillium Flow Services (UK) Ltd.)

With zero leakage flow, the pressure at D is readily calculated by considering a mass of water swirling at half the angular velocity of the rotor. Hence

$$P_D - P_C = \frac{1}{2}\rho\left(\frac{\omega}{2}\right)^2(r_3^2 - r_2^2) \tag{6.11}$$

where r_2 and r_3 are the radii at B and D, respectively, and ρ is the water density.

An additional pressure drop with a leakage flow is given by

$$\Delta P = \frac{\lambda}{d^2}Q^2\left(\frac{1}{r_2} - \frac{1}{r_3}\right) \tag{6.12}$$

where d is the axial separation between the stator and the impeller, λ the friction factor (which is a function of Reynold's number), and Q is the flow rate. This illustrates some of the complexity in an accurate assessment of the pressure distribution within the unit: the important thing is to gain an understanding of the manner in which the parameters vary.

The second leakage path is through the clearance A into chamber E where at entry to the clearance, the fluid swirls at half shaft speed. By a very similar argument, the differential pressure across B and C may be calculated. Having established both these differentials the characteristics of the sealing clearances can be calculated by an approach first presented by Black (1969). The calculation proceeds by considering Bernoulli's equation within the seal, integrating around the circumference and then transforming from a frame of reference, which rotates with the bulk of the fluid. The result is the evaluation of forces on the rotor given by

$$F_y = -\varepsilon\left[\left(\mu_0 - \frac{1}{4}\mu_2\omega^2T^2\right) + \mu_1TD + \mu_2T^2D^2\right]y - \varepsilon\omega\left[\frac{1}{2}\mu_1T + \mu_2T^2D\right]z \tag{6.13}$$

$$F_z = -\varepsilon\left[\left(\mu_0 - \frac{1}{4}\mu_2\omega^2T^2\right) + \mu_1TD + \mu_2T^2D^2\right]z - \varepsilon\omega\left[\frac{1}{2}\mu_1T + \mu_2T^2D\right]y \tag{6.14}$$

where D is the differential time operator, $T=L/V$ or the time taken for fluid to traverse the seal, L being the length of the seal under consideration and

$$\mu_0 = \frac{9\sigma}{1.5 + 2\sigma} \tag{6.15}$$

$$\mu_1 = \frac{(3 + 2\sigma)^2(1.5 + 2\sigma) - 9\sigma}{(1.5 + 2\sigma)^2} \tag{6.16}$$

$$\mu_2 = \frac{19\sigma + 18\sigma^2 + 8\sigma^3}{(1.5 + 2\sigma)^3} \tag{6.17}$$

$$\sigma = \lambda \frac{L}{y_0} \tag{6.18}$$

Finally $\quad \varepsilon = \dfrac{\pi}{6\lambda}\left[\dfrac{\sigma}{1.5 + 2\sigma}\right] R.\Delta P \tag{6.19}$

The calculations, although somewhat involved, are readily incorporated into computer models of machines. In many instances they can make significant differences to model predictions. Damping of particular vibration modes can be very sensitive to these factors.

6.3.1.1 Influence of System Pressure Distribution

The pressure distribution within the machine is complicated and it is not a simple function of shaft speed. To gain some understanding of the complexities involved, we must diverge to consider the pressure developed by a centrifugal pump. The centrifugal pump is used as an example but it is representative of a wide range of fluid handling machines. Consider a pump comprising N stages, each of which has an impeller radius r and rotating at speed Ω rad/sec. Then the maximum pressure which can be developed by such a pump is given by

$$P_{max} = \frac{1}{2}N\rho r^2 \Omega^2 \tag{6.20}$$

This, however, is only the beginning of the story and this pressure represents only a limiting value at zero flow. Any flow through the system will give rise to some dissipative drag terms, partly due to viscous wall effects, turbulent losses, and blade/diffuser entry losses, all of which can be represented by linear and quadratic terms in Q, the volumetric flow rate. Taking this into account we arrive at an expression for the pressure developed as

$$P(\Omega, Q) = P_0(\Omega) + \alpha Q - \beta Q^2 \tag{6.21}$$

Note from this expression that the maximum pressure does not necessarily occur at zero flow, although it is highly desirable that this is this case. The function shown in Equation 6.21 is known as the *characteristic* of the pump and this is key information in its operation. The characteristic in general comprises a family of curves $P(Q)$ for different values of speed. It is very desirable that α is negative: this is usually, but not always, the case.

- This may be used to establish the conditions under which the machine is operating and thereby evaluate the pressure differentials within the machine. While it may not always be necessary to evaluate detailed numerical values, it is frequently important to have an appreciation of the factors influencing both performance and vibration behavior in order to optimize plant operation. To achieve this, the properties of the system which is being fed by the pump must be considered. Let this system characteristic be represented by another curve, say, P_{sys} then the operating point is determined by the intersection of these two curves as shown in Figure 6.6. P_{sys} will, of course, not be unique but will depend on the settings of various valves within the circuit. The figure shows the operating point on the P, Q plot which becomes purely a function of rotor speed, albeit a fairly complicated one.

The pump characteristic shown in Figure 6.6 can be very helpful in analyzing the behavior of a pump; by predicting the pressure developed, the user is able to make reasonable estimates of the pressure profile within the unit and comparing measured data with model predictions can offer considerable insight into condition. Some of the work is somewhat speculative but the insight gleaned into the internal workings of the pump can be very informative.

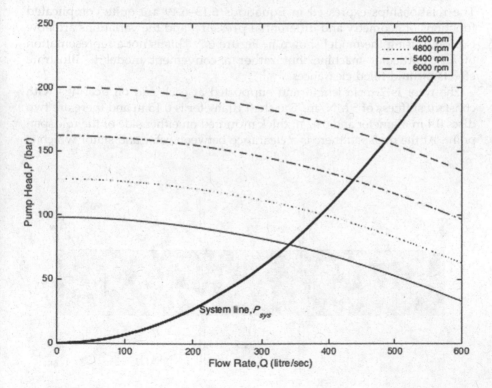

FIGURE 6.6
Typical Characteristic Curve.

To clarify this, consider the characteristic plot in a bit more detail. The declining curves show the pressure developed by the pump as a function of volumetric flow rate for different rotational speeds. The variation occurs because the fixed blade angles of the diffuser and impeller mean that the flow path will be optimized only for a single flow rate and rotational speed or, at best, only over a narrow range. In practice, machines will be required to operate rather more flexibly and impaired performance is the price paid for this. The plot shows a set of declining curves corresponding to different rotational speeds.

This leads to some intriguing possibilities: Clearly factors which modify the performance of a machine also influence the vibration characteristics. The obvious question arises of how to correlate changes in performance and vibration measurements to yield added insight into the internal condition of a machine. The complete realization of this potential is somewhat beyond current capabilities of technology but research continues to extend the available methods; some of the relevant methods are discussed in Chapter 8.

6.3.1.2 An Example of an Idealized System

The relationships expressed in Equations 6.13–6.19 are quite complicated functions of geometry and differential pressure, and the variations are now explored using the model shown in Figure 6.7. This is not a representation of any particular machine, but rather a convenient model to illustrate effects of fluid-filled clearances.

The rotor is 2 m in length and supported at each end on bearings with constant stiffness of 5 MN/m. The shaft diameter is 0.15 m and there are two discs 0.4 m diameter and 0.02 m thick mounted on either side of the midspan point. At the midspan there is a clearance between rotor and stator which is

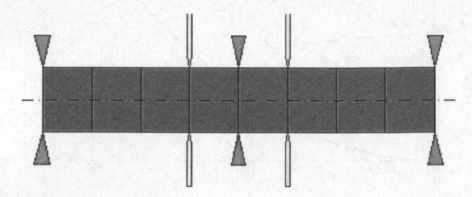

FIGURE 6.7
Simplified Rotor Model.

0.1 m in (axial) length with a radial clearance of 200 μm across which a variable pressure differential is applied. The natural frequencies and mode shapes are shown in Figure 6.8 without any constraint applied to the central bush. Note there is no damping in the bearings: this idealization was chosen in order to clarify the effect of the central sealing bush.

It is clear from the mode shapes that since the bush is placed at the midspan of the shaft, it will have essentially no effect on the "rocking" modes 2 and 4. Owing to the finite axial span of the bush, the influence is not exactly zero, but can be neglected unless pressure differential is very high. Table 6.1 shows the natural frequencies and damping coefficients of modes 1 and 3 for a range of differential pressure across the bush.

For mode 1, the "bounce" mode, as pressure increases to 5 MPa, the natural frequency increases from 26.7 to 38.6 Hz. Note that between 5 and 10 MPa, the nature of the mode changes completely and it is difficult to make general observations. An important observation, however, is the steady increase in damping as the differential pressure rises. In real plant, the first mode of pumps is often not seen in plant data owing to high damping of this mode. This means that in many cases, the lowest frequency mode observed will be the "rocking" mode, readily recognizable by the phase difference between the displacements at the bearings. In the simple test sample, the frequency of

Nat Freq = 26.7329Hz

Nat Freq = 51.1693Hz

Nat Freq = 173.5959Hz

Nat Freq = 429.9714Hz

FIGURE 6.8
Natural Frequencies and Mode Shapes.

TABLE 6.1

Effects of Bush on Natural Frequencies and Damping

Pressure Differential (MPa)	Mode 1		Mode 3	
	Frequency (Hz)	Damping Coefficient	Frequency (Hz)	Damping Coefficient
0.2	27.28	0.10	173.44	0.016
0.4	27.83	0.14	173.44	0.023
0.6	28.83	0.17	173.44	0.028
0.8	28.92	0.19	173.44	0.033
1	29.45	0.21	173.44	0.037
2	31.96	0.26	173.45	0.051
5	38.61	0.33	173.47	0.084
10	51.16	>1	173.50	0.12

mode 3 did not change, although the damping shows a steady increase. In real machines, however, the situation is considerably more complex, because there are usually a number of neck-rings and seals, each subject to pressure differentials varying with pump operation.

The influence of the bush on the forced response can be observed from Figure 6.9. An equal imbalance was applied to each of the discs so that response will be governed by the first mode. It is clear that the peak becomes heavily damped.

As illustrated in Table 6.1 the presence of the seals within a pump may have considerable influence on the overall vibrational behavior of the unit and, furthermore, this effect may vary with time owing to wear within the various clearances. While the shift in critical speeds may be modest, the crucial influence in the behavior arises from the damping. As wear progresses, damping changes can substantially alter the machine's signature. This wear pattern will also be reflected in the performance characteristics of the pump, that is, plots of developed head against flow rate. As so often, the real value of vibration signal analysis is maximized by using the measured data in conjunction with other plant information in forming an overall judgment of a machine's condition.

6.3.2 Other Forms of Excitation

The arguments set out so far outline the influence of the pumped medium on the mass, damping, and stiffness properties of the rotor system, but of course, this medium also has a major influence on the dynamic forces arising during operation. The number of veins on each impeller and the number of channels in the diffuser normally differ by one, usually six and

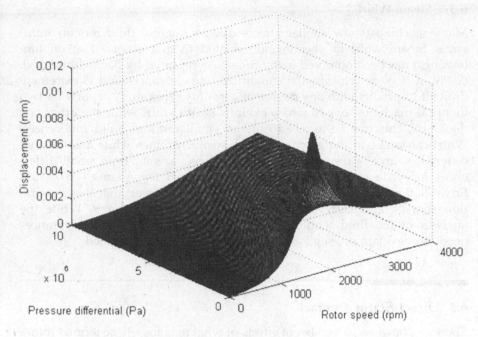

FIGURE 6.9
Response to Imbalance.

seven respectively. In a perfect (idealized) pump, each time an impeller blade passes a diffuser vane there will be some pulse of pressure, resulting in an excitation at 42 times per revolution and harmonics thereof. If the acceleration (as opposed to displacement) is monitored, then a signal is usually present at this fairly high frequency. However, in a real machine there will normally be subharmonics of this ideal frequency. For instance, some erosion or other fault on a diffuser vane would give rise to pulsations at six times per revolution, whereas an imperfection on an impeller blade will yield a frequency corresponding to seven times shaft speed.

There is a wider context to this discussion: the detailed discussion of centrifugal pumps illustrates the main (and rather obvious) point that the dynamic behavior of any piece of machinery is determined by the physics of its internal state and operation. Indeed, it is this fundamental connection which gives value to monitoring by means of vibration and performance parameters: the better the understanding of the basic physics of operation, the greater will be the information gleaned from monitoring and diagnostic procedures. The example of large centrifugal pumps has been used as an illustration, since the combination of a high energy density and a fairly light machine implies that the pumped medium has a significant influence of the vibrational behavior.

6.3.3 Steam Whirl

Many machines show similar effects due to internal fluid motion influences. Steam whirl (a phenomenon related to that observed in oil film bearings) has been observed many times. As reported by Bachschmid *et al.* (2008), it is often critically dependent on fine clearances and is extremely difficult to predict with any degree of certainty. This study reports on two identical machines, one of which exhibits steam whirl while the other one does not. This can be ascribed only to small random build differences. Whirl can occur in the glands of some steam turbines while a somewhat different source is the so-called Alford's force, arising from small differences in blade clearances and this is discussed by Adams (2001) and Friswell *et al.* (2010). It should be emphasized, however, that these excitation sources are really instabilities, rather than true forces. While the analysis of the fluid motion used for the analysis of pump clearances needs some adaption, similar physical concepts may be applied.

6.4 Direct Stator Contact

There are, however, a number of effects of what may loosely be termed rotor–stator interaction. The appropriate point at which to begin the discussion is with rubbing of a rotor against its casing during part of the cycle. This seemingly simple picture is actually very complex and covers a wide range of faults and resultant behavior. In the first instance, we consider a "light" rub in which the rotor rubs against its containing stator over a narrow angle of travel during each rotation. This is depicted in Figure 6.10, and we now consider the likely sequence of events. It is assumed that at rest, the rotor and stator are not in contact (apart, that is from in the bearings) but some whirl due to imbalance causing the rotor to rub against the stator over a limited portion of its orbit.

6.4.1 Extended Contact

6.4.1.1 Physical Effects

The possibilities are well summarized by Muszynska (1989, 2005), but in many instances, there will be insufficient information to fully explain all details and consequently it is important to be aware of the various possibilities.

As the rotor contacts the stator

a) Rubbing will cause heat generation leading to a thermal bend
b) The change in support will alter the natural frequencies and dynamic response

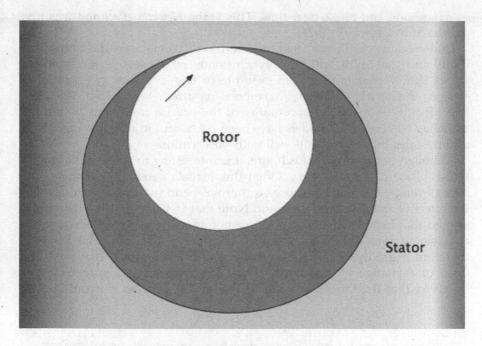

FIGURE 6.10
A Rotor–Stator Rub.

c) The impact will generate harmonics (and subharmonics)
d) Friction can give rise to torsional excitation
e) There may be a change in damping
f) Deterioration in plant performance
g) Generation of acoustic emission

6.4.1.2 Newkirk Effect

Since the whirl is synchronous, the same point on the rotor rubs on each revolution and consequently heat is generated which causes the rotor to bend.

As discussed in Section 4.3, a bend also gives rise to a synchronous vibration signal, and at constant speed, the effect of this thermal bending is to modify the effective imbalance and so the vibration will change, in general in both magnitude and phase. Without embarking on the mathematical details, it is fairly straightforward to see how a cyclic pattern of behavior develops: because the effective imbalance is changed by the thermally induced bend, the rub conditions, and consequently the heating changes. This gives rise to a range of possible scenarios which depend on the combination of physical parameters. Clearly, the vibration will change in magnitude and phase over time. This is the *Newkirk effect* and was first discussed by Newkirk (1926).

Under some conditions, the amplitude will remain within acceptable limits and the result will be a synchronous response whose amplitude varies slowly over time and an example of this is shown in Figure 6.11. Various theoretical treatments have been reported (Muszynska, 1989) but the difficulty is really the uncertainty of the precise conditions within the machine. Nevertheless, models have been successful in reproducing trends at timescales that correspond well with observations.

Although not shown on this figure, it is interesting to observe the way in which phase varies with time. Often this reveals a gradual change as the heat developed at the rub causes a thermal bend, which in turn changes the effective imbalance on the rotor. Note that in Chapter 4 it is (strongly) argued that bends and imbalances are not equivalent; nevertheless, taken at a single rotor speed they can be considered together. The result is therefore a gradually changing effective unbalance which may take the form of a change in amplitude, phase, or both. From Figure 6.11 it is observed that the timescale associated with this process is about 20 min

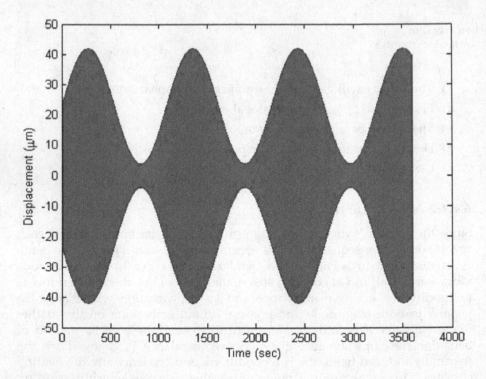

FIGURE 6.11
An Example of the Newkirk Effect.

and long periods like this are typical of large machines. This characteristic time will be determined by a variety of factors, including the dimensions of the machine, thermal conductivity of the rotor, and the rotor flexibility. Other factors such as the mass imbalance and the clearance will only influence the extent of the thermal cycling.

There are a range of possibilities covering the interaction of a rotor and its stator:

a) The rotor may rub constantly all around the circumference (although this is a rare condition).
b) The rotor may constantly rub against one part of the stator.
c) The rotor may contact the stator at one portion of its orbit on each cycle: this is the scenario which gives rise to the *Newkirk Effect* as described above.

But this seemingly simple process can give rise to a wide variety of consequences, and the parameters of a specific situation will determine which of the terms dominate. For successful diagnosis, the aim is to use a blend of all these factors to yield maximum insight into the problem.

There have been many papers on the topic of thermal rubs in rotating machines and this is not surprising in view of its importance. Treatments range from the purely empirical to detailed mathematical analyses, which is all a little frustrating in view of the limited available knowledge. It is important to appreciate that there are two quite distinct cases: If there is a short impact between rotor and casing, there will be an exchange of momentum but little heat generation. With a long rub the situation will be reversed, with little momentum exchange but considerable heat generation. In a real incident, one may observe a combination of these two extremes. Here we present a description of the rather slow rub case but even this yields a complex combination of phenomena.

To develop the discussion, consider the steady-state motion of a rotor which has both imbalance and a bend. The motion is described by

$$\mathbf{M}\ddot{y} + \mathbf{C}\dot{y} + \mathbf{G}\Omega\dot{y} + \mathbf{K}y = \mathbf{M}\Omega^2 \varepsilon e^{j\Omega t} + \mathbf{K}b e^{j\Omega t} \qquad (6.22)$$

The complication with a rub is that the bend is thermally induced and it is time dependent. In all realistic scenarios, however, the thermal timescale is much longer than that of the dynamics and so it is valid to separate the two problems. As the rotor contacts the stator, heat $\dot{Q}(t)$ will be generated and, although there will be some radiated losses, the net result will give rise to a thermal field within the rotor which will cause a bend. This time-dependent heat generation will give rise to a temperature profile within both rotor and, usually to a lesser extent, the stator. Focusing attention on the rotor, this nonuniform temperature profile gives rise to distortion, tending to bend the rotor. For a given heat input and known losses, the temperature profile of the rotor can, in principle, be calculated using standard methods, either analytical or numerical; but, of course, it is rare to have any precise knowledge of the conditions and, consequently, it is far more important to formulate a more general understanding of the processes involved.

The distribution of temperature within the rotor is easily translated into stresses and strains, which in turn result in a combination of forces and moments acting on the rotor giving rise to the bend term of Equation 6.22. This represents a complicated chain of events which is represented by the chart shown in Figure 6.12.

This process does, however, avoid one important question: does the rotor stay in contact with the stator long enough to generate significant heat, or does it simply undergo an impact and rebound? In either event, there will be significant changes to the dynamic behavior, but the two scenarios are markedly different.

In reality, any contact will give rise to both types of interaction with one dominating the other depending on the local parameters. It is worth briefly considering the physics of the interaction: as the rotor makes contact with the stator two interaction forces arise, one perpendicular to the surface and the other, a friction force, along the surface. Both these forces will be nonlinear functions of the relative displacement. The two components will remain in contact for some time, say τ, which will be determined by the local stiffness properties of both components together with the velocity of the rotor.

6.4.2 Collision and Recoil

6.4.2.1 Physical Effects

All rotating machines involve fine clearances between stationary and moving parts, and these are often filled with fluid, which may be the working fluid,

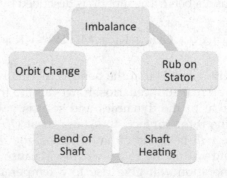

FIGURE 6.12
The Newkirk Cycle.

coolant, or lubricant. These clearances influence the dynamics of the machine in a number of ways which we can categorize into three main divisions. The classification is as follows:

a) Fluid forces predominate – the clearances may act as subsidiary bearings in the manner discussed for neck-ring seals in pumps. A very detailed discussion of the varied phenomena is given by Muszynska (2005). The overall contribution can be the addition of mass, stiffness, or damping. It can also, under certain circumstances, feed energy into the rotor, giving rise to instability in the form of whirl or whip.

b) Rubbing – leading to a thermal and dynamic cycle.

c) Impact and recoil.

A substantial number of papers on this topic take a variety of models describing the contact between stator and rotor. These involve structural parameters, surface condition, and lubrication and it is clear that the real physics of the processes involved is indeed complex. In a real machine it is highly unlikely that one would ever have the information to execute a credible calculation. But this is not to diminish the value of this research: the real value lies not in the numbers, but rather in clarifying the wide spectrum of effects that can arise from rotor–stator contact.

Edwards *et al.* (1999) were the first to discuss quantitatively the generation of torsional excitation from rotor–stator contact and this is, perhaps, a little surprising. It reflects a widespread tendency to neglect torsion and this is understandable in view of the difficulties of accurate measurement of torsional motion. As discussed elsewhere, the effort involved in monitoring torsion cannot be justified in many cases, but it remains important to gain an understanding of the phenomena involved.

As the rotor makes contact with the stator, the dynamics of the system is fundamentally changed. The stiffness and damping parameters alter in addition to the introduction of impulsive forces. If the time in contact is short, then these impulsive forces play an important role, whereas for extended contact, discussed in 6.4.1, thermal processes assume greater importance. As an illustration of the complexity which can arise with impulsive contacts, an idealized system is now considered in a single direction.

6.4.2.2 Simulation

Now return to the issue of contact between rotor and stator. In addition to the thermal aspects (discussed in section 6.4.1) of the contact, there is impulse exchange which may occur sufficiently rapidly so that the rebound of the rotor is far more significant than the distortion of the motion caused by heat generation. Clearly, the determining factor in this process is the time over which rotor and stator remain in contact and this in turn depends on the relative stiffness terms as well as velocities.

Once a rotor comes into contact with its stator, the effective stiffness will change. This is indeed a complicated scenario involving impact, torsional excitation, and heat generation. Taking just the simplest case, it is instructive to model this as a mass on a spring which is impacting on a second spring as shown in Figure 6.13. At rest the clearance between the mass and the second spring is δ.

Free vibration is considered first; for small amplitudes the displacement will be given by

$$x_1(t) = x_0 \sin \omega t \tag{6.23}$$

where $x_0 \leq \delta$ and $\omega = \sqrt{k_1/m}$.

Now consider the case when $x_0 > \delta$. Then contact will occur at time

$$\tau = \omega^{-1}\sin^{-1}\left(\delta/x_0\right) \tag{6.24}$$

After the impact has occurred, the displacement is given by

$$x_2(t) = A \cos \omega_2(t - \tau) + B \sin \omega_2(t - \tau) \tag{6.25}$$

where $\omega_2 = \sqrt{(k_1 + k_2)/m}$. By equating position and velocity at the instant of contact

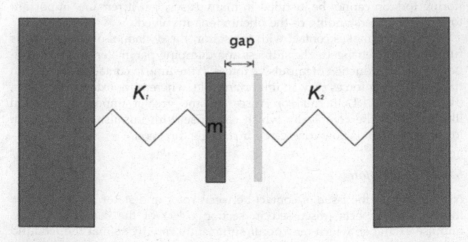

FIGURE 6.13
Idealized One-Dimensional Impacting.

$$A = \delta$$
$$\omega_2 B = x_0 \omega \cos \omega \tau \qquad (6.26)$$

Hence

$$B = \frac{\omega}{\omega_2} \sqrt{x_0^2 - \delta^2} = \sqrt{\frac{k_1 (x_0^2 - \delta^2)}{k_1 + k_2}} \qquad (6.27)$$

This study can now be continued to establish the "effective frequency" of this nonlinear cycle by determining the time at which the oscillation returns to the zero point. Note that Equation 6.25 can be expressed as

$$x_2(t) = \sqrt{A^2 + B^2} \cos \left[\omega_2(t - \tau) - \tan^{-1} \left(\frac{B}{A} \right) \right] \qquad (6.28)$$

Then the maximum deflection occurs at time

$$t_{\max} = \tau + \frac{1}{\omega_2} \tan^{-1} \left(\frac{B}{A} \right) \qquad (6.29)$$

This is illustrated in Figure 6.14. In this slightly artificial example, the base frequency is 1 Hz while the stiffness $k_2 = 4 \times k_1$, the vibration has an initial

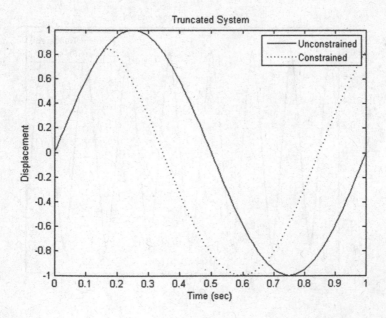

FIGURE 6.14
System with an Impact.

velocity corresponding to unit amplitude, and contact occurs at an amplitude of 0.8.

Hence zero displacement will occur after time $2 \times t_{max}$ and the cycle will be complete at time given by

$$T = \frac{2}{\omega_2} \tan^{-1}\left(B/A\right) + 2\tau + \frac{\pi}{\omega_1} \tag{6.30}$$

The effective frequency is $\Omega = {2\pi}/{T}$, although this terminology must be used with some care: Ω does indeed specify the fundamental frequency but this will be accompanied by harmonics since the displacement function is no longer sinusoidal. This simple model is reflected in the machine with rotor–stator contact which generates harmonics of the shaft rotation frequency. More generally this is the case with all forms of nonlinear behavior.

This study of the free vibration gives some insight into the mechanics but the analysis of the forced response is a little more complicated. This is because transients arise each time the support conditions change and while an analysis is in principle tractable, the algebra rapidly becomes too complicated to yield a meaningful physical picture. It is, therefore, more convenient to resort to a numerical simulation. Figure 6.15 compares the "unconstrained" free vibration with that with impacting. It is clear that the

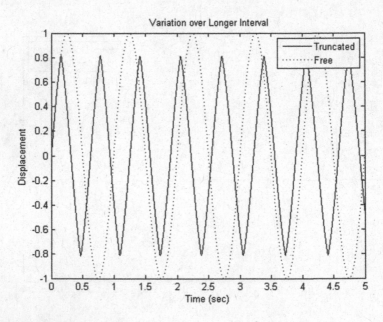

FIGURE 6.15
Free and Constrained Motion.

cycles are shorter but the motion is no longer sinusoidal and contains a number of harmonics arising from the impacts.

Referring back to Figure 6.13, the forced equation of motion is given by

$$K(x)x + C\dot{x} + M\ddot{x} = F \sin \omega t \qquad (6.31)$$

In this case damping has been included since there is no great simplification in ignoring it. Note however that in this case a sinusoidal form for x has not been assumed since nonlinearities are present giving rise to harmonics.

$$
\begin{aligned}
K(x) &= k_1 && \text{for } x < \delta \\
&= k_1 + k_2 && \text{for } x > \delta
\end{aligned}
\qquad (6.32)
$$

If the sample is subject to forcing, a plot of velocity against displacement reveals the harmonic content and the gradual approach to chaos as δ is reduced. Figure 6.16 shows plots of the response of this example to a unit force at 1.5 Hz with

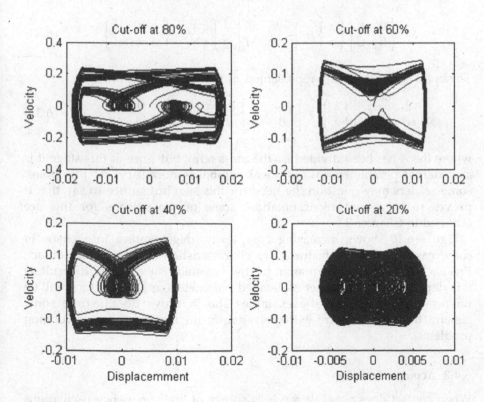

FIGURE 6.16
Phase Plot with Impacting.

gap values of 80%, 60%, 40%, and 20% of the excited amplitude. This provides a very graphic view of the extent of the nonlinear behavior.

A detailed study of the situation by Edwards *et al.* (1999) makes it clear that for some combinations of speed and geometric parameters, the motion can become chaotic. Muszynska (1989) gives an extensive discussion of the analysis which is indeed rather complex. It confirms that for certain speed and imbalance combinations chaotic motion can result. Since in the present volume the primary emphasis is on the recognition of fault conditions, attention is restricted to the basic phenomena.

At the most basic level, we consider the motion of a rotor with a single degree of freedom. Note that in the case $x_{max} > \delta$ the simple sinusoidal solution cannot be applied since the stiffness varies over a cycle. At first sight, one may consider that the two regimes could separately be solved in the frequency domain, but if this route were taken due allowance would be needed of the transients arising as the mass passes between the two regimes. It is therefore most appropriate to consider this problem numerically and this can be achieved using state-space notation. Equation 6.31 is rewritten in the form

$$\begin{bmatrix} 0 & K \\ K & C \end{bmatrix} \begin{Bmatrix} x \\ \dot{x} \end{Bmatrix} + \begin{bmatrix} -K & 0 \\ 0 & M \end{bmatrix} \frac{d}{dt} \begin{Bmatrix} x \\ \dot{x} \end{Bmatrix} = \begin{Bmatrix} 0 \\ A \sin \omega t \end{Bmatrix} \tag{6.33}$$

This may be conveniently represented in nondimensional units as

$$\begin{bmatrix} 0 & \omega_n^2 \\ \omega_n^2 & 2\zeta\omega_n \end{bmatrix} \begin{Bmatrix} x \\ \dot{x} \end{Bmatrix} + \begin{bmatrix} -\omega_n^2 & 0 \\ 0 & 1 \end{bmatrix} \frac{d}{dt} \begin{Bmatrix} x \\ \dot{x} \end{Bmatrix} = \begin{Bmatrix} 0 \\ \frac{A}{M} \sin \omega t \end{Bmatrix} \tag{6.34}$$

where the A has been divided by the mass term, but since at this stage it is an external parameter no additional symbol is necessary. At this point, some readers may question the need for this step but suffice to say that it proves to be a convenient notation: some of the reasons for this are outlined in Chapter 3.

Figure 6.16 shows increasing complexity, degenerating into chaos in some cases, and these features are characteristic of rotor–stator contact. Precise theoretical representation of the dynamics is extremely difficult as it is dependent on a number of detailed parameters, many of which will be unknown. Nevertheless, the examples shown above do illustrate some general trends which are extremely helpful in the diagnosis of rubbing problems.

6.4.3 Acoustic Emission

When contact does occur, it can be a source of high-frequency oscillations which are somewhat loosely termed "acoustic emission" by many authors.

As rotor and stator impact there will indeed be a range of high-frequency modes excited and the precise mechanism bringing this about will be dependent on the stiffness profile of the two bodies. Furthermore, the process will be nonlinear involving something approaching Hertzian contact (albeit involving cylindrical, rather than spherical, geometry). In addition to the set of high-frequency modes excited, as the surfaces make contact, surface asperities will distort and break and it is these processes which give rise to elastic waves in both rotor and stator. The effective forcing function is wide band and consequently the observed AE frequencies will reflect the resonances of the various components. Price *et al.* (2005) were among the first to use the frequency content of AE signals: a more common approach is simply to monitor the number of "events" as a measure of acoustic activity. Through this route, the monitoring of AE has become a valuable tool in the monitoring of static structures, but it is only in recent years that its applicability to rotating plant has become appreciated.

Figure 6.17 shows a typical signal of the high-frequency vibration. Note these signals are very low in amplitude and need normally 40–60 db of amplification. The figure shows a typical transient event, which may be characterized by a number of parameters which are illustrated in the figure. Conventionally, studies in AE have focused on counting the number of so-called "events," the number of time the signal exceeds some predetermined threshold value. While this approach has proved very successful in the monitoring of static structure, such as bridges, it is the author's belief that AE technology has considerably more potential for greater use in the future.

True AE signals differ from normal vibration measurements in that they arise as a result of mechanisms within the material, such as the breaking of molecular bonds. They occur at high frequency, normally in the range 25 kHz–1 MHz, but these frequencies do not correspond to any on the molecular scale as is sometimes thought (erroneously!). As outlined in Chapter 1, the frequencies corresponding to molecular bonds are at least six or seven orders of magnitude above those of AE and therefore it is worth considering the origin of the signals observed. As any event on a molecular scale, such as the breaking of bonds, occurs suddenly on the timescales being considered, the effect on the body is effectively an impulse. This means that a full spectrum of frequencies is generated. Consequently, a stress wave travels through the body, some undergoing reflections and absorption. The components of the wave that are reinforced by the various reflections correspond to the resonant frequencies of the body, or, rather, the partitions of the body between reflecting components. Hence the frequency composition of AE signals reflects the dimensions of the component within which the acoustic activity is occurring. The analysis of the complete AE waveform is a relatively new use of acoustic emission but it is an area which can bring substantial benefits in the study of rotating machinery.

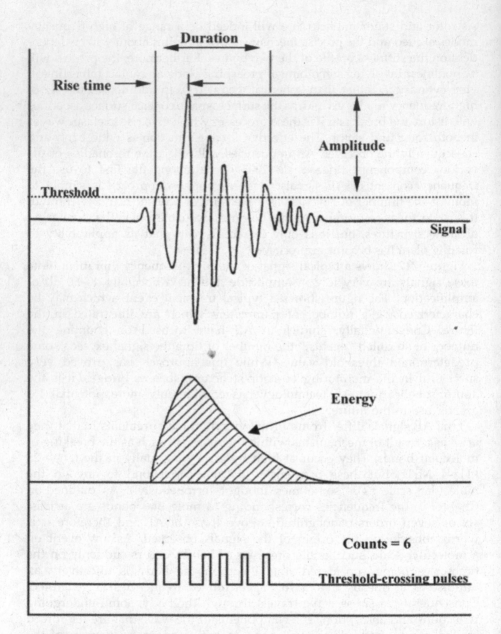

FIGURE 6.17
Typical Acoustic Emission Signal.

(Reproduced with permission from Proceedings of the Institution of Mechanical Engineers, Part J, 219(2), pp. 89–98, 2005)

Consider now the processes which may give rise to AE activity: these may be enumerated as follows

1. Initiation or growth of cracks
2. Plastic deformation
3. Rubbing surfaces (with deformation and breaking of asperities)
4. Turbulent flow in bearings
5. Effects of the working fluid

Several papers have reported the use of AE for the detection of rotor rub and this provides a sensitive and useful monitor of rotor rubbing. Similarly, studies have been undertaken to detect component cracking in ball bearings (Price *et al.*, 2005).

Any form of rubbing between rotor and stator may be expected to produce both high-frequency vibration and acoustic emission, noting here the fundamental difference between the two. The AE is produced as surface asperities distort and break, sending a stress wave through the solid. Leahy et al. (2006) have applied AE techniques to the detection of rubbing phenomena.

6.5 The Morton Effect

A superficially similar effect can occur in rotors with substantial overhung components (such as compressors), yet the phenomenon is rather different in that there are no harmonics or AE generated but simply a cyclic synchronous term. This is known as the *Morton Effect* and was first observed in 1976. Again, the root cause of the fluctuating synchronous vibration is the formation of a thermal bend, but in this case, no contact occurs between rotor and stator. Heat is generated not by friction but by the shearing of the oil film at the point of minimum film thickness and, consequently, maximum shear. Because there is no contact, there is no change in dynamic properties and no torsional excitation. Simulations of this effect have been undertaken by Keogh and Morton (1993), Kirk and Guo (2005), and others, but it remains a difficult effect to predict with any accuracy, since it is dependent on a fine balance of conflicting factors.

De Jong (2008) reports work on a large compressor which exhibited the Morton effect. He discusses a range of corrective actions ranging from substantial design changes to simply changing the lubricating oil. Unlike the Newkirk effect, that of Morton is associated only with film bearings, whereas an actual rub can occur on any rotating machine. More recently the same effect has been discussed in gas bearings (Tong and Palazzola, 2018).

6.6 Harmonics of Contact

The essential point about rotor motion within a clearance is that the effective stiffness of the rotor increases sharply as the clearance is taken up. That is, the effective stiffness is displacement dependent and the restoring force is a non-linear function of the displacement. An idealized form of this is expressed in Section 5.6.

In reality, nothing is truly linear but linear analysis helps give consider-able insight and is very much simpler that a full nonlinear treatment. Following the logic of Section 5.6, we take an equation of the form

$$m\ddot{u} + k_0 u - k_1 u^3 = Fe^{i\omega t} \tag{6.35}$$

where, for the sake of clarity, damping has been neglected.

Letting

$$u(t) = \sum_{n=-\infty}^{\infty} a_n e^{in\omega t} \tag{6.36}$$

Then substituting this expression into Equation 6.35 gives

$$\sum_{n=-\infty}^{\infty} \left(-a_n mn^2\omega^2 + a_n k_0\right)e^{in\omega t} - k_1 \sum_{n=-\infty}^{\infty} a_n^3 e^{3in\omega t} = Fe^{i\omega t} \tag{6.37}$$

While this is moderately complicated, it is clear that the coefficients a_n are interrelated. For all values of $n \neq 1$

$$\left(-a_{3n} m 9n^2\omega^2 + a_{3n} k_0\right) - k_1 a_n^3 = 0 \tag{6.38}$$

Hence various harmonics of the forcing frequency arise; the precise details of the amplitudes are not of concern here. However, the presence of the harmonics is of great interest (and often concern) in the monitoring of rotating machinery. In the case presented, imbalance will give rise to a synchronous term plus a series of harmonics as given by Equation 6.37.

This can often be detected graphically in rotor orbits: consider the plots shown in Figure 6.18. The first figure shows a simple orbit which in this case is singular. With the presence of a second harmonic, as depicted in 6.18(b), a loop appears, the size of which reflects the amplitude of the second harmonic term. Figures (c) and (d) show that further loops arise with higher harmonics. This extremely simple view of the orbit provides a sensitive

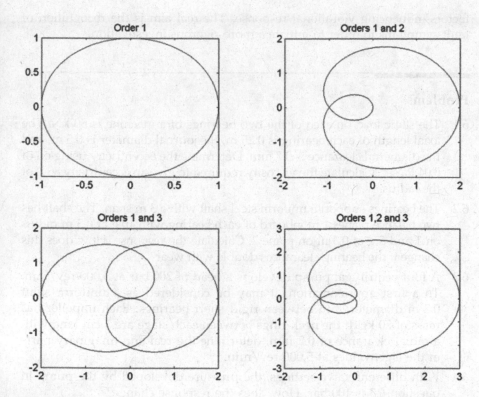

FIGURE 6.18
Effects of Harmonics.

indicator of nonlinearity: the nonlinear terms give rise to harmonic excitation, which in turn generate a looped structure in the rotor orbits.

6.7 Concluding Remarks

The interaction between rotor and stator are complex and often involve details of the machine's operation. In this chapter, the phenomena have been categorized as outlined in the introduction and in many respects the behavior observed will be entirely satisfactory. It is in the small minority of cases where deviations from normal behavior give indications as to the internal operation and condition of the machine. It is in such cases where an understanding of the internal operation is invaluable to fault diagnosis. Although several analyses are offered here, almost always, there will be little, if any, precise data on which meaningful calculations can be based. The real value of the modeling studies is to gain an appreciation of sensitivities of factors influencing vibrational response. The real aim is the recognition of fault symptoms in order to initiate a more rigorous investigation.

Problems

6.1 The static load on each of the two bearings of a machine is 50 kN. The axial length of each bearing is 0.25 m, the journal diameter is 0.5 m, and the diametral clearance is 0.7 mm. Determine the eccentricity using oil of 0.04 Pa sec. Calculate the viscosity required to give an eccentricity 80% of the radius.

6.2 The bearings support a uniform steel shaft with a 3 m span. The shaft has two balance planes 1 m inboard of each bearing with 10 g at 0.3 m radius on 1 and 5 g at 0.3 m on plane 2. Calculate the response. How does this change if the bearing clearances double with wear?

6.3 A four centrifugal pump develops a head of 200 bar at 5,000 rev/min. To a first approximation, it may be considered as a uniform shaft 0.2 m diameter, 3 m between rigid short bearings. Each impeller has mass of 50 kg. If the neck-rings between each stage are 5 cm long with a radial clearance of 0.2 mm, determine the real and imaginary parts of the eigenvalues at 5,000 rev/min.

6.4 With different valve settings, the pressure developed by the pump in question 6.3 is 100 bar. How does the response change?

6.5 A machine shaft passes through a clearance of 0.5 mm. The calculated natural frequency is 38 Hz but it is observed to be 40 Hz. Estimate the response within the clearance.

References

Adams, M.L., 2001, *Rotating Machinery Vibration: From Analysis to Troubleshooting*, Marcel Dekker, New York.

Bachschmid, N., Pennacchi, P. and Vania, A., 2008, Steam-whirl analysis in a high pressure cylinderof a turbo generator, *Mechanical Systems and Signal Processing*, 22, pp. 121–132.

Black, H.F., 1969, Effects of hydraulic forces in annular pressure seals on the vibration of centrifugal pump rotors, *Journal of Mechanical Engineering Science*, 11, pp. 206–213.

Bonello, P., 2019, The extraction of Campbell diagrams from the dynamical system representation of a foil-air bearing rotor model, *Mechanical Systems and Signal Processing*, 120, pp. 502–530.

Childs, D.W., 1993, *Turbomachinery Rotordynamics: Phenomena, Modeling and Analysis*, Wiley, New York.

De Jong, F., 2008, The synchronous rotor instability phenomena – Morton effect, *Proceedings of the 37th Turbomachinery Symposium*.

Edwards, S., Lees, A.W. and Friswell, M.I., 1999, The influence of torsion on rotor-stator contact in rotating machinery, *Journal of Sound and Vibration*, 137, pp. 463–481.

Friswell, M.I., Penny, J.E.T., Garvey, S.D. and Lees, A.W., 2010, *Dynamics of Rotating Machines*, Cambridge University Press, New York CambCNe.

Hamrock, B.J., Schmid, S.R. and Jacobson, B.O., 2004, *Fundamentals of Fluid Film Lubrication*, Marcel Decker, NJ.

Keogh, P.S. and Morton, P.G., 1993, Journal bearing differential heating evaluation with influence on rotordynamic behaviour, *Proceedings of the Royal Society: Mathematical and Physical Sciences*, 445, pp. 527–548.

Kirk, R.G. and Guo, Z., 2005, Morton effect analysis: Theory, program and case study, *3rd International Symposium on Stability Control in Rotating Machinery*, Cleveland, OH.

Kramer, E., 1993, *Dynamics of Rotors and Foundations*, Springer-Velag, Berlin, Germany.

Leahy, M., Mba, D., Cooper, P., Montgomery, A. and Owen, D., 2006, Experimental investigation into the capabilities of acoustic emission for the detection of shaft-to-seal rubbing in large power generation turbines: A case study, *Proceedings of the Institution of Mechanical Engineers Part J – Journal of Engineering Tribology*, 220, pp. 607–615.

Muszynska, A., 1989, Rotor to stationary element rub-related vibration phenomena in rotating machinery: Literature survey, *Shock and Vibration Digest*, 21, pp. 3–11.

Muszynska, A., 2005, *Rotordynamics*, CRC Press, Taylor & Francis, Boca Raton, FL.

Newkirk, B.L., 1926, Shaft rubbing: Relative freedom of rotor shafts from sensitiveness to rubbing contact when running above their critical speed, *Mechanical Engineering*, 48, pp. 830–832.

Price, E.D., Lees, A.W. and Friswell, M.I., 2005, Detection of severe sliding and pitting fatigue wear regimes through the use of broadband acoustic emission, *Proceedings of the Institution of Mechanical Engineers Part J – Journal of Engineering Tribology*, 219, pp. 85–98.

Schwietzer, G., 2005, Safety and reliability aspects for active magnetic bearing applications – A survey, *Proceedings of the Institution of Mechanical Engineers Part I – Journal of Systems and Control Engineering*, 219(16), pp. 383–392.

Tong, X. and Palazzola, A., 2018, Tilting pad gas bearing induced thermal bow rotor instability (Morton Effect), *Tribology International*, 121, pp. 269–279.

7
Machine Identification

7.1 Introduction

This chapter is a little different from earlier chapters in that it reports the use of novel techniques which have appeared over recent years. Whilst several of them have been well validated in both laboratory and operational plant, their complexity has meant an inevitable time-lag in becoming widely used by the community rather than being restricted to researchers in the field. Nevertheless, these ideas indicate the direction in which machine diagnostics may progress in the foreseeable future. The bases of the key procedures are outlined and the reader is guided to the references for further details.

7.2 Current State of Modeling

In the previous three chapters, methods are discussed aimed at analyzing vibration signals and the signature types which are generated by a range of basic faults are reviewed. Also discussed are various techniques of machine modeling and in situations where the models are sufficiently accurate, it is clear that such models may prove very useful in mapping symptoms to causes, hence providing a useful diagnostic tool. Even prior to the advent of digital computers, some simple models provided significant insight into the dynamics of machinery. This trend can be traced back to the work of Jeffcott (1919) and Stodola (1927). In the last 50 years, more sophisticated numerical models have been developed. The earlier forms were based on the transfer matrix concept, but in recent years the finite element approach has become predominant. The transfer matrix method, while leading to a wildly non-linear computation of critical speeds, had the virtue of making very efficient use of computer storage.

Most modern approaches are based on the finite element method, an approach applied to rotors by several authors in the 1970s (see e.g. Nelson, 1980, Nelson and McVaugh, 1976). Since the introduction of these techniques, increasingly sophisticated models have been applied to understand observed plant behavior and a number of highly beneficial applications have been reported. There are, however, a number of limitations in current models and some of the reasons for this are discussed in this chapter.

At this point some general points on modeling are appropriate. As a generic term, modeling may cover a wide variety of tools ranging from a statistical digest of the operational behavior of a machine to a physics-based mathematical description. In essence, it is a construct to aid under-standing of plant operation. The finite element method has been applied to rotor systems for many years to gain understanding and the location of some faults and a basic formulation is given in Chapter 3 and discussed more fully by Friswell *et al.* (1996). While the statistical approach will be expected to cover the range of operation with some degree of accuracy, it is limited in terms of predictive capability: expressed in more mathemati-cal terms, it can interpolate but not extrapolate, and indeed this is the case with all empirical models, including artificial neural networks.

Physics-based (almost invariably FE) models do, in principle, have pre-dictive capability. The important question is their fidelity to the real plant and the extent to which fine details of construction are adequately reflected in the models. In many cases, the required details are not known. This variability can be seen in instances where nominally identical machines operate; because they are built to the same drawings, the models for these machines are identical. However, it is almost invariably the case that although the machines show strong similarities in their behavior, there are subtle differences. In Section 7.4 the sources of these variations are dis-cussed, and in the remaining part of the chapter some strategies for over-coming these difficulties are outlined.

It should be stressed at this point that this is an important practical issue. All machine operators have a vast amount of data relating to steady on-load running. As described in Chapter 2, this data usually reveals variation with time owing to minor changes in various parameters but very often these variations are of the same order of magnitude (or lower) than the uncer-tainty in modeling and consequently little meaningful information can be gleaned from this significant resource of data.

A number of difficulties have been outlined which explain the lack of agreement between models derived purely from engineering drawing and the observed behavior of actual plant. This leads to the conclusion that, given the benefits of accurate models, the monitored plant behavior must play some part in the development of these models. There are, however, a number of ways in which this problem may be addressed.

One approach would be to develop a purely empirical model to relate applied forces to response. For a machine, this may take the form of the response to a series of imbalance distributions but such an approach would involve considerable expense. This could be made somewhat more practical by utilizing the approach described in Section 7.5.1 to calculate the forces, then calculate the parameters to give optimum fit between model and observations. This fitting may then be obtained by an ARMA (autoregressive moving average) approach (see e.g. Worden and Tomlinson, 2001).

Used correctly, this approach can yield effective models, but it is difficult to relate individual terms to physical parameters. This can be overcome by using a physics-based approach in which a model is developed in which specific parameters are adjusted to give best fit – although, as discussed in the next two sections, even this is open to differing interpretations.

Section 7.5 gives, in some detail, the basic method proposed by the author and his colleagues, with subsequent sections giving a number of refinements. In this method, a good model for the rotor is assumed while the stiffness, mass, and damping matrices representing the foundation are to be determined. This means that each element of these three matrices is a parameter to be determined by the least square (LS) process. As described, the LS optimization is applied in a single step. Other authors have applied Kalman filtering to this problem as discussed in Section 7.6.3. The use of the filtering approach has many attractions, but has some disadvantages, and it is discussed by Provassi *et al.* (2000).

7.3 Primary Components

To address the issue of model uncertainty and possible tuning models to specific plant items, consider the factors involved in the representation of a machine. It is convenient for our discussion to divide the system into three main components, although it is perhaps more accurate to term theses as groups of components. These three are:

1) The rotating element
2) The bearings and other interactions between rotor and stator
3) The stator and its supporting structure

Let us now consider each of these in turn. The complete rotating element will, in the case of most turbo-machinery, be a (fairly complex) beamlike structure. This may be represented in terms of an FE model, by some type of beam elements and discs. Appropriate gyroscopic terms are readily included. Experience has developed over the years to the point where rotors may be represented with a reasonable degree of accuracy. There are a number of points to take into consideration, such as the effects of shrink fit stresses, but the experienced analyst can produce good rotor models. Furthermore, these models may be readily compared with numerical model, by suspending the rotor in a sling and measuring the response of the rotor to either a shaker or impact test. This technique has been reported by Mayes and Davies (1976) and Lees and Friswell (2001), among others. The response measured in a sling gives a very good approximation to the free–free modes of the rotor, thus providing a point of reference to the rotor model.

Unfortunately, this type of independent check is not readily available in the case of the other two components of our global model. In considering the interaction with the stator, the overall phenomena will, of course, be highly dependent on the type of machine in question. The one ever-present contribution will be that of the bearings and indeed these will normally represent the dominant force in locating the rotor. Seals and neck-rings also have an important contribution and these have been outlined in Chapter 6, and the book by Childs (1993) has a much more extensive discussion on these effects. In discussing large, heavy machinery, the bearings will usually be oil journal bearings, although there are now some large compressors being mounted on active magnetic bearings. These journal bearings are nonlinear in their properties and there is some uncertainty as to their behavior. However significant strides have been made in recent years and the performance of journal bearings for small oscillation about the equilibrium position is adequately understood. The one limitation here is that the bearing stiffness and damping coefficients depend strongly on the static load on the bearing. With a simple two-bearing machine, this presents no problems, but in a complex, multibearing situation, there is currently no means of knowing the load on each bearing.

While this limits the knowledge of the dynamics, it is by no means the major source of uncertainty. Unless the machine in question is misaligned to a dramatic extent, it is known that the bearing will give a positive direct stiffness at a given speed, and the magnitude of this together with cross terms can be evaluated, albeit approximately. However, the force transmitted via the bearings is dependent on the supporting structure and in this instance, the analyst has little or no idea what the dynamic properties are. These supporting structures often have their own dynamic characteristics and may show resonances below shaft running speed; therefore, it is unknown whether the structure will give a positive or negative restoring force at any given speed. For turbo-generators, this problem has come to the fore in the last 50 years as machines have dramatically increased in size. Foundation structures were at one time high tuned; that is, they had no resonances below running speed. As the machines became bigger, it became increasingly difficult to retain adequate rigidity and there was a move toward low-tuned, steel foundations for large machines and these offered a number of practical and financial attractions. It should be noted also that large concrete foundations may also be low tuned. Provided adequate precautions are taken in the avoidance of resonances (or, more precisely, critical speeds) close to operating speeds, such machines are very successful, but there is added difficulty in producing an adequate machine model and in interpreting the machine vibration behavior. In fact, the representation of the support structure is now believed to be the most significant defect in the modeling of large turbo-machinery.

For a time, it was thought that this problem could be overcome by modeling the support structure using FE techniques. In the period 1970–1990, several groups around the world attempted this, but success was somewhat limited. The extent of agreement was fairly poor. While it may be argued that the models have predicted the overall trends correctly, it is clear that the agreement is nowhere near good enough for the development of a usable model.

The use of computer methods for the study of machine foundations was reviewed by Lees and Simpson (1983). Their paper concluded that although FE studies could be helpful in the design process, by identifying global trends, far greater precision would be required if a model were to be use as a part of monitoring/fault diagnosis. A few years later Pons (1986) reported an extensive study on concrete foundations in which extensive measurements were made during power station construction. Once again poor correlation was reported between measured FRF and model predictions. Furthermore, the acquisition of plant data in this way is extremely costly, not only in terms of the effort involved, but, more significantly, in terms of the loss of production time on the plant. The situation is exacerbated by the subtle but significant differences between nominally identical units.

7.4 Sources of Error/Uncertainty

While, in principle, the effects due to the supporting structure are easily included in a model, in practice it is impossible to obtain adequate data to formulate the model. The supporting structures often include bolted and welded joints, sliding keyways and expansion joints, multivarious pipework, and variety of ancillary equipment. In addition to these factors, there are minor build differences which lead to marked changes in vibrational behavior. A simple example of this may be illustrated by considering the modal behavior of a portal frame such as that shown in Figure 7.1. Portal frames such as this form part of many machine support structures, but there are a number of decisions to be made as to the modeling of it: some of these are:

1) Can a beam representation be used or is solid modeling required? If the latter, the effort required to generate the model is significantly increased.

2) If beams are used, where is the neutral axis (the usual assumptions of center lines may distort rotational effects at the joints)?

3) How are the joints treated?

Hirdaris and Lees (2005) discuss the problems of modeling a portal frame and Oldfield *et al.* (2005) give a more general discussion of the modeling of joints.

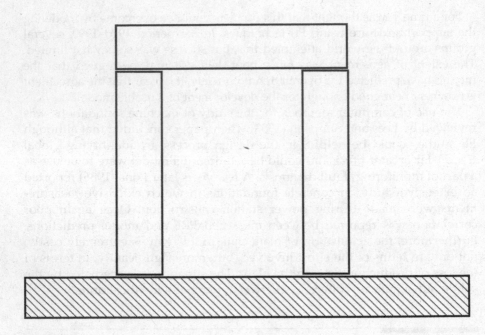

FIGURE 7.1
Portal Frame Structure.

As with any large, complex structure there are numerous difficulties in modeling the structure. The conventional modeling units, beams and plates, are in themselves idealizations of practical geometries. Yet fully three-dimensional elements present a prohibitive problem of data preparation and computing. However, the real problem goes much deeper than this; in virtually all practical cases there is insufficient data to specify the structural properties. Other problems arising for the modeler may be listed as follows

- Appropriate representation of individual components
- The influence of machine attachment points, including keyways
- The contribution to stiffness of machine casings
- The influences of ancillary machinery attached to the support structure

By reference to Figure 7.1, some of the difficulties of modeling become readily apparent. For a slender structure, the frame may be adequately represented by three beams. The corner joints are represented as constraints to a 90° angle with the vertical and horizontal members. With such a model, the natural frequencies and mode shapes are readily calculated.

As the thickness of the beams becomes greater in proportion to their length, the modeling options become much less clear. While it is probably acceptable to make representation in terms of beams on the center-line, it is clear that problems arise over the modeling of the corners. Firstly, the mass distribution is inappropriate, and secondly, perhaps more important, the corners have now some shear characteristics and the joint can no longer be represented as simply two beams meeting at right angle.

The portal frame shown in Figure 7.1 is typical of many which form part of the supporting structure of turbo-alternators. In this instance, the top was taken to be rigid and of significant mass but the flexural rigidity of the two vertical columns differ by 1%. Such a discrepancy is encountered in real plant; indeed it would be difficult to ensure consistency to within such a small tolerance. To put this in context, assuming that the columns are a rectangular box section, the wall thickness must be consistent to 1% or the box width to 0.33% – a demanding specification.

7.5 Model Improvement

There are very real practical difficulties in obtaining the required data on turbine foundations. A very convenient way of measuring the dynamics of many structures is through the use of a calibrated hammer which applies an impulse to the structure under study. Provided that the impact is sharp enough, this form of excitation provides a wide spectral range. While there are a number of practical problems associated with the technique, it is generally convenient for site use and usually more convenient (although somewhat less accurate) than, for example, electromagnetic shakers, the problems being associated with frequency resolution and signal/noise ratio.

In the case of large turbo-set foundations, however, the usefulness of the impact approach was somewhat limited. While there has been little detailed study of this issue, Lees and Simpson (1983) reported difficulty in obtaining adequate signal to noise ratios to enable the determination of cross receptances, particularly in the lower part of the frequency range. This, added to the large modal density, dictates that any detailed study would have to be performed using a shaker. However, this implies further problems. Clearly a large exciter, applying sinusoidal forces of order 10 kN, is required for large power turbines, and this implies a test which is laborious to plan and execute. This creates a further difficulty in the time requirements: It is precisely because the plants have a very high earning capacity that accurate models are required for the rapid diagnosis of faults. But this also means that time on the plant is at a premium and it will rarely be cost beneficial to carry out extensive vibration surveys.

A further difficulty arises because the plant will never be in the ideal state for testing. To examine the dynamics properties of the supporting structure, the requirement is for the machine to be complete with all its casings but no rotors in place. This is important since the upper casings can contribute significantly to the overall stiffness of the assembly. This is a configuration which will seldom/never arise during normal operation and furthermore it would prove extremely expensive to provide these conditions. To put these aspects into correct perspective, it is important to observe that the large efficient machines may have very high replacement costs; their high value, on the one hand, dictates the importance of having accurate models but also severely limits the availability for plant testing.

Kramer (1993) has suggested that the influence of foundations may be adequately modeled within an overall system but the difficulty lies in the choice of appropriate parameters. With relatively stiff concrete foundations, reasonable answers may be obtained but for low-tuned steel structures, the problems in obtaining an accurate representation are very much greater.

The high cost of plant implies that the only real option to obtain the required data is to access it while the machine is in operation, but this poses a number of questions as to the appropriate way forward. One might consider the application of an external force to the machine during operation, but few plant managers would sanction such a procedure. The remaining possibility is the use of plant data directly. There are two closely related approaches to this which will now be discussed.

7.5.1 System Identification

At this point it is important to distinguish between direct and inverse problems. To use a relevant example, consider the equation of motion of a system under the action of a sinusoidal force or set of forces. This can be expressed by the usual equation (omitting gyroscopic terms for clarity)

$$\mathbf{K}\mathbf{x} + j\omega\mathbf{C}\mathbf{x} - \mathbf{M}\omega^2\mathbf{x} = \mathbf{F}_0 e^{j\omega t} \tag{7.1}$$

Clearly, if the three matrices are known, then for any set of forces $\mathbf{F}_0$ the resulting displacement vector is readily calculated. This is a direct problem and is reasonably straightforward. This is so because usually it is clear from the physics of the situation that there will be a clear and unique answer.

The inverse problem to that just posed would be the determination of the matrices $\mathbf{K}$, $\mathbf{C}$, $\mathbf{M}$, given the measured response to a given set of forces. In contrast to the direct (or forward) problem, it is not clear that there is a unique answer, and in general, the approach to inverse problems is somewhat more involved. It is worth noting that in the direct problem there are n parameters to be determined (just the number of degrees of freedom) whereas in the inverse problem, the number of unknowns is $3^* n^* (n+1)/2$ and so clearly several frequency components must be considered together. But, even given this, the problem is not straightforward. Simply considering a range of frequencies in order to yield the problem theoretically soluble may not be sufficient in a realistic case in which the signal (vibration) is contaminated with measurement noise. It is therefore appropriate to utilize all the information available to yield the most accurate result.

To do this, Equation 7.1 is rewritten in the rather different form

$$\mathbf{WS} = \mathbf{F}_0 \tag{7.2}$$

In this equation, the matrix $\mathbf{W}$ contains all the measurement and frequency terms, while the vector $\mathbf{S}$ comprises all the unknown matrix elements in some predetermined order. To clarify this, consider the two degrees of freedom case. Note there will be a total of nine unknown parameters, namely $\{K_{11}, K_{22}, K_{21}, M_{11}, M_{22}, M_{21}, C_{11}, C_{22}, C_{21}\}$. The ordering can be chosen and the reasons for this particular choice will be outlined in due course. The immediate task is to express the equations of motion in the form of Equation 7.2. For each (discrete) frequency component, we may write

$$
\begin{bmatrix}
x_{01} & 0 & x_{02} & -\omega^2 x_{01} & 0 & -\omega^2 x_{02} & j\omega x_{01} & 0 & j\omega x_{02} \\
0 & x_{02} & x_{01} & 0 & -\omega^2 x_{02} & -\omega^2 x_{01} & 0 & j\omega x_{02} & j\omega x_{01}
\end{bmatrix}
\begin{Bmatrix}
K_{11} \\ K_{22} \\ K_{21} \\ M_{11} \\ M_{22} \\ M_{21} \\ C_{11} \\ C_{22} \\ C_{21}
\end{Bmatrix}
=
\begin{Bmatrix}
F_{01} \\ F_{02}
\end{Bmatrix}
\tag{7.3}
$$

There is one such equation for each of the N_f frequencies and these can be concatenated to give a single equation of the form 7.2 in which the matrix $\mathbf{W}$ has the dimension $(2 \times N_f) \times 9$.

With a large number of frequencies measured, the problem is over-specified, but since the measurements are contaminated with noise, this is desirable and allows the analyst to obtain a "best estimate" and to assess the errors involved. In general, $\mathbf{W}$ will be rectangular and hence an appropriate way forward is to multiply 7.2 by the transpose of $\mathbf{W}$ to give

$$\mathbf{W}^T \mathbf{W} \mathbf{S} = \mathbf{W}^T \mathbf{F}_0 = \mathbf{A} \mathbf{S} \tag{7.4}$$

Since $\mathbf{A}$ is now a square matrix, we may write

$$\mathbf{S} = (\mathbf{W}^T \mathbf{W})^{-1} \mathbf{W}^T \mathbf{F}_0 \tag{7.5}$$

This is called the *Moore–Penrose* solution and it is easily shown that this is equivalent to the LS solution of the over specified set of equations.

At this stage, the problem is, in principle, solved but there are some difficulties to be resolved. These problems center on the conditioning of the pseudo-inverse in Equation 7.5. This is closely associated with the level of noise in the measurements: the dichotomy is in seeking to fit measurements to a model, generally increasing the number of degrees of freedom will improve the agreement between model and measurement; too many degrees of freedom will result in the modeling of the noise. This leads to a physically unrepresentative model which, although reproducing the particular data used, has little or no predictive capability.

7.5.2 Error Criteria

In all identification procedures, it is important to include all available data and because of inevitable measurement and other errors, the data will contain some inconsistencies. The problem becomes one of minimizing some measure of the errors, usually the sum of the squares of the individual errors. But even then, there is a choice between the *equation error* and the *output error*. From an analytic standpoint, these two at first sight seem equivalent but this is not the case as the errors are weighted rather differently.

Consider first the errors in Equation 7.1. At each frequency considered the equation error can be defined as

$$\varepsilon(\omega) = \mathbf{F}(\omega) - \mathbf{K}\mathbf{x} - j\omega\mathbf{C}\mathbf{x} + \omega^2\mathbf{M}\mathbf{x} \qquad (7.6)$$

This is a vector quantity and a total error parameter can be defined as

$$E = \sum \varepsilon^T(\omega_i)\varepsilon(\omega_i) \qquad (7.7)$$

where the summation includes all frequencies in the transient being studied. The quantity E is called the *equation error*. This approach has the great benefit that the resulting LS calculation in linear in all the unknown elements of the matrices $\mathbf{K}$, $\mathbf{M}$, and $\mathbf{C}$, but unfortunately it can lead to errors because it can be sensitive to noisy data. This can be appreciated physically by observing that at some frequencies, there may be high forces acting on the foundation but with little response. As displacement is the quantity measured, this may be subject to substantial noise, which can give rise to bias in the estimated parameters. The way around this is to express the error in terms of the output, which is, of course, the measured quantity. This gives

$$\varepsilon_2(\omega_i) = \mathbf{x}(\omega_i) - \frac{F(\omega_i)}{K + j\omega_i C - \omega_i^2 M} \qquad (7.8)$$

Then

$$E_2 = \sum \varepsilon_2{}^T(\omega_i)\varepsilon_2(\omega_i) \qquad (7.9)$$

E_2 is called the *output error*. Note that for exact data, the two formulations are equivalent but they weight noise very differently. Output error leads to much better estimates of the unknown parameters because of the more appropriate weighting on interference in the data. The substantial penalty paid for this is the mathematical complication arising from the fact that Equation 7.9 is nonlinear in all the unknown parameters. In practice, this means that an iterative approach is required to obtain the solution. One approach is to use the equation error method to give an initial solution, then to use this as a starting point in an iterative scheme.

7.5.3 Regularization

The process of improving the quality of the results from an inverse problem is known as "regularization" and this consists of a range of methods which used in combinations to suit the particular problem aim to produce a unique, physically meaningful solution.

There are a number of techniques employed to achieve a good result and in this section three of the most important approaches are described. The methods used in regularization can be conveniently divided into three:

a) Appropriate scaling of parameter and weighting of measurements to minimize errors.
b) The elimination of unimportant variables by using any available knowledge of the underlying physics of the problem.
c) Use of a stable numerical procedure, such as the singular value decomposition, which is outlined in Section 7.5.4 and discussed further in Chapter 8.

The first step in seeking to optimize the estimation procedure is to ensure the matrix elements in Equation 7.3 are all of a similar order of magnitude. At first sight, this is problematic simply because the scale of stiffness and mass terms is very different, but this can be resolved simply by rescaling the frequency scale. Each frequency ω is divided by the first natural frequency ω_c and then the unknown parameters of Equation 7.3 are transformed to become

$$v = \left(K_{11}\ K_{22}\ K_{12}\ C_{11}\omega_c\ C_{22}\omega_c\ C_{12}\omega_c\ M_{11}\omega_c^2\ M_{22}\omega_c^2\ M_{12}\omega_c^2\right)^T \qquad (7.10)$$

The important point to note here is that all the nine unknowns are of a similar order of magnitude. A very approximate value for the natural frequency ω_c will suffice for this scaling. Equation 7.3 then is rewritten as

$$\begin{bmatrix} x_{01} & 0 & x_{02} & -\left(\frac{\omega}{\omega_c}\right)^2 x_{01} & 0 & -\left(\frac{\omega}{\omega_c}\right)^2 x_{02} & j\frac{\omega}{\omega_c}x_{01} & 0 & j\frac{\omega}{\omega_c}x_{02} \\ 0 & x_{02} & x_{01} & 0 & -\left(\frac{\omega}{\omega_c}\right)^2 x_{02} & -\left(\frac{\omega}{\omega_c}\right)^2 x_{01} & 0 & j\frac{\omega}{\omega_c}x_{02} & j\frac{\omega}{\omega_c}x_{01} \end{bmatrix} \{v\}$$

$$= \left\{\begin{matrix} F_{01} \\ F_{02} \end{matrix}\right\} \qquad (7.11)$$

Having scaled the parameters in this way, the next step is "weighting" or simply recognizing that in practice, some readings are more reliable than others. Typically, at frequencies close to resonance, high vibration levels will result which far exceed any background noise and consequently, these readings will be relatively accurate and should be assigned a high weighting. Mathematically, this is achieved by returning to Equation 7.2 and multiplying by a diagonal matrix, Π. The equation becomes

$$\Pi W S = \Pi F_0 \qquad (7.12)$$

Following the discussion above, the solution is given by

$$S = \left(W^T \Pi^T \Pi W\right)^{-1} W^T \Pi^T F_0 \qquad (7.13)$$

This can prove an effective step in improving the accuracy of the estimation procedure, but all available steps are often required to yield accurate results. A most important step is to ensure that an appropriate number of degrees of freedom are selected. It is obvious that to fit any given curve, increasing the number of degrees of freedom in the "trial function" will enhance the degree of fit. This, however, is not the crucial issue: The problem is that if the fitting curve is given too much freedom, the result will reflect the noise in the measured. To some degree, choosing the correct combination of parameters is an iterative process.

Care is needed to ensure that the appropriate number of variables is included for identification and this in itself has significant discussion in the literature (Smart *et al.*, 2000 for example). This is a problem for both physics-based models and the more statistically based methods. The problem is that increasing the number of variables will always give an improved fit to the experimental data, but the derived model will have little or no predictive capability. This is because the parameters are "used" to model the data noise rather than reflecting the true structure of the machine. An effective way around this difficulty is simply to constrain certain variables to be zero and several instances of this are mentioned in the study on machine foundations. This constraint is simply applied by deleting the appropriate rows and columns of the matrices and vectors. With these considerations, Equation 7.13 can be used to obtain appropriate estimates, but care is needed to minimize the effects of measurement errors. Rather than directly proceeding to use the pseudo-inverse matrix as in Equation 7.13, it is preferable to use the Singular Value Decomposition. This is discussed in many textbooks (see for example Golub and Van Loan, 1996) and just the basic outline is given in the next section.

7.5.4 Singular Value Decomposition

Using this technique enables the analyst to solve Equation 7.13 as a series of contributions from the set of modal terms. Errors arise in the treatment of modes with very small singular values. This is because the small singular value means a contribution will be magnified and secondly combinations of variables with small singular values have little influence on the errors in the set of equations. The way around this is to truncate the set of singular values at some minimum value, setting all subsequent terms to this value. To solve the problem by this method, consider Equation 7.12 and decompose the matrix $[\Pi W]$ which has dimensions $m \times n$ where $m \gg n$, as

$$\Pi W = U A V^T \tag{7.14}$$

where U has dimensions $m \times m$, A has $m \times n$, and V has $n \times n$. Both U and V are orthogonal matrices, while A contains the singular values on its diagonal, with zeros in all other elements. Hence rewriting Equation 7.12

$$U A V^T S = \Pi F_0 \tag{7.15}$$

Then, using the orthogonality properties of U and V, the solution may be recovered as

$$S = V A^{-1} U^T \Pi F_0 \tag{7.16}$$

Although A is a rectangular matrix, it does have an inverse because of its particular structure:

$$
\mathbf{A} = \begin{bmatrix} \lambda_1\ 0\ 0\ 0\ 0\ 0 \\ 0\ \lambda_2\ 0\ 0\ 0\ 0 \\ \cdots\cdots\ \ddots\ \cdots\cdots \\ 0\ 0\ 0\ \lambda_n\ 0\ 0 \end{bmatrix} \Rightarrow \mathbf{A}^{-1} = \begin{bmatrix} \frac{1}{\lambda_1}\ 0\ 0\ 0 \\ 0\ \frac{1}{\lambda_2}\ 0\ 0 \\ \cdots\cdots\ \ddots\ \cdots \\ 0\ 0\ 0\ \frac{1}{\lambda_n} \\ 0\ 0\ 0\ 0 \\ 0\ 0\ 0\ 0 \end{bmatrix} \qquad (7.17)
$$

This allows the solution of the inverse problem to be expressed in terms of the singular values and their corresponding eigen-functions. More importantly, by substituting low eigenvalues with some specified minimum value, the contribution of these terms to numerical errors can be minimized. This is a useful technique for solving large linear problems applied in many fields of which system identification is just one. Further details are discussed in Chapter 8.

7.6 Application to Foundations

It has been appreciated for a considerable time that the supporting structures play a significant role in determining the dynamic properties of large machines and this increased in significance with the trend toward flexible low-tuned supports, that is, structures with resonances below the machine's running speed. From an academic point of view, the obvious approach to this problem would be the performance of a full modal test but this is not feasible for commercial reasons. An additional issue is that such an extensive series of tests would give information on a single machine whereas it is now appreciated that even nominally identical machines in fact display subtle but important differences in behavior. A series of detailed tests were carried out during the construction of Bugey Power Station as reported by Pons (1986), and these results have helped insight into the modeling of these structures.

7.6.1 Formulating the Problem

Recognizing the problems discussed in the preceding section, attention has inevitably turned to the use of operational plant data to infer system properties. The problem of applying system identification techniques to a rotating machine has received attention over recent years and several approaches have been developed by groups around the world. In determining the dynamic properties of structures modal analysis (see e.g. Ewins, 2000) has become the predominant approach. In this method a known force is applied to a structure at one or more locations and a highly developed methodology is employed to determine modal parameters. Some elements of this approach have been applied to the identification of machine foundation particularly by Provassi *et al.* (2000) and the subsequent work by Pennacchi *et al.* (2006). Their general approach is to tune models using a Kalman Filtering approach (basically and iterative LS method) to refine an initial modal representation. A number of difficulties arise, but substantial progress has been made in recent years.

In using the operating machine as the test-bed, the force levels exciting the structure have to be assessed and there are two main ways of doing this: the first option is to measure the relative motion across each of the bearings and then use a bearing model of some form to formulate the force transmitted to the structure (Vania *et al.*, 1996). The alternative route, suggested by Lees (1988), is to use a validated model of the rotor and then calculate the force arising from a known imbalance, considering the dynamics of the (fully determined) shaft. While this route demands some additional calculation, it does have some advantages. Later work by Lees and Friswell (1997) demonstrated that the imbalance could be regarded as an additional parameter (or set of parameters) to be identified.

The other main choice to be made is the choice of parameters for the model: should the choice be physical parameters or modal ones? Again there are strong arguments on both sides. Modal models typically have fewer degrees of freedom to deal with, but the equations are inherently nonlinear, implying the need for some form of iterative solution. This imposes the necessity of an initial FE model of structure in order to provide a starting point for the full calculation. Because each bearing is implicitly treated separately in the calculation no means of estimating the overall imbalance has yet been suggested using the modal picture.

7.6.2 Least Squares with Physical Parameters

The use of physical coordinates is in some ways rather clumsy, involving a larger number of unknowns but offering a clear physical picture with some advantages. Although it becomes necessary to perform a nonlinear estimation to refine the answers, the approach leads to a linear estimation which is then used in the iterative loop. Lees (1988) suggested using plant data by considering the rotor itself to be the source of excitation. In this approach, the forces were calculated by relying on a good rotor model and then making the assumption that a bearing model is available. A modified form of this approach was later developed as discussed below. The calculation seeks values for the stiffness and mass matrices of the support on the assumption that machine damping is dominated by the bearings.

The calculation was originally developed for use on older machines where data on absolute shaft location is not readily available. The distinguishing feature of the method of Lees (1988) and the subsequent papers, culminating in Smart *et al.* (2000), is the use of a detailed rotor model to infer the magnitude of forces acting. The basic concept of the method is that the rotor motion is produced by the action of two sets of forces, each acting at a discrete set of axial locations.

a) A set of bearing forces, which are unknown, acting at the bearing location.

b) A set of unbalance forces acting at a predetermined set of balance planes. These forces may be considered as known by considering the difference between two balance runs. (Note, however, that later work [Lees and Friswell, 1997] shows that this restriction can be removed.

Because the shaft motion, or, rather, the related pedestal motion, is measured, the bearing forces can be expressed in terms of the unbalance.

Since the forces at the bearings on the rotor are now determined, the forces acting on the structure are equal and opposite, and their effect is measured as the pedestal vibration. The dimension of the vector $\{F\}$ is the number of bearings (times two if both perpendicular directions are considered). The vector $\{F\}$ represents the force acting on the rotor from the bearing reaction, hence the force acting on the supporting structure is just $-\{F\}$.

In the case of a foundation with many modes within the running frequency range, the structural matrices should vary. This may be treated by subdivision of the frequency range, resulting in a larger LS problem, but no additional difficulty in principle.

This approach will now be considered in some detail. With the usual assumption of linearity, the equation of motion can be written as

$$\mathbf{K}x + \mathbf{C}\dot{x} + \mathbf{M}\ddot{x} = \mathbf{F} \qquad (7.18)$$

Written in this form, the equation is not very helpful in as much as there is only incomplete knowledge of the stiffness and mass matrices. Consider now purely sinusoidal excitation: now Equation 7.18 can be expressed as

$$(\mathbf{K} + j\omega\mathbf{C} - \omega^2\mathbf{M})\mathbf{X}_0 e^{j\omega t} = \mathbf{Z}(\omega)\mathbf{X}_0 e^{j\omega t} = \mathbf{F}_0 e^{j\omega t} \qquad (7.19)$$

where $\mathbf{Z}$ is the dynamic stiffness, part of which is unknown. However, the parts which are known can be utilized to infer the remaining terms and to achieve this, after canceling the sinusoidal terms in 7.19, the dynamic stiffness is partitioned and the equation is rewritten in the form

$$\begin{bmatrix} Z_{R,ii} & Z_{R,ib} & 0 \\ Z_{R,bi} & Z_{R,bb}+Z_B & Z_B \\ 0 & Z_B & Z_B+Z_F \end{bmatrix} \begin{Bmatrix} X_{R,i} \\ X_{R,b} \\ X_{F,b} \end{Bmatrix} = \begin{Bmatrix} F_u \\ 0 \\ 0 \end{Bmatrix} \qquad (7.20)$$

The dynamic stiffness terms Z_R, Z_B, and Z_F refer to the rotor, bearing, and foundation respectively. Experience has shown that the rotor can be modeled with good accuracy and, furthermore, the adequacy of any rotor model may be verified by impulse or shaker tests in works; therefore Z_R can be regarded as known. The subscripts i,b refer to internal and bearing locations. The bearing may also be modeled and, although some difficulties arise, it can be shown that the uncertainties have only a limited influenced on the overall result, namely the estimate of the foundation dynamic stiffness, Z_F.

There are several ways to proceed from Equation 7.20, depending on what plant measurements are available. Here the case considered relies solely on measurements of bearing pedestal vibration; while this is simplest from a measurement perspective, it does involve a little more mathematical manipulation. This is because the terms $X_{R,i}$ and $X_{R,b}$ have to be eliminated from 7.20. These two parameters can be expressed as

$$X_{R,b} = -\left(Z_{R,bb} + Z_B\right)^{-1}\left(Z_{R,bi}X_{R,i} + Z_BX_{F,b}\right)$$
$$X_{R,i} = Z_{R,ii}^{-1}\left(F_u - Z_{R,ib}X_{R,b}\right) \tag{7.21}$$

With these substitutions, the third line of 7.20 can be expressed as

$$-Z_B\left(Z_{R,bb} + Z_B\right)^{-1}\left(Z_{R,bi}\left(Z_{R,ii}^{-1}(F_u - Z_{R,ib}X_{R,b})\right) + Z_BX_F\right) + (Z_B + Z_F)X_F = 0 \tag{7.22}$$

This equation however contains $X_{R,b}$ which is unknown. But this can be eliminated using the third equation in 7.20 and writing

$$X_{R,b} = -Z_B^{-1}(Z_B + Z_F)X_F \tag{7.23}$$

Substituting 7.23 in 7.22 yields and equation in which the only unknown quantities are Z_F and F_u.

At first sight, this looks fairly formidable, but it contains only two terms which are unknown, Z_F the foundation dynamic stiffness and F_u, the imbalance. In fact, working with the difference between two imbalance runs, it is possible to regard F_u as known. Lees and Friswell (1997) show that the imbalance may simply be treated as a set of additional parameters to be identified and this is briefly discussed in 7.7. Later work (Sinha *et al.*, 2002) demonstrates that the bearings may also be treated in similar manner. By defining some new terms, Equation 7.20 can be expressed as

$$PF_u + Q + SZ_F = 0 \tag{7.24}$$

where the matrices $\mathbf{P}$ and $\mathbf{S}$, together with the vector $\mathbf{Q}$, are rather complicated but well-defined functions of $\mathbf{Z}_R$, $\mathbf{Z}_B$ which are both known and the vector $\mathbf{X}_F$ which is measured. This equation is now recast in the form of 7.2 for the identification process. The foundation matrix $\mathbf{Z}_B$ is expanded to its constituent stiffness mass and damping terms and the equation becomes

$$\mathbf{P}_\omega \mathbf{F}_{u\omega} + \mathbf{Q}_\omega + \mathbf{S}_\omega v = 0 \tag{7.25}$$

where, as discussed in 7.5.1, the matrix S_ω contains the measured vibration readings, as does the vector Q_ω and the vector $v^T = \{k_{11} \;\cdots\; k_{1n} \; m_{11} \;\cdots\; m_{1n} \; c_{11} \;\cdots\; c_{1n}\}^T$ represents the parameters to be determined. The subscript ω is a reminder that this equation is formulated for each frequency of the excitation and hence all these results in an overspecified problem of the form

$$\mathbf{S}v = -\mathbf{P}\mathbf{F}_u + \mathbf{Q} \tag{7.26}$$

This formulation is suitable for machines on which only measurements of bearing vibrations are available, but, as shown, some modeling of the bearings is needed, although the results are not very sensitive to bearing errors. On more modern machines, however, measurements of the rotor vibration are often available using proximity probes. In this case, a different, and rather simpler, formulation is used and this is discussed in 7.6.2.1. Not only is this simpler but the availability of rotor data facilitates access to some further insights into the machine's condition.

7.6.2.1 Formulation with Shaft Location Data

For the case in which shaft motion is also monitored, a rather simpler formulation may be derived from Equation 7.19 to give

$$\begin{bmatrix} Z_{R,ii} & Z_{R,ib} & 0 \\ Z_{R,bi} & Z_{R,bb} & 0 \\ 0 & 0 & Z_F \end{bmatrix} \begin{Bmatrix} r_{R,i} \\ r_{R,b} \\ r_F \end{Bmatrix} = \begin{Bmatrix} f_u \\ f_b \\ -f_b \end{Bmatrix} \tag{7.27}$$

At this stage, of course, the force exerted on the shaft at the bearings, f_b is unknown, but it is easily eliminated and recovered from the calculation later. By adding the second and third rows

$$\begin{bmatrix} Z_{R,ii} & Z_{R,ib} \\ Z_{R,bi} & Z_{R,bb} \end{bmatrix} \begin{Bmatrix} r_{R,i} \\ r_{R,b} \end{Bmatrix} + \begin{bmatrix} 0 & 0 \\ 0 & Z_F \end{bmatrix} \begin{Bmatrix} 0 \\ r_F \end{Bmatrix} = \begin{Bmatrix} f_u \\ 0 \end{Bmatrix} \tag{7.28}$$

It is worth considering each of the terms in some detail: all the values of *r* are known, since these are the measured quantities. In the first matrix, all four terms Z_R are known, since this is just the model of the rotor which (it is assumed) has been validated separately. Z_F is just the model of the foundation which is to be determined. The first row of Equation 7.28 is now used to eliminate the internal rotor degrees of freedom using dynamic condensation. The imbalance force may be either known (by taking the difference of two balancing runs) or treated as an additional unknown vector to be determined.

In general, the more measurements available, the better, so the acquisition of rotor information allows the extraction of information about the bearings themselves. For instance, if there is data on both rotor and pedestal position, the calculation can, in principle, give information on forcing and relative displacement hence stiffness and damping information can be inferred.

7.6.2.2 *Applying Constraints*

As shown by the SVD (discussed in 7.5.4 and further in Chapter 8) the range of the singular values of the matrix $[\Pi W]$ determines the number of parameters which can be extracted with reasonable accuracy. Equivalently, the criterion can be based on the eigenvalues of the regression matrix used for the Moore–Penrose inverse $[W^T \Pi^T \Pi W]$.

Too great a range of the eigenvalues is a source of ill-conditioning in the regression equation. Physically, this means that there is insufficient information to discriminate all the unknown parameters and the main techniques for dealing with this issue are discussed in 7.5.3.

While it is important to assign enough freedom to a model to obtain a good correspondence with measured data, too much freedom will result in the modeling of noise terms, and although this may improve the fit, it will reduce the predictive capability of the derived model. It is important to emphasize that the crucial test of a derived model is its predictive capability. The decision as to the requisite number of degrees of freedom is a difficult one, which may on occasion require some iteration. As a consequence, it is often desirable to reduce the number of unknowns in a model and this may be based on physical reasoning.

Consider the case of a machine with 14 bearings. For the sake of this argument, we consider only motion in a single direction.

Then the stiffness matrix **K** is 14×14 and has 105 (=14×15/2) independent terms. However, this could be reduced significantly if we assume that each bearing can only interact with its three nearest neighbors. If this is the case, then the number of terms in **K** will be 50 = (14+13+12+11), more than halving the size of the matrix. The matrix will now be of the form

$$
\mathbf{K} = \begin{bmatrix}
x & x & x & x & 0 & 0 & 0 & 0 & 0 & 0 & 0 & 0 & 0 & 0 \\
x & x & x & x & x & 0 & 0 & 0 & 0 & 0 & 0 & 0 & 0 & 0 \\
x & x & x & x & x & x & 0 & 0 & 0 & 0 & 0 & 0 & 0 & 0 \\
x & x & x & x & x & x & x & 0 & 0 & 0 & 0 & 0 & 0 & 0 \\
0 & x & x & x & x & x & x & x & 0 & 0 & 0 & 0 & 0 & 0 \\
0 & 0 & x & x & x & x & x & x & x & 0 & 0 & 0 & 0 & 0 \\
0 & 0 & 0 & x & x & x & x & x & x & x & 0 & 0 & 0 & 0 \\
0 & 0 & 0 & 0 & x & x & x & x & x & x & x & 0 & 0 & 0 \\
0 & 0 & 0 & 0 & 0 & x & x & x & x & x & x & x & 0 & 0 \\
0 & 0 & 0 & 0 & 0 & 0 & x & x & x & x & x & x & x & 0 \\
0 & 0 & 0 & 0 & 0 & 0 & 0 & x & x & x & x & x & x & x \\
0 & 0 & 0 & 0 & 0 & 0 & 0 & 0 & x & x & x & x & x & x \\
0 & 0 & 0 & 0 & 0 & 0 & 0 & 0 & 0 & x & x & x & x & x \\
0 & 0 & 0 & 0 & 0 & 0 & 0 & 0 & 0 & 0 & x & x & x & x
\end{bmatrix} \tag{7.29}
$$

Now let us consider how to eliminate the unwanted degrees of freedom: note that this is the same as constraining them to be zero. Returning to Equation 7.2, suppose that one or more components of $\mathbf{S}$ are constrained to take specified values $\mathbf{S}_c$, then dividing the vector $\mathbf{S}$ into free and constrained elements

$$
\begin{bmatrix} \mathbf{W}_f & \mathbf{W}_c \end{bmatrix} \left\{ \begin{matrix} \mathbf{S}_f \\ \mathbf{S}_c \end{matrix} \right\} = \{ \mathbf{F}_0 \} \tag{7.30}
$$

which may be expanded to give

$$
\mathbf{W}_f \mathbf{S}_f = \mathbf{F}_0 - \mathbf{W}_c \mathbf{S}_c \tag{7.31}
$$

In the particular, but very common, case of constraining a parameter to zero, this is achieved simply by deleting the column of the matrix $\mathbf{W}$ corresponding to the degree of freedom in question. For nonzero constraints, the forces term is also modified as illustrated in Equation 7.31.

For the sake of illustration, consider a machine with two bearings; then in identifying full stiffness mass and damping matrices there will be 9 degrees of freedom. If the three cross terms are to be eliminated, using the ordering convention outlined above the vector of unknowns becomes

$$
a = \{\, k_{11} \quad k_{22} \quad 0 \quad m_{11} \quad m_{22} \quad 0 \quad c_{11} \quad c_{22} \quad 0 \,\} \Rightarrow \{\, k_{11} \quad k_{22} \quad m_{11} \quad m_{22} \quad c_{11} \quad c_{22} \,\}
$$

This means that columns 3, 6, and 9 of the $\mathbf{A}$ matrix are deleted, and so if the measurements cover N frequency components, then the modified $\mathbf{A}$ matrix has the dimensions $N{\times}6$. This is, perhaps, a rather arbitrary example, but now consider a more realistic scenario. It is reasonable to assume that a force at the high-pressure turbine end of the machine will not influence the exciter; that is, there will be no significant coupling between bearing 1 and bearing 14, assuming the case of a 14-bearing turbo-generator. So the stiffness, mass, and damping matrices have a banded structure, the extent of which is, a priori, again a matter of judgment which may be aided by considering the singular values of the measurement matrix. Recalling that the unknown terms in the vector are just the elements of the system matrices, the ordering is arbitrary and may be chosen for computational convenience.

It is less obvious how to deal with the damping in the system; there is certainly no justification to assume classical (proportional) damping in a rotating machine and this might lead to the representation of the damping as a full matrix. However, this is not necessary and indeed it is undesirable, potentially leading to significant errors. There are two important reasons for this: firstly, the measured data of any system will have only a significant information content of the damping in a relatively restricted frequency range around the natural frequencies. Hence identifying the damping in this way introduces a significant number of variables for which there is limited information. A second related issue for rotating machines is that, in the case of oil journal bearing mounts, about 99% of the total damping arises in the bearings, and this dissipative force is already represented within the calculated bearing forces. In principle, therefore it would be physically reasonable to set the foundation damping to zero, but experience has shown that better numerical stability is obtained if there is a damping term within the identified model.

Although the unknown parameters can be identified together, the size of the problem can be reduced by careful study of the physical behavior of a real machine. The problems of machine diagnostics are most acute in the case of large machines mounted on low-tuned (usually steel) foundations and these machines are almost invariably supported on oil film journal bearings. This being the case, the overwhelming proportion of damping in the system takes place in the journal oil film and consequently, the damping properties of the structure may be represented with simply a diagonal **C** matrix. It should, perhaps, be emphasized that some damping terms have been found necessary to obtain convergence in identification.

An interpretation of this is that in the experimental results being analyzed, the noise level is of a similar order to those terms necessary to resolve differences between possible solutions. Figure 7.2 shows a comparison of measured and calculated response on a large laboratory test-rig subjected to a known imbalance. The figure shown is typical of several from the experiments as reported by Smart *et al.* (2000).

Note that at all frequencies above about 15 Hz the agreement is excellent in terms of both magnitude and phase. The reason for this poor performance at low frequencies is unclear, but it is of little practical importance on real plant.

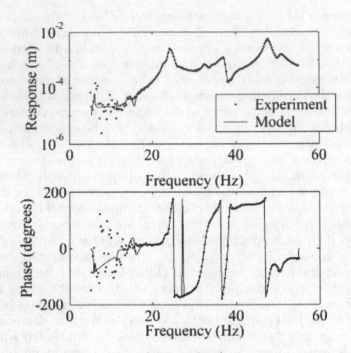

FIGURE 7.2
Comparison of Measured and Calculated Response.

(Reproduced with permission from Proceedings of the Royal Society, A, 456, pp. 1583–1607, 2000)

7.6.3 Modal Approach with Kalman Filters

The alternative to the identification of matrices described above is the use of a modal model and this has been pioneered by several authors in Italy (Provassi *et al.*, 2000, Vania, 2000, Zanetta, 1992). Once modal behavior has been determined, then clearly all dynamic behavior can be recovered. The big advantage of using the modal approach is a significant reduction in the number of parameters to be identified but a penalty is paid for this insofar as the equations are nonlinear in the unknown terms.

The force is inferred from the bearing properties and hence in contrast to the Lees (1988) approach, measurements of rotor position, in addition to pedestal motion, are required. Feng and Hahn (1995), retain physical, rather than modal, parameters, but in their initial study, they too use the calculated bearing coefficients, with nominal static loadings to calculate the bearing dynamic forces. In later work, the same authors (Feng and Hahn, 1998) reported the measurement of bearing oil pressure circumferentially, integrating to obtain the force exerted on the rotor. There were, understandably, significant practical difficulties in doing this. While this is a very interesting laboratory investigation, it is not seen as a practical scheme for operational plant.

For several years, there has been study of this problem in Italy, involving a number of workers. Vania (2000) and Zanetta (1992) have played major roles in the work of this group. Their approach differs from that followed by the Swansea team in three respects, two of which are of physical importance, while the third is, perhaps, more a mathematical detail. Having said that, at the time of writing the whole problem is not understood to the extent of making categorical statements about the relative importance of different aspects.

The three respects distinguishing the work of Zanetta and Vania *et al.* are:

a) A modal model, rather than a physical model, is used

b) Forces are calculated using calculated bearing parameters

c) Identification is done by Kalman filtering

It is the author's belief that each of these points has advantages and disadvantages, some of which raise fundamental issues as to the eventual use as to these methods. It must be considered highly likely that some approach will emerge which incorporates elements of both approaches. At present, however, the route of such a synthesis is far from clear.

In expressing the model in modal terms, two advantages are immediately introduced. Firstly, the dimensionality of the problem may be significantly reduced, and secondly, the behavior of each bearing pedestal may be written independently. The price paid for this is that the whole problem becomes very nonlinear with the important implication that any feasible solution scheme must be iterative. All such methods, including the Kalman filter approach, require an initial trial state, from which the final solution can be developed. This may be a substantial disadvantage in practice and it is well known that the absolute convergence and solution efficiency of such algorithms is strongly dependent on the closeness of the initial state to the final solution.

There is an inherent assumption of linearity in reducing the system to a modal model. At present this may not be too restrictive an assumption since the models used by all authors in this area are currently linear. However, retaining the model in physical space rather than modal space preserves the option of introducing nonlinearities at a later stage. The price paid for this luxury is a much larger, coupled problem. It results however in a linear, although large, problem.

The response at each location satisfies the expression

$$z_i = \sum_{r=1}^{n} \psi^i{}_r \frac{\psi^T{}_r F(\omega)}{(\omega^2{}_r - \omega^2 + 2j\xi_r\omega_r\omega)} + \psi^i{}_l \frac{\psi_l{}^T F(\omega)}{-\omega^2} + \psi^i{}_n \frac{\psi_n{}^T F(\omega)}{\omega_n{}^2} \qquad (7.32)$$

Note that the last two terms on the right-hand side of this equation are there to allow for the influence of modes below and above the frequency range of the study. This is a device commonly used in modal analysis (see Ewins, 2000). The forces are calculated from bearing properties and the differential motion. Others (e.g. Feng and Hahn, 1995) have relied on a solution of Reynold's equation to achieve this. This involves some uncertainty with regard to both bearing internal clearances and static load, which, of course, can vary with operational conditions. Notice that Equation 7.32 is inherently of the output error type and hence the conditioning will tend to be superior. It is however nonlinear, thus introducing some mathematical complexity. The parameters to be determined are

$$x = \left\{ \psi^T{}_1, \cdots, \psi^T{}_r, \cdots, \psi^T{}_n, \cdots, \xi_1, \cdots \xi_r, \cdots, \xi_n, \omega_1, \cdots, \omega_r, \cdots \omega_n, \psi^T{}_n \psi^T{}_n \right\}^T$$

$$(7.33)$$

To give this some context, consider a machine with 14 bearings, restricting attention to vibration in a single direction. Then for n modes the number of variables (or degrees of freedom) will be $(14 * n + 2n + 14 * 2)$.

Applying nonlinear corrections is relatively straightforward. In the analysis presented above, the derived model is purely linear, but there is nothing in the fundamental approach which limits attention to a linear model. Nonlinear bearing models could be introduced into this scheme, although the sensitivity of results to input uncertainties has not been assessed to date. In this regard the input would include a detailed nonlinear bearing model. This dependence on a bearing model may, however, be removed by measuring the instantaneous position of the rotor at each journal together with the bearing pedestal motion. Indeed, these measurements are readily available on some newer turbo-generators.

At worst, the approach used in ordinary balancing procedures may be employed: all conventional balancing techniques are based on the assumption of linearity. In practice, this may lead to an iterative approach, in which successive refinements to the state of balance are made using essentially a quasilinear approach at each stage.

While shaft measurement is often found on newer turbo-generator units, many machines do not have this facility and the provision and maintenance of the instrumentation would not be cost beneficial, or at least it is not currently deemed to be so.

A more onerous requirement is an assessment of the static load on the bearings. This will rarely, if ever, be the case using currently accepted approaches. Multirotor systems are aligned under cold conditions and corrections made for anticipated thermal movements. However, there is little information on the alignment of machines under operating conditions. This is important because the dynamic characteristics of oil journal bearings are strongly dependent on the static load carried by the bearing. In some current methods this difficulty is circumvented by the use of nominal values in the calculation. It is important, however, to appreciate that this is a source of error in the overall estimation of parameters.

7.6.4 Essentials of Kalman Filtering

Kalman filtering can be regarded as a continuous form of LS analysis which is extensively used in control applications. Here, a brief discussion is given but further insight can be gained from numerous textbooks. A very clear explanation is given by Noton (1972) and an interesting application to structures is given by Simonian (1979), although both these references are now rather dated. Where in the standard LS method a single calculation is made which minimizes the sum of the squares of the error terms, with the filtering approach, an initial estimate (i.e. guess!) of the answer is modified as each set of data is added. As the calculation evolves, estimates of the unknown parameters and their weightings are formed and the calculation is self-consistent in so far as the appropriate weightings are used. In this respect, filtering has an advantage over the straightforward application of the Moore–Penrose inverse calculation. In this type of calculation, each new piece of data is used in turn to give a current fit. At stage i for example, the parameter estimate is given in the form,

$$y_i = \alpha_i y_{i-1} + (1 - \alpha_i)x_i \tag{7.34}$$

where x is the estimate calculated using the current set of data. In simple terms, at each stage of the calculation, the overall estimate is formed as a weighted mean of the existing estimate and that derived from the "new" set of data. When compared with the straightforward LS approach, the filtering method does avoid the very large matrix operations by splitting the calculation into a set of rather smaller calculations. However, the calculations are far from trivial and the effective number of variables is considerable and it is worth considering the reasons for this. In the case of the stiffness, mass, and damping matrices each having n degree of freedom, the total number of variables will be $3*n*(n+1)/2$, but the weighting terms in Equation 7.34 is dependent on the variance matrices of each of the three variables. This will involve a total of $3 \times (n^2 + n) \times (n^2 + n + 1)$ coefficients; hence, these calculations are demanding in terms of storage requirements. Efficient algorithms are available and these large correlation matrices can be calculated directly from the updated parameter estimates.

7.7 Imbalance Identification

One advantage of the direct modeling approach (Smart *et al.*, 2000) is that it is a straightforward matter to add the eccentricity vector to the list of variables for identification. This is discussed in detail by Lees and Friswell (1997), but actually the method is a straightforward development of the method discussed in the previous section. Writing the basic equation of motion in its simplest form (following Smart *et al.*, 2000)

$$\mathbf{Z}_F r_{F,b} = f_{F,b} \tag{7.35}$$

The bearing force $f_{F,b}$ can be calculated as

$$f_{F,b} = \mathbf{Z}_1 r_{F,b} + \mathbf{Z}_2 f_u \tag{7.36}$$

where

$$\mathbf{Z}_1 = \mathbf{Z}_B \left(\mathbf{P}^{-1}\mathbf{Z}_B - \mathbf{I}\right) \qquad \mathbf{Z}_2 = \mathbf{Z}_B \mathbf{P}^{-1}\mathbf{Z}_{R,bi}\mathbf{Z}^{-1}_{R,oo} \tag{7.37}$$

and

$$\mathbf{P} = \mathbf{Z}_{R,bb} + \mathbf{Z}_B - \mathbf{Z}_{R,bi}\mathbf{Z}^{-1}_{R,ii}\mathbf{Z}_{R,ib} \tag{7.38}$$

While these formulae are complicated, their physical meaning is straightforward, the matrices used involve only known or measured quantities, and they are readily calculated. Using this notation, the equation can be cast in form for identification as

$$\mathbf{W}\nu = f_{F,b} = \mathbf{Z}_1 r_{F,b} + \mathbf{Z}_2 \omega^2 e \tag{7.39}$$

where e is the unknown mass eccentricity vector (defined at each of the preselected balance planes). This equation is easily rearranged as

$$\left[\mathbf{W} - \mathbf{Z}_2\omega^2\right]\begin{Bmatrix} \nu \\ e \end{Bmatrix} = \mathbf{Z}_1 r_{F,b} \tag{7.40}$$

The estimation of imbalance is a valuable device and, in contrast to the foundation parameters, estimates can be readily verified using balancing techniques. Table 7.1 shows the results of a series of tests on a large

TABLE 7.1

Comparison of Estimated and Actual Imbalance

Test No.	Bal. plane	Magnitude (kgm) Actual	Phase (Deg) Actual	Magnitude (kgm) Estimated	Phase (Deg) Estimated
1	3	0.160	45	0.1678	44.4
2	1	0.160	195	0.1797	188.2
	3	0.160	45	0.1606	37.5
3	3	0.160	45	0.1715	48.4
	4	0.160	150	0.2108	149.1
4	1	0.160	195	0.1737	195.9
	5	0.160	0	0.1811	−0.02
5	1	0.160	195	0.1624	182.3
	3	0.160	45	0.1318	42.9
	4	0.160	150	0.1941	154.2
6	1	0.160	195	0.1852	195.7
	3	0.160	45	0.1829	42.2
	5	0.160	0	0.1830	2.77

four-bearing laboratory rig. Six imbalance configurations were tested and the magnitude and phase of the estimates are shown.

Having established validity of this method in the laboratory, it was applied "blind" to a 350 MW turbo-alternator (overall length about 40 m) which had suffered a blade loss, the details of which were not revealed to the team at Swansea University. Calculations indicated high imbalance at the two balance planes on the HP rotor. This was equated to a single imbalance by considering moments and the estimates were compared with the measurements taken by the company involved. The imbalance estimate was 15% low and the estimated position along the rotor was 500 mm away from the location of the actual damage blade. This is considered very acceptable in view of the fact that the actual fractured portion of blade was an irregular shape. Some error was found in the phase estimate and this has never been resolved fully; possibly there were phase discrepancies in the signals but a full investigation of this was not possible. Nevertheless, rig results together with plant experience suggest this is a useful approach.

The method has also been to recover details of shaft bends. As discussed in Chapter 4, bends give rise to synchronous vibration but it is important to distinguish them from imbalance. The extension to the calculation is straightforward and is not discussed here. The interested reader can find a full discussion in Edwards *et al.* (1998).

7.8 Extension to Misalignment

The inclusion of rotor motion data, where available, is discussed in Section 7.5 and as with all additional data, if used correctly, it can enhance the quality of the estimations and consequently understanding of the machine' operation. However, it also opens two additional facilities to the overall technique. The first effect is very simple in physical terms; because there is, in effect, a monitor of bearing forces together with the differential motion between shaft and stator, the two may be used to assess bearing condition. The properties of the bearing, at any given configuration, depends on the static load carried and consequently, given the properties, a calculation can be set up whereby the load is determined by comparing the measured stiffness with an idealized short bearing (see Hamrock *et al.*, 2004), but any other bearing model could be used if deemed more appropriate for a particular case. This calculation is inherently nonlinear, so an output error is appropriate. The vector of unknowns is now $\theta = \{\nu \, e \, F_1 \cdots F_n\}^T$. An iterative scheme is used to reach the solution. The calculation is slightly more involved than those presented in this chapter but full details are given by Sinha *et al.* (2004).

7.9 Future Options

Most of the studies discussed in this chapter have emerged in recent years and considerable progress has been reported on the identification of large machines and indeed, inverse problems in general. It is a prime example of an area where measured data and modeling expertise complement each other, to yield an enhanced understanding of the operation of plant. As discussed, several approaches to the problems have been presented and are now at a usable stage. Currently, though, these techniques do add some complexity to the task of data analysis and there is a need for some degree of automation. While this will demand some effort, it is not difficult to envisage a future system which will activate automatically on each rundown and produce estimates of the mass and stiffness terms at each bearing. In addition, the system would identify the unbalance profiles of the rotor and information on any bending, although hopefully this would be fairly rare. Extending the system to give an estimate of the misalignment could also prove invaluable in plant operation. The net result of all this would be a trend toward saving the physical parameters of a system as they develop over time, rather than the current practice of storing vast quantities of vibration data, which require considerable expert interpretation.

Looking further ahead, there are some intriguing possibilities. After a model has been derived following a rundown, it could be fine-tuned using a filtering approach. The long-term aim must be to gain maximum insight from data of machines on-load. This data is plentiful but currently of limited use because the information content is much lower than transient data. However, in a regime of fine-tuned models, the analyst may be able to examine the consequences of subtle on-load changes and hence gain further insight.

7.9.1 Implementation

Although this procedure might appear to some fairly complex, it is purely algorithmic and, in time, can be readily automated. This would allow significant information gain following each machine rundown. This sequence is readily followed:

1) Rundown automatically processed as described above

   There is a rather modest overhead in doing this: the rotor model would need to be set up for each class of machines. On a power station with say four machines, there will probably be a single rotor model. This point has been discussed in several papers (e.g. Lees, 1988). The variations that exist between nominally identical machines are functions of their supporting structure and these differences can be treated. Ideally, rotor models should be validated using a free-free test.

2) Estimated unbalance trended against previous values (classified among hot and cold rundowns)

3) Foundation parameters compared with previous values

   With the automated calculations, separate records could be assembled to monitor foundation parameters, state of balance, rotor bend, bearing (static) forces, and hence alignment.

4) Alignment changes monitored

5) If proximity probe data is available on the unit, the bearing characteristics may be inferred and may be indicative of wear within the bearing.

   With a separate treatment, a full record of bearing properties can be assembled. This is again made possible by the derivation of the forces at the bearings by use of a good rotor model.

Given such a system, a great deal of insight into a machine's condition can be gained with virtually no additional equipment or effort about that which is expended currently. Basic research has now developed techniques for the calculation of the parameters mentioned above, although there is still discussion among several research groups with regard to the most efficient methods. When these issues have been resolved, significant software development will be required to implement these approaches. However, prototype schemes could be installed in the very near future.

A further intriguing possibility arises; because after analysis we have full knowledge of the rotor's position in the bearings and the unbalance forces, full information may be recovered using the model of the rotor to yield the deflected shape of the rotor, thus highlighting any areas of rubbing.

7.9.2 Benefits

By pursuing this analysis, what is being done is at all times seeking to fit the observed behavior and the numerical model. Regarding the model and the plant data as sources of information, a judicious combination of the two will yield the maximum information about the state of the plant. This idea may be surprising to some who might argue that the plant data is "real" while the model is a mere figment of imagination. This, however, is not the case. Any piece of information which enhances plant understanding is to be valued. Information from plant and model should be viewed as complementary evidence as to the overall state and can operate as a cross-check on each other.

7.10 Concluding Remarks

This chapter is rather different from others in this book in that a new approach to monitoring has been outlined. Trials so far by several groups around the world have shown success and hopefully they will develop further. Despite some initial complexity, it is believed that the methods outlined here offer some very considerable benefits.

Problems

7.1   Table 7.2 shows the response of a two-spring, two-mass system in the range 5–100 Hz. The system is excited by a unit force acting on mass 2. Formulate the regression equation to determine the stiffness mass and damping parameters.

7.2   Repeat the calculation of Problem 7.1, using only data up to a frequency of 50 Hz. Explain the difference between in the derived estimates and those obtained using the complete frequency range.

7.3   Using the data in Table 7.2, estimate the system parameters **K**, **M**, and **C**, on the assumption the damping matrix is diagonal. Compare the estimated model's displacement with those measured in the table.

7.4 Calculate the singular values of the regression matrix of the data shown in Table 7.2 and show how this varies as data is added to the system.

7.5 A uniform rotor of diameter 350 mm has an overall length of 3 m. At one end is a fan which may be represented as a steel disc of diameter 600 mm and thickness 150 mm. The two bearings are at the drive end of the shaft and 2 m along the rotor. The imbalance on the fan is 0.0003 kgm. At full speed, the shaft motion at the bearing locations is measured as

	x-Direction	y-Direction
Bearing 1	6×10^{-6} m @30°	4×10^{-6} m @−20°
Bearing 2	8×10^{-6} m @20°	5×10^{-6} m @70°

Determine the dynamic bearing forces.

TABLE 7.2

Input data for Problems 7.1–7.4

	Displacement 1		Displacement 2	
Frequency (Hz)	Magnitude(μm)	Phase (°)	Magnitude(μm)	Phase (°)
10	1.00	−1	2.43	−1
20	1.15	−11	2.71	−8
30	1.74	−28	3.79	−24
40	3.09	−87	5.84	−80
50	1.74	−149	2.59	−138
60	1.13	−168	1.11	−150
70	1.1	179	0.46	−136
80	2.19	144	1.13	−67
90	0.86	28	1.13	−161
100	0.27	12	0.61	−170

References

Childs, D., 1993, *Turbomachinery Rotordynamics: Phenomena, Modeling and Analysis,* Wiley, New York, NY.

Edwards, S., Lees, A.W. and Friswell, M.I., 1998, The identification of a rotor bend from vibration measurements, *16th International Modal Analysis Conference,* pp. 1543–1549, Santa-Barbara, CA.

Ewins, D.J., 2000, *Modal Testing: Theory, Practice and Application*, Second Edition, Wiley-Blackwell, Oxford, UK.

Feng, N.S. and Hahn, E.J., 1995, Including foundation effects on the vibration behaviour of rotating machinery, *Mechanical Systems and Signal Processing*, 9(3), pp. 243–256.

Feng, N.S. and Hahn, E.J., 1998, Vibration analysis of statically indeterminate rotors with hydrodynamic bearings, *Journal of Tribology – Transactions of ASME*, 120(4), pp. 781–788.

Friswell, M.I., Lees, A.W. and Smart, M.G., 1996, Model updating techniques applied to turbogenerators mounted on flexible foundations, *NAFEMS Conference, Structural Dynamics Modelling: Test Analysis and Correlation*, pp. 461–472, Cumbria, UK.

Golub, G.H. and Van Loan, C.F., 1996, *Matrix Computations*, The John Hopkins University Press, Baltimore, MD.

Hamrock, B.J., Schmid, S.R. and Jacobson, B.O., 2004, *Fundamentals of Fluid Film Lubrication*, Marcel-Dekker, NJ.

Hirdaris, S.E. and Lees, A.W., 2005, A conforming unified finite element formulation for the vibration of thick beam and frames, *International Journal for Numerical Methods in Engineering*, 62(4), pp. 579–599.

Jeffcott, H.H., 1919, The lateral vibration of loaded shafts in the neighbourhood of a whirling speed: The effects of want of balance, *Philosophical Magazine*, 6(37), pp. 304–314.

Kramer, E., 1993, *Dynamics of Rotors and Foundations*, Springer-Verlag, Berlin, Germany.

Lees, A.W., 1988, The least square method applied to identify rotor foundation parameters, *Institution of Mechanical Engineers Conference 'Vibrations in Rotating Machinery'*, pp. 209–215, Edinburgh.

Lees, A.W. and Friswell, M.I., 1997, The evaluation of rotor imbalance in flexibly mounted machines, *Journal of Sound and Vibration*, 208(5), pp. 671–683.

Lees, A.W. and Friswell, M.I., 2001, The vibration signature of chordal cracks in asymmetric rotors, *19th International Modal Analysis Conference*, Kissimmee, FL.

Lees, A.W. and Simpson, I.C., 1983, Dynamics of turbo-alternator foundations, *Institution of Mechanical Engineers Conference*, London, paper C6/83.

Mayes, I.W. and Davies, W.G.R., 1976, The behavior of a rotating shaft system containing a transverse crack, *Institution of Mechanical Engineers Conference on Vibrations in Rotating Machinery*, Cambridge, UK, pp. 53–64.

Nelson, H.D., 1980, A finite rotating shaft element using Timoshenko beam theory, *Journal of Strain Analysis for Engineering Design*, 102, pp. 793–803.

Nelson, H.D. and McVaugh, J.M., 1976, The dynamics of rotor-bearing systems using finite elements, *Journal of Engineering for Industry*, 98, pp. 593–599.

Noton, M., 1972, *Modern Control Engineering*, Permagon Press, Oxford, UK.

Oldfield, M., Ouyang, H. and Mottershead, J.E., 2005, Simplified models of bolted joints under harmonic loading, *Computers & Structures*, 84(1–2), pp. 25–33.

Pennacchi, P., Bachschmid, N., Vania, A., Zanetta, G.A. and Gregori, L. et al., 2006, Use of modal representation for the supporting structure in model-based fault identification of large rotating machinery – Part 1: Theoretical remarks, *Mechanical Systems and Signal Processing*, 20(3), pp. 662–681.

Pons, A., 1986, Experimental and numerical analysis on a large nuclear steam turbo-generator, *IFToMM Conference on Rotordynamics*, Tokyo, p. 269.

Provassi, R., Zanetta, G.A. and Vania, A., 2000, The extended Kalman filter in the frequency domain for the identification of mechanical structure excited by multiple sinusoidal inputs, *Mechanical Systems and Signal Processing*, 14(3), pp. 327–341.

Simonian, S.S., 1979, System identification in structural dynamics: An application to wind force estimation, Ph.D. Thesis, University of California, University Microfilms International.

Sinha, J.K., Friswell, M.I. and Lees, A.W., 2002, The identification of the unbalance and the foundation model of a flexible rotating machine from a single rundown, *Mechanical Systems and Signal Processing*, 16, pp. 255–271.

Sinha, J.K., Lees, A.W. and Friswell, M.I., 2004, Estimating the static load on the fluid bearings of a flexible machine from run-down data, *Mechanical Systems and Signal Processing*, 18, pp. 1349–1368.

Smart, M.G., Friswell, M.I. and Lees, A.W., 2000, Estimating turbogenerator foundation parameters – Model selection and regularisation, *Proceedings of the Royal Society, A*, 456, pp. 1583–1607.

Stodola, A., 1927, *Steam and Gas Turbines*, Translation by S. Loewenstein, McGraw-Hill Book Co., Inc., New York.

Vania, A., 2000, On the identification of a large turbogenerator unit by the analysis of transient vibrations, *Institution of Mechanical Engineers Conference Vibrations in Rotating Machinery*, pp. 313–322, Nottingham, UK.

Worden, K. and Tomlinson, G.R., 2001, *Non-linearity in Structural Dynamics*, 2001, IOP Publishing, Bristol, UK. Later re-published by CRC Press, Boca Raton, FL.

Zanetta, G.A., 1992, Identification methods in the dynamics of turbogenerator rotors, Paper C432/092, *Institution of Mechanical Engineers Conference Vibrations in Rotating Machinery*, pp. 173–182, Bath, UK.

8

Some Further Analysis Methods

8.1 Introduction

In the text so far, a number of machine faults have been discussed and the main methods of modeling and data analysis have also been explained. In this chapter some further approaches to data analysis are explored. Before proceeding to more sophisticated techniques, the next section examines some very basic approaches to data which very often will provide an effective prelude to more advanced approaches.

8.2 Standard Approaches

There are several distinct aspects to the data processing involved in assessing the condition of a piece of machinery ranging from purely empirical to more physics-based approaches. Workers in the field tend to concentrate on certain aspects within this broad range of activities, but few would doubt the need for this broad-ranging approach to the problems.

The overall aim is to extract information from data. Of itself, a vibration reading is of little value unless translated into context and hence given some physical meaning but some ways of doing this form the subject of this chapter. This conversion of data into information is by no means trivial and warrants some study. At one end of the spectrum would be a purely statistical approach. For instance, a user may build up a large bank of vibration and other machine parameters and suggest that any marked change in the pattern signifies some change in the machine which should then be investigated. This is indeed a simple model but it may be appropriate to some relatively low-value plant where the user may specify that a given change in some features of the parameter should trigger removal of the machine concerned from service.

In dealing with vibration data appropriate functions might be the root mean square (RMS) vibration or other functions such as the Kurtosis, which basically measures the spikiness of a signal and gives a good indication of impacts within the system. Slightly more complex systems would involve measuring vibration in two planes and recording on a scatter diagram as discussed in

Chapter 2. As shown there, some statistical tests are readily introduced to assess the point at which a change of significance has taken place. Of course, the data recorded in such a system will usually have more than just vibration data but will also incorporate records of speed, load, and performance.

Whilst this simplest level system may be adequate to guide the safe operation of plant, for more complex (and valuable) machines, the requirements will often be rather more sophisticated. On the more critical items of plant rather more is required from a monitoring system. It is common in the literature to classify monitoring schemes into three categories as follows:

1. Detection of a fault
2. Diagnosis and localization of a fault
3. Correction or mitigation of the effects of a fault

Each of these stages includes preceding steps. Category 1 has already been discussed and the crucial requirement at this level is just some criteria to alert an alarm or shutdown signal. This can be done on purely statistical grounds, but insight may be required to identify which parameter should be used as a guide to the condition. As described above one very simple measure is the RMS value of vibration. Let $y(t)$ be the measured vibration, then

$$y_{rms}(t) = \frac{1}{T} \int_{t-\frac{T}{2}}^{t+\frac{T}{2}} y^2(t)dt \tag{8.1}$$

where the period T is chosen to suit the problem in hand. For machines which operate under nominally stationary conditions a fairly long time window can be used, whereas a machine in transient conditions will require a rather narrower time window. This is a useful concept even in cases where there are multiple components of vibration components, but it may not fully reflect the important features of a problem. For instance, consider the monitoring of a rolling element bearing. Of particular interest is the presence and magnitudes of impacts between different components of the bearing. The Kurtosis is a measure of the "spikiness" of a signal and hence can be helpful in highlighting important changes. The Kurtosis is defined as

$$K(t) = \frac{\int_{t-\frac{T}{2}}^{t+\frac{T}{2}} y^4(t)dt}{\left(\int_{t-\frac{T}{2}}^{t+\frac{T}{2}} y^2(t)dt\right)^2} \tag{8.2}$$

where T is selected with the same considerations as outlined above. However, the use of this measure as a monitoring parameter relies on some, albeit limited, insight into the operation of the machine involved.

Figure 8.1 shows a signal which is heavily contaminated with Gaussian random noise. The issue here is how the engineer can unequivocally discriminate this from a signal comprising several sinusoidal components. In fact, there are a number of approaches to this, but the examination of Kurtosis is possibly the most convenient.

Other purely statistical approaches include the use of the auto- and cross-correlations and their respective spectral representations. These methods are useful in cases which have a significant random component to the measurement as might be the case on a fluid handling machine for example. 'n' Give a signal with a substantial random element, a common approach to dealing with this is to calculate the autocorrelation of the signal, which in physical terms expresses how often the signal repeats itself.

Given a measurement $y(t)$ the autocorrelation function is calculated as

$$R(\tau) = \lim_{T \to \infty} \frac{1}{T} \int_0^T y(t)y(t+\tau)dt \qquad (8.3)$$

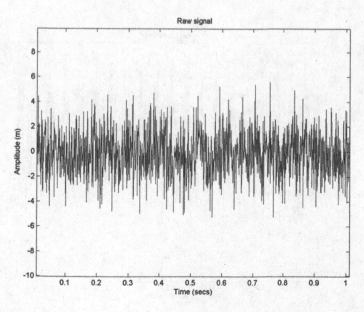

FIGURE 8.1
A Signal Heavily Contaminated with Noise.

In practice, of course, T will be finite, but the longer it is, the better will be the averaging out of random terms to reveal the underlying structure of the signal. This presupposes that the signal is stationary, meaning that although there are substantial fluctuations, the statistical properties of y do not vary with time. Note that R is a function only of the interval between the readings and not of the absolute time.

Let us look at these in an example in which the signal has very significant noise. Part of the raw time signal is shown in Figure 8.1. The figure shows the first second of the raw data and it is immediately apparent that any analytic form is heavily submerged in noise. Because there is so much noise considerable averaging is required to obtain the true signal. In Figure 8.2, the autocorrelation is shown after averaging over 1,000 sec: of course, there is a substantial assumption here that the system is constant for such a length of time – a condition that is rarely satisfied in rotating machinery. This is discussed in Section 8.6.5. For the present discussion it is assumed that a suitable period of steady operation is available. The variation of the autocorrelation is shown over a period of 1 sec and, when compared with Figure 8.1, it is immediately obvious how the periodic structure emerges. Note that although there is still significant noise, underlying patterns are clearer.

The extent of noise introduced in this example far exceeds that which would normally be observed in machinery but it has been exaggerated to

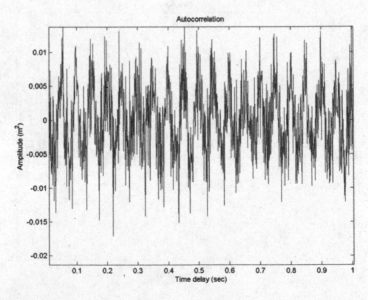

FIGURE 8.2
Autocorrelation Function.

illustrate the point. Although Figure 8.2 is still noisy, the structure is clear and its Fourier transform, the power spectral density is shown in Figure 8.3. The two frequency components in the original signal are immediately apparent at 20 and 40 Hz, respectively.

In this case, these two components can also be identified by performing an FFT directly on the original signal, but the noise level is rather higher and the component at 40 Hz becomes partially submerged. This is clear from Figure 8.4, which shows the FFT for a period of 1,000 sec.

The progression through these stages has gradually clarified the under-lying structure of what, at first, appeared to be a purely random signal. It is difficult to present a precise recipe for processing data in this way, but there are standard tools at the disposal of the analyst, some of which have been applied in this case. While some of the standard techniques will often be applied, some prior knowledge of the operation of the machine produc-ing the data in question is often a distinct advantage.

The important point to note is that all the basic signal processing approaches are purely statistical in nature. The basic role of the analyst is to seek to infer a physical interpretation from the available data. This will often involve a combination of techniques, the choice of which will be dependent on the nature of the investigation.

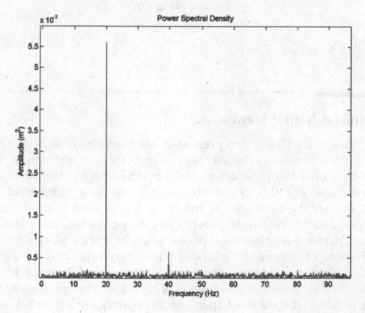

FIGURE 8.3
The Power Spectral Density.

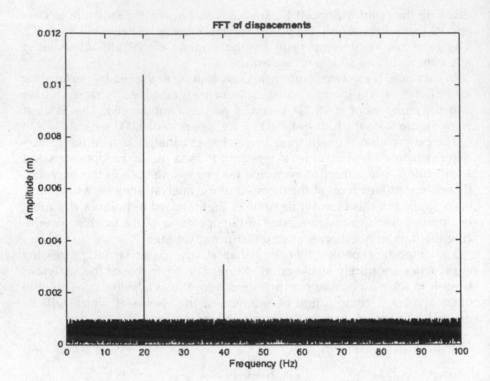

FIGURE 8.4
FFT over 1,000 Sec of Data.

8.3 Artificial Neural Networks

The majority of the discussion presented has been deterministic in nature. Deterministic, physically based representations are extremely helpful because they offer the prospect of real physical insight into the operation of the machine. But this is not always possible as a number of factors which influence the operation of machines are beyond current modeling capabilities and in these instances different approaches are needed. For instance, Chapter 2 discusses the presentation of scatter plots which show the variation of vibration levels (both magnitude and phase) over a considerable period of time. The example shown in Figure 2.6 illustrates how the vibration varies over a period of time even though the machine is running at nominally constant speed and operators are led to ask why the machine behaves in this way.

At one level, the answer is obvious: although the machine is operating under nominally constant conditions, numerous parameters are constantly changing, but the control system is making corrections in order to hold the output constant. For instance, although load and speed are (nominally) constant, the rotor current, steam pressure, and condenser levels are constantly varying. Consider the case of a turbine/generator having six bearings: In addition to the two components of vibration at each bearing together with the shaft speed, 54 additional parameters were monitored, mainly comprising temperatures and pressures at various points around the circuit. While these parameters were ideally constant, they all show some fluctuation. These fluctuations have an effect on the vibration levels but in a rather complicated way and in order to give an accurate assessment of a machine's condition a method is needed to assess these variations. Since this is beyond the capability of current direct modeling methods, a more empirical approach is required. One such route is the use of artificial neural networks (ANNs).

An ANN is a method by which a set of inputs is associated with a set of output by adjusting a set of weight parameters. In the present example, the input comprises the 54 parameters of the machine (implying that in the network there will be 54 nodes in the input layer). The output layer in this instance will have 12 nodes yielding the vibration levels at each bearing (or, more generally, each measurement point) which may be expressed as either magnitude and phase or real and imaginary parts. Between the input and output layers there will be one or more intermediate layers and the number of nodes in each of these layers is dependent on the problem. The situation is reminiscent of many identification problems: too few nodes will give an inadequate fit, while too many will lead to the modeling of noise. The 54 input parameters are the various operational parameters such as pressures, temperatures, and rotor current, whilstL the 12 output parameters will be the (complex) bearing vibration levels.

All of the nodes in each layer are linked to each node in the next layer and there is a weighting associated with each link. These weights are adjusted to give optimum fit between the inputs and outputs. This process is called "training." The trained network can then be used to predict the vibration given any set of parameters, provided, of course, that there has been no underlying change in the machine's condition.

This may be regarded as a form of interpolation. The most common form of ANN used in condition monitoring applications is the multilayer perceptron and its structure is shown in Figure 8.5.

The figure shows just one hidden layer but in principle there may be an arbitrary number. As shown in Figure 8.5, each node is connected to each node of the preceding layer and it passes an output to each node of the subsequent layer. The forward signal at each stage is given by

$$y = f\left(\sum w_i x_i + w_0\right) \tag{8.4}$$

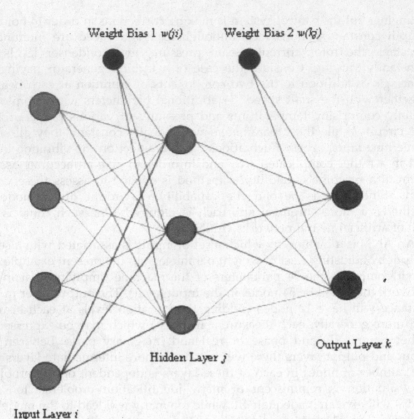

Weight Bias 1 $w_{(ji)}$ Weight Bias 2 $w_{(kj)}$

Output Layer k

Hidden Layer j

Input Layer i

FIGURE 8.5
The Multilayer Perceptron.

The function f can take several forms, but a common choice is the sigmoid, that is,

$$f(x) = \frac{1}{1 + e^{-x}} \tag{8.5}$$

This is an appropriate function because it has a continuous derivative which vanishes at high positive and negative values. The weighting functions w are obtained by optimizing the agreement between input and output for the set of training data. Because the function f (called the activation function) is nonlinear it can be shown that any set of data may be represented by the network.

There are, of course, some pitfalls in the application of these techniques. Because the inputs may cover a wide variety of physical measurements ranging from displacements of order 10^{-6}m to temperatures of hundreds of degrees, it is important to normalize all inputs to a range of -1 to $+1$, to ensure all measurements are given due significance in the calculation. This is easily achieved by simply rescaling each input node on a scale -1 to 1. The other difficulty is the choice of network parameters to use; too few hidden layers will restrict the ability to fit to data but an excessive number will result in the modeling of noise. This dilemma is encountered in many problems of system identification, as discussed in Chapter 7.

Note that this fit is obtained by adjusting the weights, this process may be considered as an advanced form of curve fitting and this may be regarded as both its strength and weakness: using an appropriate network represents a convenient means of obtaining some functional representation of a set of data, but this is just a curve fit and does not necessarily relate to any physical parameters of the system, in marked contrast to the work presented in Chapter 7. A full discussion of the training of ANNs is beyond the scope of this chapter and the interested reader is referred to Worden and Tomlinson (2001) for example. There are a number of user-friendly routines available for the application and manipulation of ANNs; see for example Nabney (2003).

Of greater importance to the chapter is a discussion of the usefulness in the context of machine health monitoring. Being a sophisticated form of curve fitting, information may be interpolated but not extrapolated. This is simple because the data is simply fitted rather than associated with any underlying physical equation. However, used appropriately, an ANN can help give considerable physical insight into machine behavior. We illustrate this point with the following example associated with data in the scatter diagrams of Chapter 2.

The data shown in the scatter diagram of Figure 2.6 is a plot of the vibration at one of the bearings over a period of 11 days. There are 8,600 individual measurements, and each of these depends on 54 parameters such as rotor speed, steam temperature, and pressure together with a range of other measurements which, although nominally constant, do vary due to continual adjustments in the control system to meet varying load conditions. The overall task is to determine how the bearing temperature depends on each of the parameters. There is no reason to presume that this dependence is linear but we can gain some understanding of some important issues by initially assuming a linear relationship. Although one might suspect that not all of the 54 variables are of equal importance, there is a need to establish an approach to assess the relative importance of the variables: in other words, which terms are driving the observations.

Given this rather complicated set of data, a primary requirement is the determination of the relative importance of the different parameters. If the system is linear, an effective analysis can be achieved using the singular value decomposition (SVD), mentioned briefly in Section 7.5.4 and discussed more fully in Section 8.5. Without assuming linearity, however, the data can be effectively studied using ANNs.

A proportion of the data, maybe half, can be used to train the network. Ideally the data chosen for the training should cover the full span of various parameter values, then having trained the network; the remaining data is used to test the system and gain insight into its structure. For instance, whereas the data presented has 54 input variable and a single output, the trained network can give the effect of varying each parameter individually. Hence the interactions of the parameters can be analyzed and the linearity of the dependence on each parameter can be established. These ideas are not applied to machine vibration data with an aid to diagnosis, and early example is the work of Futter et al. (1991).

8.4 Merging ANNs with Physics-Based Models

ANNs are nonlinear tools which can provide an excellent set of tools for fitting observational data. In this respect they are much more powerful than linear regression but this "flexibility" can also be a disadvantage in some circumstances. One benefit of using a regression technique is that the engineer can fit data to a given equation form and thereby impose some constraints on the underlying physics. There would be significant advantages to find some form of network, which could constrain the fit in this manner. Although this concept may appear somewhat counterintuitive, it can be an advantage to seek a less optimum fit. In Chapter 7, the identification of some foundation models is improved by choosing the appropriate degrees of freedom to evaluate. Typically, it is found that fitting to a large number of degrees of freedom will invariably improve the "fit," but the derived model has little or no predictive capability. The same argument applies to ANNs: constraining the form of the derived model may lead to improved capability but it is far from obvious how this might be achieved directly. Hopefully, some method of applying these ideas will emerge in due course.

There are, however, indirect ways in which the accuracy and flexibility of ANNs can be merged with regression approaches. In Section 8.2 the influence of the pseudo-static parameters can change the overall behavior and ANNs can be used to separate the effects of the different parameters. Once this has been achieved, then a separate regressive form can be developed.

8.5 Singular Value Decomposition

In discussions elsewhere (see Section 7.5.4) use has been made of the SVD, but in view of its usefulness some further discussion of the technique is appropriate. SVD is a direct extension of eigenvalue analysis to matrices which are not square. Consider a rectangular matrix **A** with dimensions $m \times n$.

The case $m > n$ is of particular interest as this corresponds to an over-specified problem for which an approach such as least squares must be applied to reach a satisfactory solution. All matrices $\mathbf{A}$ may be decomposed as the product of three matrices

$$\mathbf{A} = \mathbf{U} \Lambda \mathbf{V}^\mathrm{T} \tag{8.6}$$

where $\mathbf{U}$ has dimensions $m \times m$, $\mathbf{V}$ has dimensions $n \times n$, and Λ has the dimensions of $\mathbf{A}$. The matrices $\mathbf{U}$ and $\mathbf{V}$ are orthogonal, that is,

$$\mathbf{U}^\mathrm{T}\mathbf{U} = \mathbf{I}_m \qquad \mathbf{V}^\mathrm{T}\mathbf{V} = \mathbf{I}_n \tag{8.7}$$

where $\mathbf{I}_n$ is the unit (diagonal) matrix of dimension n. The matrix Λ is given by

$$\Lambda = \begin{bmatrix} \lambda_1 & 0 & 0 & 0 \\ 0 & \lambda_2 & 0 & 0 \\ 0 & 0 & \ddots & 0 \\ 0 & 0 & 0 & \lambda_n \\ 0 & 0 & 0 & 0 \\ 0 & 0 & 0 & 0 \end{bmatrix} \tag{8.8}$$

From Equations 8.7 and 8.8 two equations may be developed to yield

$$\mathbf{A}\mathbf{V}^T = \mathbf{U}\Lambda \qquad \mathbf{U}^T\mathbf{A} = \Lambda\mathbf{V} \tag{8.9}$$

Note here that if $\mathbf{A}$ is square, then $\mathbf{U}$ and $\mathbf{V}$ are identical and Equations 8.9 become the standard eigenvalue problem.

Now consider the meaning of all this using an over specified problem of the form

$$\mathbf{A}\mathbf{s} = \mathbf{f} \tag{8.10}$$

This can be reexpressed as

$$\mathbf{U}\Lambda\mathbf{V}^\mathrm{T}\mathbf{s} = \mathbf{f} \tag{8.11}$$

Therefore, using the orthogonality relationships

$$\Lambda\mathbf{V}^\mathrm{T}\mathbf{s} = \mathbf{U}^\mathrm{T}\mathbf{f} \tag{8.12}$$

If we now examine the pth column of matrix $\mathbf{U}$, then if the corresponding singular value (λ_p) is low (relative to the rest of the set) this means that the combination of parameters $(\mathbf{U}^T\mathbf{f})$ contribute little to the overall equation, but the corollary of this is that this low sensitivity can lead to significant errors. The way around this is a truncation of the series of eigenvalues as used in Section 7.5.4. It is then a straightforward matter to invert Equation 8.12. Although the matrix Λ is rectangular, an inverse can be established because of its particular form.

Often this is indicating that the model has too many parameters for the data available and this is a useful check in many investigations. The SVD gives good insight into the structure of a dataset. A full account of the mathematical details is given in a number of textbooks (e.g. Lawson and Hanson, 1974, republished 1995). From a user's standpoint SVD is a very easy means of analysis via, for example, MATLAB.

With this approach, reconsider the fluctuations of vibration shown in Figure 2.6. If it is assumed to be linear, then the SVD approach can be applied and the linearity can be easily verified using ANNs. Let θ represent the temperature at each time, P is the array of parameters, and R contains the unknown coefficients. Then

$$\theta = \mathbf{P}R \tag{8.13}$$

In this equation, since the displacement at the measured position depend on 54 parameters, θ is a vector of length 8,600 is this example (that is, the number of time steps), $\mathbf{P}$ is a matrix 8,600×54, and R is a vector of length 54. The Moore-Penrose inverse may be applied to this problem to give

$$R = \left(\mathbf{P}^T\mathbf{P}\right)^{-1}\mathbf{P}^T\theta \tag{8.14}$$

Unfortunately, this answer will be unreliable as it is subject to considerable noise. To shed some light on this problem we modify the approach and analyze the matrix $\mathbf{P}$ using SVD. It is known that

$$\mathbf{P} = \mathbf{U}\Sigma\mathbf{V}^T \tag{8.15}$$

Σ is a diagonal matrix containing the 54 singular values in nonascending order: $\mathbf{U}$ and $\mathbf{V}$ are matrices with orthonormal columns and where $\mathbf{U}^T\mathbf{U} = \mathbf{V}^T\mathbf{V} = \mathbf{I}_n$. There are several algorithms available to achieve this decomposition (see e.g. Golub and van Loan, 1996), including a standard function in MATLAB. The mechanics of these algorithms need not be covered here: more important is its usefulness. Manipulating Equations 8.13–8.15 yields

$$\Sigma^{-1}\mathbf{U}^T\theta = \mathbf{V}^TR \tag{8.16}$$

A formal solution for R may be obtained by expressing it in terms of the columns of $\mathbf{V}$; hence,

$$R = \sum_{j=1}^{54} a_j v_j \tag{8.17}$$

From this we obtain

$$a_j = \frac{u_j^T \theta}{\sigma_j} \tag{8.18}$$

It is apparent here that terms for which σ_j is small make a significant difference to the calculated answer, but this is of concern for reasons which can be appreciated by returning to Equation 8.13. Combining Equations 8.15 and 8.17 shows that

$$\mathbf{P}v_j = \sigma_j u_j \tag{8.19}$$

where u_j is the jth column of the matrix $\mathbf{U}$.

This means that there are some vectors which simply produce noise in the system and the way to avoid this noise is simply to truncate the range of singular values which are considered in the solution, in effect setting all the rest to zeros. Actually, this type of analysis can lead to considerable insight. This will now be illustrated by returning to the example of the data with 54 variables discussed in the previous section.

The two graphs shown in Figure 8.6 are only marginally different. They differ only insofar as some clearly erroneous readings have been removed from the "modified" data. The figure shows the distribution of singular values and it is very clear that the rate of changes with respect to index (column) number changes markedly around the value $x=12$, but clearly the precise location is somewhat debatable. At higher indices, owing to the low singular values, higher terms make little contribution, yet, if included, these terms would severely distort the solution. Hence terms in these higher modes are regarded as noise and ignored.

8.6 Other Useful Techniques

The various techniques described so far find regular application in the analysis of machine behavior. In this section some rather more specialized approaches are discussed which prove useful in the study of some problems.

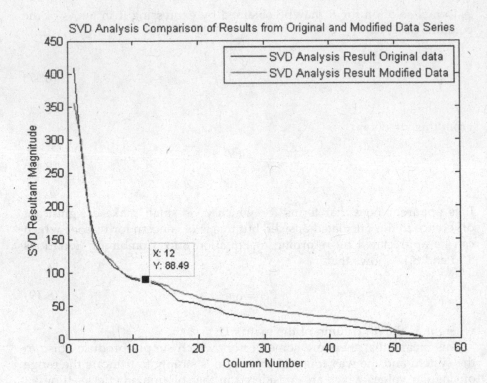

FIGURE 8.6
Singular Values of Bearing Data.

8.6.1 The Hilbert Transform

Whilst the Fourier and Laplace transforms project a function into a different domain (e.g. time to frequency), the HT remains in the same space and hence there is a fundamental distinction between this and some on the other common transformations. However, the HT gives some deep insights into the nature of a signal. Here only a brief discussion is given on the nature and main uses of the transform is described. Readers with more specific requirements are referred to the more extensive discussions by Worden and Tomlinson (2001) or Feldman (2011).

The definition of the transform is

$$H[x(t)] = \tilde{x}(t) = \pi^{-1} \int_{-\infty}^{\infty} \frac{x(\tau)}{t - \tau} d\tau \qquad (8.20)$$

This straightforward definition in itself gives little insight into the application of the transform. It has the effect of "smearing out" the time variation of the function. The full meaning of the transform is in fact rather deep and there are a number of ways of approaching it. In the parlance of signal processing, HT may be regarded as a linear filter in which all amplitudes remain unchanged, yet the phases are shifted by $-\pi/2$. The transform has a range of properties, the details of which can be found in a number of sources (see for instance Feldman, 2011 or Worden and Tomlinson, 2001). Owing to the difficulty in forming the integral owing to the singularity, it is taken as the Cauchy integral or principal value.

The focus here, however, is how HT can be useful in the analysis of signals from machinery. The two most important applications are the derivation of a signal envelope and the detection of nonlinearity: both of these depend on the so-called analytic signal, again a function of time but now complex. This is defined as

$$X(t) = x(t) + j\tilde{x}(t) \tag{8.21}$$

where $\tilde{x}(t)$ is the HT of $x(t)$.

Treating this as any other complex function, one may immediately write

$$A(X) = \sqrt{x^2(t) + \tilde{x}^2(t)} \tag{8.22}$$

Now consider as an example the decaying sine wave. Let

$$x(t) = Ce^{-at}\sin\omega t \tag{8.23}$$

Then (as shown below)

$$H(x(t)) = \tilde{x}(t) = Ce^{-at}\sin\omega t \tag{8.24}$$

Therefore

$$A(t) = |X(t)| = Ce^{-at}\sqrt{\sin^2\omega t + \cos^2\omega t} = Ce^{-at} \tag{8.25}$$

Hence a simple application of the transform provides a means of extracting the envelope of a signal. This is often the most salient feature of a signal rather than the more rapid variations that can mask overall features. In addition to the envelope amplitude, Equation 8.25 also yields a value for α and the time derivative of this yields the instantaneous frequency.

Individual HTs can prove tricky to calculate: the good news is that tables exist so that full calculations are rarely, if ever, needed in practice. As an example, consider the important function $\cos \omega t$. For the HT

$$H\{\cos \omega t\} = \frac{-1}{\pi}P \int\limits_{-\infty}^{\infty} \frac{\cos \omega(y + t)}{y} dy \tag{8.26}$$

where the last expression is obtained simply by changing variable. This is now expanded to give

$$H\{\cos \omega t\} = \frac{-1}{\pi} \left\{ \cos \omega t \, P \int\limits_{-\infty}^{\infty} \frac{\cos \omega y}{y} dy - \sin \omega t \, P \int\limits_{-\infty}^{\infty} \frac{\sin \omega y}{y} dy \right\} \quad (8.27)$$

The first integrand on the right-hand side is an odd function and hence the integral becomes zero. The evaluation of the second term is somewhat more involved. It may be evaluated by considering the contour integral

$$I = \oint \frac{e^{j\omega z}}{z} dz \quad (8.28)$$

with $\omega > 0$ and the contour encircles the upper half plane, but circumvents the origin. The overall integral is divided into four sections corresponding to two parts of the real axis, then two semicircle sections, one infinitely large and one infinitesimally small. By Cauchy's theorem it is known that the total contour integral is zero and from this it can be deduced that

$$\int\limits_{-\infty}^{\infty} \frac{\sin \omega y}{y} dy = \pi \quad (8.29)$$

Using this it follows that

$$H\{\cos \omega t\} = \sin \omega t \quad (8.30)$$

An important feature of signals from mechanical systems is causality. A system cannot respond until an excitation has been applied to it. For the purpose of this discussion it suffices to define a causal function as one which has zero value for negative time and clearly this is a prerequisite for any description of a physical system. An elegant discussion of this issue is given by Worden and Tomlinson (2001) and only a brief summary is given here.

For any function of time $g(t)$ there is a decomposition:

$$g(t) = g_{even}(t) + g_{odd}(t) \quad (8.31)$$

But if $g(t)$ is causal,

$$g_{even}(t) = g(|t|)/2, \qquad \text{all } t \quad (8.32)$$

And

$$\begin{aligned} g_{odd}(t) &= g(|t|)/2, \qquad t > 0 \\ &= -g(|t|)/2, \qquad t < 0 \end{aligned} \quad (8.33)$$

But, as observed by Worden and Tomlinson (2001), these two equations may be written more elegantly (and meaningfully!) as

$$g_{even}(t) = g_{odd}(t)\varepsilon(t)$$

(8.34)

$$g_{odd}(t) = g_{even}(t)\varepsilon(t)$$

where $\varepsilon(t)$ is the signum function (often denoted sgn(t)) given by

$$\varepsilon(t) = \begin{cases} 1, & t>0 \\ 0, & t=0 \\ -1, & t<0 \end{cases}$$

(8.35)

Noting that the Fourier transform of this function is $\frac{j}{\pi\omega}$ the relationship with the HT emerges and Equation 8.20 may be reexpressed as

$$Re(G(\Omega)) = -\frac{1}{\pi} \int\limits_{-\infty}^{\infty} \frac{Im(G(\Omega))}{\Omega - \omega} d\Omega$$

(8.36)

$$Im(G(\Omega)) = \frac{1}{\pi} \int\limits_{-\infty}^{\infty} \frac{Re(G(\Omega))}{\Omega - \omega} d\Omega$$

(8.37)

The gives a succinct but very profound conclusion: The Fourier transform of a causal function is invariant under HT, that is,

$$G(\omega) = \tilde{G}(\omega) = H\{G(\omega)\}$$

(8.38)

These relationships can also be utilized to assess the linearity (or otherwise) of a signal.

In summary the HT has three major roles in the area of machine monitoring:

1) The determination of envelope curves
2) The detection of nonlinearity
3) Determination of causality

8.6.2 Time–Frequency and Wigner–Ville

Section 2.2.4 discusses the two methods of analyzing data from a machine during transient operation. This is important because, although it presents a number of analysis challenges, the data obtained during transients is rich in information as discussed in Chapter 7. In Chapter 2, one approach was to impose a window in time which moves as the transient progresses. The general term for this approach is the short term Fourier transform (STFT) and this proves a very useful tool despite some difficulties. The problem arises from the difficulty of defining frequency and time simultaneously, and this is not merely a mathematical issue but rather a matter of examining the underlying concept. The frequency of a sinusoid is perfectly defined, but a sine wave is not limited in time. On the other hand, a time signal that is localized in time contains a spread of frequencies. In the case of the STFT the transform of the signal can be written as

$$S(t, \omega) = \int x(\tau)w(t - \tau)e^{j\omega\tau}d\tau \qquad (8.39)$$

where the window function is given by

$$w(t) = 1 \qquad 0 < t < \delta$$
$$w(t) = 0 \qquad \text{otherwise}$$

where δ is the width of the window. As discussed in Chapter 2 the choice of this width is a compromise between the resolution of the FFT and the acceleration/deceleration of the rotor.

Insight into the development of the spectrum in time can be gained by analogy with the way in which stationary signals may be analyzed. A useful measure is the autocorrelation function of a stationary signal that is given by

$$R(\tau) = \int x(t)x(t + \tau)dt \qquad (8.40)$$

This will reveal the periodicities by showing the relationship of two signals separated by a time τ. For nonstationary signals, the analogous quantity will also be a function of time and hence this cannot be integrated. So that we may write

$$R(t, \tau) = x^*(t - {}^\tau/_2)x(t + {}^\tau/_2) \qquad (8.41)$$

where suffix * denotes complex conjugate.

And the Wigner–Ville distribution is the Fourier transform of this quantity and this gives a view of the spectral content of the signal as a function of time:

$$S_w(t, \omega) = \int x^*(t - \tau/2)x(t + \tau/2)e^{-j\omega\tau}d\tau \tag{8.42}$$

It may be observed that the expression for the STFT shown in Equation 8.42 is actually imprecise – its actual form is dictated by the window function w. While the STFT has some intuitive credibility, there is some lack of precision in the inability to define the instantaneous frequency.

The issues raised in this discussion are very relevant to several subsequent points. Indeed, the concept of time and frequency is central to a number of issues in machine monitoring. The formalism expressed here is helpful in outlining other sections.

8.6.3 Wavelet Analysis

It is true to say that the Fourier transform and its digital descendants (DFT, FFT) form the backbone of signal processing as it is applied to rotating machinery monitoring and diagnostics. It is indeed a convenient approach and has the immense mathematical advantage that it converts a differential equation of motion into an algebraic equation. Provided that the system is linear, or at least predominantly so, then this proves an extremely convenient approach, so much so that it has become embedded in the engineer's way of thinking; frequencies are thought of as independent entities. All this is very attractive in many instances, and yet there are fundamental conflicts when transient conditions are to be considered. By its very nature, a sinusoid acts over all time; any pulse or transient behavior must be represented as a combination of frequency components and this leads to significant complication. It is the case that most methods for the analysis of machinery are based on the assumption of stationarity. For example, in examining the transient operation of turbo-machines, it is common to consider them to be nearly stationary use the so-called short-term Fourier transform. Indeed, it is conceptually difficult to imagine precisely what is meant by a frequency of, say, 50 Hz, if it persists for only a very short period of time.

This brings the discussion into the realm of time–frequency problems, an area with its own very substantial literature. Among a wide class of techniques one of the most prominent is the *wavelet transformation*. Conceptually the distinction between a Fourier resolution and one involving Wavelets is that in the latter, the basic unit is localized in time unlike the sinusoid function. A common Wavelet function is the Morlet function given by

$$\Psi(t) = c_\sigma \pi^{-1/4}e^{-t^2/2}\left(e^{i\sigma t} - e^{-\frac{1}{2}\sigma^2}\right) \tag{8.43}$$

where c_σ is for normalization. There are several forms of wavelet in common usage but the essential point is that they are functions which are localized in time. A full discussion of the requirements of a wavelet is beyond our present scope but the interested reader may refer to the wide-ranging review of methods by Feng *et al.* (2013).

Given a signal $x(t)$, whereas the STFT proves a useful tool in many cases, the discussion of preceding sections illustrates some of the shortcomings, all related to the time/frequency conflict. One possible way of treating transient problems effectively is by means of the wavelet transformation. This takes the form

$$W_x(a, b; \psi) = a^{-1/2} \int x(t)\psi^* \left(\frac{t-b}{a} \right) dt \tag{8.44}$$

where * denotes the complex conjugate. Ψ is the wavelet chosen and so W is a transformation of x involving two variables a and b, in just the same way as the STFT is a transform involving two variables, in that case t and ω .

A wide range of functions may be chosen as wavelets; essentially they are pulses which are limited in time. The two main requirements are that the wavelet has zero mean

$$\int \psi(s)ds = 0 \tag{8.45}$$

and the so-called admissibility condition

$$\int \frac{\psi^2(s)}{|s|} ds < \infty \tag{8.46}$$

Wavelet decomposition offers advantages over STFT analysis and has been applied to many cases by researchers. As yet however, it has had limited impact on the industrial community. This relatively slow take can probably be ascribed to two causes: a) the slightly bewildering choice as to wavelet form and b) confusion over the physical interpretation of results. Computation is not a problem as commercial packages (e.g. LABVIEW) have well-developed facilities.

It is the author's view that interpretation is the main obstacle to progress; in STFT the meaning of time and frequency is clear despite the difficulties outlined in the previous section. In contrast, the wavelet transform yields parameters a,b. The first represents scale while b is a dilation term, but the physical meaning is less clear. These issues are discussed by Peng *et al.* (2005) who provide an extensive literature on the application of wavelets to machine diagnostics.

8.6.4 Cepstrum

Many practitioners regard Cepstrum as a "novel" technique even though it dates back to the early 1960s and predates the FFT. One of the coauthors of the paper which introduced Cepstrum was J.W. Tukey who two years later was coauthor of the FFT algorithm (Cooley and Tukey, 1965). This first paper (Bogert *et al.*, 1963) defines the Cepstrum as "the power spectrum of the logarithm of the power spectrum." The history of the technique has been admirably described by Randall (2013).

In mathematical terms the Cepstrum is given by

$$C_p(\tau) = F^{-1}\left(\log(F_{xx}(\omega))\right) \tag{8.47}$$

where $F_{xx}(f)$ is the power spectrum, and F^{-1} denotes the inverse Fourier transform. Despite this apparently contorted definition, the logarithmic form is extremely useful. In the study of gears, for instance, the logarithmic scale reveals sideband structure which is often not apparent in the direct calculation on a linear scale. The Cepstrum gives insight into sideband structure and this has proved particularly useful in the analysis of gearing problems. Other notable studies have analyzed steam turbines with damages blades.

Another significant feature of the Cepstrum is that, because it is logarithmic, the force terms and the impulse response properties are additive rather than multiplicative, and this can be used to simplify some analyses substantially. Hence in general we may write

$$x(t) = \int_0^t p(t')h(t - t')dt' = p(t) * h(t) \tag{8.48}$$

where $p(t)$ is the applied force and $h(t)$ is the impulse response and $*$ denotes the convolution operation. Taking the Fourier transform and using the convolution theorem give

$$X(\omega) = P(\omega) \times H(\omega) \tag{8.49}$$

where X, P, and H are the respective Fourier transforms of x, p, and h respectively. Hence taking logs

$$\log(X(\omega)) = \log(P(\omega)) + \log(H(\omega)) \tag{8.50}$$

This means an effective separation of force and system terms, a feature of considerable help in identifying sources.

Although most engineers working in the area of machine condition monitoring are aware of cepstrum, it remains something of a specialty. It is an extremely powerful tool that, perhaps, warrants wider usage.

8.6.5 Cyclostationary Methods

The development of cyclostationary analysis is rather more recent than many methods in current use. The approach is a sophisticated signal analysis approach to the study of systems which are not exactly periodic, yet exhibit some cyclic behavior. A good example of this, indeed a problem to which it has been successfully applied, is the motion of rolling element bearings. Naively one might expect that such a bearing would show truly periodic behavior, but this is not the case for a variety of reasons. Random slipping between the elements, possible speed fluctuations, and variations of axial and radial loads impose slight variations on the cycle time. Although these variations may be only slight, Antoni (2007) shows that such a variation can destroy the harmonic nature of the response.

The problem is resolved by observing that the variations in frequency are statistical and the variance of the speed fluctuations is periodic, albeit at an as yet unknown frequency. Denoting the Fourier transform of $x(t)$ as $X(\omega)$ then

$$S_x(\omega; \alpha) = \lim_{L \to \infty} \frac{1}{L} E\left\{ X\left(\omega + \frac{\alpha}{2}\right) \times X^*\left(\omega - \frac{\alpha}{2}\right) \right\} \qquad (8.51)$$

where L is the interval over which the signal is sampled. $S_x(\omega; \alpha)$ is called the cyclic power spectrum. Maxima of this quantity indicate the frequency content of the original signal.

This technique involves some sophisticated signal processing, but when applied to appropriate problems has proved capable of revealing spectral features not apparent using more conventional approaches. Once the precise frequencies have been established in this way, any modulation caused by the variation from true periodicity can be corrected.

8.6.6 Higher-Order Spectra

A variety of techniques are described in this text, most of which treat the vibration spectra in various ways in order to detect and/or diagnose the machine's condition. As shown elsewhere, in many cases substantial progress can be made by assuming the system under study is linear even though it is known that this is rarely the case in any strict sense. Indeed, it can often be a valuable step in diagnosis to attempt an evaluation of the extent of nonlinear effects and there are several ways of approaching this, one being the use of so-called higher-order spectra (HOS). In examining

HOS, the analyst seeks to evaluate not only the distribution of frequency components (as in most studies), but also the relationships between them as this can yield considerable insight.

As a starting point, consider the power spectral density of a time series signal τ. This is computed as

$$S_{xx}(f_k) = \frac{1}{n} \sum_{r=1}^{n} X_r(f_k) X_r^*(f_k) \tag{8.52}$$

and gives a measure of the energy in a frequency component. This notion may be generalized to give the bispectrum which indicates the relationship between two frequency components, and their sum, for instance f_k, f_l, and f_{k+l}. The bispectrum is defined mathematically as

$$B(f_k, f_l) = \frac{1}{n} \sum_{r=1}^{n} X_r(f_k) X_r(f_l) X_r^*(f_{k+l}) \tag{8.53}$$

A bicoherence may also be defined (Fackerell *et al.*, 1995) as

$$b^2(f_k, f_l) = \frac{\left| \sum_{r=1}^{n} X_r(f_k) X_r(f_l) X_r^*(f_{k+l}) \right|^2}{\sum_{r=1}^{n} \left| X_r(f_k) X_r(f_l) \right|^2 \sum_{r=1}^{n} \left| X_r(f_{k+l}) \right|^2} \tag{8.54}$$

The bicoherence at any frequency pair f_k, f_l can be interpreted as the fraction of power at f_{k+l} that is phase coupled to the two components.

Of course, this just defines the first term of a set of HOS, but to some extent the first few of the set are the most valuable. The next in the series is the trispectrum and this is given by

$$T(f_k, f_l, f_m) = \frac{1}{n} \sum_{r=1}^{n} X_r(f_k) X_r(f_l) X_r(f_m) X_r^*(f_{k+l+m}) \tag{8.55}$$

For the effective use of these tools there are a range of considerations with regard to sampling rate, sample size, and windowing techniques that are discussed by Fackerell *et al.* (1995). Here attention is focused on why these techniques are useful. A complete discussion would be ambitious indeed, but it is relatively straightforward to give an indication based on physical principles. Collis *et al.* (1998) also give some useful insight.

Consider a typical dynamic equation describing the behavior of a machine. In matrix formulation this may be written as

$$Kx + K_n x^2 + C\dot{x} + M\ddot{x} = f(t) \qquad (8.56)$$

Here a nonlinear stiffness term has been included, K_π. There is a slight difficulty here in rigorously defining the meaning of the squared term in a vector sense, but it can be taken as individual components for the purposes of the present discussion. Now assume that the forcing term comprises two components with angular speeds ω_1, ω_2, and suppose that, in some sense, the nonlinear terms are small, then the problem of solving 8.56 may be approached by first obtaining a linear solution and then using this to apply a correction. In this way an iterative procedure can be developed, but only the first-order correction is required for the current argument. Now let $x = x_L + x_n$, for the linear

$$x_L = \left[K + j\omega_1 C - \omega_1^2 M\right]^{-1} f_1 e^{j\omega_1 t} + \left[K + j\omega_2 C - \omega_2^2 M\right]^{-1} f_2 e^{j\omega_2 t} \qquad (8.57)$$

which may be written as

$$x_L = A e^{j\omega_1 t} + B e^{j\omega_2 t} \qquad (8.58)$$

The second-order term in the solution can now be calculated using Equation 8.58 to give

$$Kx_n + C\dot{x}_n + M\ddot{x}_n = -K_n x_L^2 \qquad (8.59)$$

The essence of the argument is that the forcing terms of the nonlinear correction will contain terms $\left(A e^{j\omega_1 t} + B e^{j\omega_2 t}\right)^2$ which may be expanded to give $\alpha e^{2j\omega_1 t} + \beta e^{2j\omega_2 t} + \gamma e^{j(\omega_1+\omega_2)t}$ and so it is clear that components of the response at frequency $\omega_1+\omega_2$ are produced as a direct consequence of the system nonlinearity. This can be utilized as a test for system nonlinearity and in many instances this in itself is a good condition indicator.

The simple physical result is that if a linear system is excited simultaneously with forcing at frequencies f_1 and f_2, then there will be excitation at these two frequencies and only these two frequencies. However, if the system has some nonlinearity, there will also be response at $(f_1 \pm f_2)$; hence, this provides an indicator of nonlinear behavior. Similarly, excitation at a single frequency will produce a range of harmonics. This is closely related to the discussion of Section 5.5 and the rather complicated behavior can be presented graphically by plotting the bispectrum against to frequency ranges.

Yunusa-Kaltungo and Sinha (2014) have discussed the use of bispectrum and trispectrum for the diagnosis of faults in rotating machine and simulated faults on a laboratory rig. They studied misalignment, shaft cracks, and shaft rub in their tests, and different patterns emerge for each category. This is physically reasonable since the type of nonlinearity is somewhat different for each case. Their summary results suggest considerable potential for condition monitoring. The paper emphasizes the need for both bispectrum and trispectrum results to distinguish between different faults. Further work is needed in terms of both experimental work and analytic studies on the types of nonlinearity arising in different faults on a range of machines. Nevertheless, this is a developing area with some potential.

8.6.7 Empirical Mode Decomposition

A number of sections in the chapter discuss aspects of the analysis of nonstationary and nonlinear data, and no discussion of this topic would be complete without inclusion of the empirical mode decomposition, also known as the Huang–Hilbert transform. The technique was first published in 1998 (Huang *et al.*, 1998) and has been applied widely for the analysis of nonstationary and nonlinear data. Data from machinery is, of course, often stationary being dominated by rotational speed effects, but there are instances in which full nonstationary analysis is an advantage.

The term "mode" in the name of the technique is somewhat deceptive in so far as it has nothing to do with the analytic mode shapes corresponding to eigenvectors of matrix models. The "modes" used in this technique are just shapes which prove convenient for the data decomposition.

The technique can be applied to any signal, but it is advantageous only in the case of data which is nonstationary and/or nonlinear. The approach is to resolve the complete data set into a linear combination of so-called intrinsic mode functions (IMFs). The real relevance of this is that these intrinsic mode functions all have well-defined HTs and this leads to the ability to define instantaneous energy and frequency since the data is expressed as a sum of IMFs.

An IMF can be any function which satisfies two conditions:

a) In the whole data set, the number of extrema and the number of zero crossings must be either equal or differ by one.

b) At any point, the mean value of the envelope defined by local maxima and that defined by local minima is zero.

The decomposition is largely intuitive and is derived through an iterative process, the steps of which are as follows:
Let the input data be $x(t)$.

1) All maxima are identified and joined with an envelope (e_{max}) using a cubic spline fit.

2) All minima are identified and joined with an envelope (e_{min}) using a cubic spline fit.

3) The mean of mean of these two envelopes is calculated at every point as $m(t) = \frac{e_{max}(t) + e_{min}(t)}{2}$.

4) If $m(t) = 0$ the function is an IMF; otherwise, subtract this mean function from the input and repeat the process.

5) This cycle is repeated until, ideally, (4) is satisfied or some tolerance is met. The first IMF, say, IMF1, is defined.

6) Subtract this function from the original data and repeat the whole process for $r(t) = x(t) - IMF1$.

7) The whole process is repeated until some error criterion is met.

For each IMF, the HT will give the instantaneous amplitude and frequency.

The real power of this approach is in its treatment of nonstationary data. It is still relatively new when compared to other techniques such as FFT but clearly has a role to play. It has been applied to some investigations on subharmonic vibration on gas blowers (Wu and Qu, 2009) and to some interesting gearbox diagnostics (Ma *et al.*, 2015).

8.6.8 Kernel Density Estimation

One of the objectives of this chapter in the presentation of a range of analysis techniques which either have proved useful or, in the case of newer developments, are considered promising.

Such an approach is kernel density estimation which may be used for the estimation of probability distribution functions from measured data. Given a limited number of measurements of a variable x_j the problem is to estimate the probability distribution, $p(x)$. Of course, if there were an infinite number of measurements this would be a trivial exercise, but in reality some subtle approach is required particularly if there is more than a single variable. The first step is fairly obvious – one simply divides the range into a series of bins. Mathematically this may be expressed as

$$p(x) = \frac{1}{n} \sum_{j=1}^{n} \frac{1}{h} w\left(\frac{x - x_j}{h}\right) \tag{8.60}$$

where n is the number of data points and h is the window width, x_j is the sampled data and

$$\begin{aligned} w(r) &= \frac{1}{2} && |r| < 1 \\ w(r) &= 0 && |r| \geq 1 \end{aligned} \tag{8.61}$$

While this gives some idea of the distribution, it is rather crude and is discontinuous. This is a particular problem where the number of samples, n, is small in some sense. This distribution is sometimes called the naïve estimate. Basically, this has been constructed by placing a "box" of width h and the resulting so-called naïve estimate is shown in Figure 8.7.

An approximate probability distribution can be formed by taking n overlapping bins of width $2h$ and height $(2nh)^{-1}$ in the appropriate position for each measurement then summing. It is implicitly assumed that each measured value it at the center of a bin.

The situation can be substantially improved simply by inserting a Gaussian curve instead of a rectangular box for each measurement point, or "atom" as they are (rather confusingly) called. The improvement stems from the fact that the Gaussian and its derivative are well behaved in the sense of being continuous and single valued in contrast to the top hat function. Hence, we can form an estimate as

$$\hat{f}(X) = \frac{1}{n}\sum_{i=1}^{n}\frac{1}{h}K\left(\frac{X - x_i}{h}\right) \tag{8.62}$$

where K can, in principle, be any well-behaved distribution but the Gaussian is normally used so that

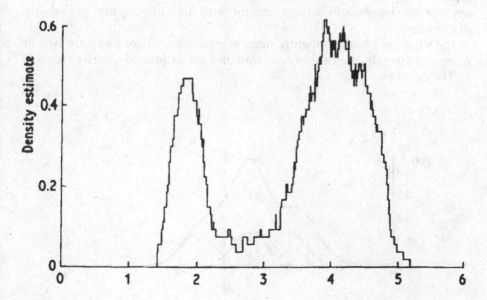

FIGURE 8.7
Naive Estimate.

$$K(x) = \frac{1}{\sqrt{2\pi}} \exp\left(-\frac{x^2}{2}\right) \tag{8.63}$$

Inserting the Gaussian into Equation 8.67, the only parameter to be chosen is h, the so-called bandwidth. This is chosen to minimize the mean integrated square error (between the derived distribution and the actual readings). Silverman (1986) describes how this is used to derive the optimal choice for the bandwidth and his result is

$$h = 0.9An^{-\frac{1}{5}} \tag{8.64}$$

where A is min(standard deviation, interquartile range/1.34), this "interquartile range" being the difference between the locations of quartiles 2 and 3. In the literature there are several (minor) variations in the magnitude of this expression. In principle the precise optimum will depend on the shape of the (unknown) distribution. One can, of course, formulate an iterative procedure, but in reality, errors from this source will be fairly small. The way in which smooth curves (Gaussian being the most common) are used as components to form distribution curves is illustrated in Figure 8.8.

The choice of h, the bandwidth, or sometimes called the "smoothing parameter," is, of course, crucial. The effect of changing this parameter is shown in Figure 8.8. These figures show the importance of choosing the appropriate bandwidth to get insight into the underlying probability distribution.

The situation becomes slightly more complicated where there are two or more variables. In the case of two variables the estimated density function will be given by

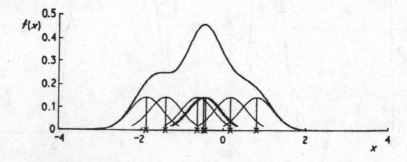

FIGURE 8.8
Composite Distribution Curve.

$$\hat{f}(v) = \frac{1}{n}|H|^{-1/2} \sum_{i=1}^{n} K\left(H^{-1/2}[v - X_i]\right) \tag{8.65}$$

where

$$H = \begin{bmatrix} h_1^2 & h_3 \\ h_3 & h_2^2 \end{bmatrix} \tag{8.66}$$

h_1, h_2 are the smoothing parameters for the x and y values, respectively, and h_3 controls the orientation. It is assumed that

$$h_3 = \rho h_1 h_2 \tag{8.67}$$

with $0 < \rho < 1$. As one might anticipate the situation becomes rapidly more complicated for higher dimensions.

So, how can these techniques be used to advantage in condition monitoring? While only limited use has been made to date, it would appear that there is significant potential. Baydar *et al.* (2001) reported the application of KDE techniques in the study of gearbox wear. It was shown that the improved representation of the probability distribution could be used to detect and incipient fault at an earlier stage than was possible with more conventional approaches.

This reasoning would also suggest that this approach can also benefit studies in remaining life prediction of machine components. The author, however, is not aware of any work to date in this area.

8.7 Concluding Remarks

A broad range of techniques are presented in this chapter. The varying methods will be useful in different circumstances and hence there is no golden rule on which approach to use when. The suitability will depend on the type of data and the features to be extracted.

Problems

8.1 A uniform steel rotor, 1 m in length, is supported at each end on bearings, each of which has constant stiffness 5×10^5 N/m with damping od 150 Nsec/m . There are two discs mounted 250 mm from either end and both of these are steel with thickness 20 mm and diameter 100 mm. The first disc has an imbalance of 0.04 kg mm, while the second has an imbalance of 0.02 kg mm, both at zero phase. The shaft passes through a clearance of 10μm at midspan. Calculate the response and determine the speed at which the shaft will contact the stator.

8.2 A time signal has the form

$$y(t) = (5e^{-0.05t} \sin(100\pi t) + 3e^{-0.08t} \cos(60\pi t))e^{-(t-25)^2}/100$$

Determine the maximum instantaneous frequency and velocity, and the time at which these occur.

8.3 A vibration trace has the form

$$y(t) = 5\sin(40\pi t) + 2\sin(100\pi t) + \text{noise}$$

The white noise has amplitude of 10. Determine the autocorrelation function and identify the main frequencies.

8.4 A time-limited signal has the form

$$y(t) = \frac{t(100 - t)\sin\left(\frac{\pi t}{2}\right)}{20} + t$$

For 0 < t < 300 sec.

Determine an empirical mode decomposition for this signal. (Note this may be achieved either manually or with freely available MATLAB routines.)

8.5 Taking the matrix

$$U = \begin{bmatrix} 1 & 2 & 3 \\ 4 & 5 & 6 \end{bmatrix}$$

determine the SVD of both U and its transpose and explain the relationship between the two.

<hr>

References

Antoni, J., 2007, Cyclic spectral analysis of rolling-element bearing signals: Facts and fictions, *Journal of Sound and Vibration*, 304, pp. 497–529.

Baydar, N., Chen, Q., Ball, A. and Kruger, U., 2001, Detection of incipient tooth defect in helical gears using multivariate statistics, *Mechanical Systems and Signal Processing*, 15(2), pp. 303–321.

Bogert, B.P., Healey, M.J.R. and Tukey, J.W., 1963, The quefrequency analysis of time series for echoes: Cepstrum, psuedo-autocvariance, cross-cepstrum and saphe cracking, *Proceeding of the Symposium on Time Series Analysis*, pp. 209–243, Providence, RI ProvidePP.

Collis, W.B., White, P.R. and Hammond, J.K., 1998, Higher order spectra: The bi-spectrum and tri-spectrum, *Mechanical Systems and Signal Processing*, 12(3), pp. 375–394.

Cooley, J.W. and Tukey, J.W., 1965, An algorithm for the machine calculation of complex Fourier series, *Mathematics of Computation*, 19(90), pp. 297–301.

Fackerell, J.W.A., White, P.R., Hammond, J.K. and Pinnington, R.J., 1995, The interpretation of the bispectra of vibration signals – 1. Theory, *Mechanical Systems and Signal Processing*, 9(3), pp. 257–266.

Feldman, M., 2011, Hilbert transform in vibration analysis, *Mechanical Systems and Signal Processing*, 25, pp. 735–802.

Futter, D.W., Hewitt, G. and Mayes, I.W., (1991). Remote monitoring and processing of large quantities of on-line vibration data by an expert system, *Institution of Electrical Engineers, Colloquium on Advanced Condition Monitoring Systems for Power Generation*, pp. 1011–1018, IEEE, London.

Golub, G.H. and van Loan, C.F., 1996, *Matrix Computations*, The John Hopkins University Press, Charles Village, Baltimore, MA.

Huang, N.E., Shen, Z., Long, S.R., Wu, M.C., Shi, H.H., Zheng, Q., Yen, N.-C., Tunf, C.C. and Liu, H.H., 1998, The empirical mode decomposition and the Hilbert Spectrum for nonlinear and non-stationary time series analysis, *Proceedings of the Royal Society, A*, 454, pp. 903–995.

Lawson, C.L. and Hanson, R.J., 1974, *Solving least square problems*, Prentice-Hall Inc., Eaglewood Cliffs, NJ. Republished in 1995 by the Society for Industrial and Applied Mathematics.

Ma, H., Pang, F.R., Song, R. and Wen, B., 2015, Fault feature analysis of cracked gear considering the effects of the extended tooth contact, *Engineering Failure Analysis*, 48, pp. 105–120.

Nabney, I.T., 2003, *Netlab, Algorithms for Pattern Recognition*, Springer-Verlag, London.

Peng, Z.K., Tse, P.W. and Chu, F.L., 2005, A comparison study of improved Hilbert-Huang transform and wavelet transform: Application to fault diagnosis for rolling bearing, *Mechanical Systems and Signal Processing*, 19, pp. 974–988.

Randall, R.B., 2013, A history of Cepstrum analysis and its application to mechanical problems, *Conference 'Surveilance 7'*, Chatres, France, (October).

Silverman, B.W., 1986, *Density Estimation for Statistics and Data Analysis*, London, Chapman and Hall.

Worden, K. and Tomlinson, G.R., 2001, *Non-Linearity in Structural Dynamics*, CRC Press, Boca Raton, FL (originally published by IOP Publishing, London, UK).

Wu, F. and Qu, L., 2009, Diagnosis of subharmonic faults of large rotating machinery, *Mechanical Systems and Signal Processing*, 23, pp. 467–475.

Yunusa-Kaltungo, A. and Sinha, J.K., 2014, Coherent composite HOS analysis of rotating machines with different support flexibilities, *Vibration Engineering and Technology of Machinery, Proceedings of VETOMAC-X*, pp. 145–153, Manchester, UK.

9

Case Studies

9.1 Introduction

A collection of the principal tools available for the monitoring and fault diagnosis of vibration problems in rotating machinery is described in the earlier chapters of this book. Each investigation takes on its own characteristics and there is no such thing as a standard investigation. There are nevertheless some standard approaches and, more importantly, disciplines to be adhered to. In this chapter, five differing studies are described in some detail in order to convey an overview of the application of the techniques employed. Four of the studies report investigations successfully completed, whilst the fifth briefly describes an on-going project related to a topic of concern in many countries at the present time.

9.2 A Crack in a Large Alternator Rotor

The events reported in this section occurred many years ago, in 1981 at a time when measurement, data processing, and computing facilities were very much less developed than those currently available. Nevertheless, the diagnosis of the machine fault provides an admirable example of the use of vibration data in machine fault diagnosis. In May of that year, a large turbo-alternator returned to service following a repair on the alternator which involved refitting of the outboard end-bell. These are heavy components which cover electrical connections on the alternator rotor and it is quite common for them to cause some imbalance problems following a refit. Experience shows however that there is a tendency to bed-in. On its return to service, the vibration levels on the run-up were remarkably consistent with measurements taken over a year before. There were, however, some changes in the response to loading.

The machine was a large 500 MW unit comprising high-pressure (HP) turbine, intermediate-pressure (IP) turbine, three low-pressure (LP) turbines, the alternator, and the exciter. The vibration changes were most pronounced at bearing 11, the drive end of the alternator rotor.

The amplitude and phase of the synchronous and twice per revolution components for this bearing is shown in Figure 9.1. As shown in the figure, the vibration levels at that date were remarkably similar to those monitored 18 months earlier. It is not clear if the sensitivity to load was caused by a thermal process or whether it was an influence of the cracking which was subsequently diagnosed. The unit ran until 22nd July, during which the synchronous vibration on bearing 11, after an initial increase, settled to the low level of 1.6 mm/sec, while the twice per revolution showed a very slight gradual increase from 2.1 to 2.3 mm/sec. The situation, however, was confusing, as these particular machines are known to exhibit load sensitivity in the 2/rev component.

The machine was then taken off load for refueling and returned to service eight days later. The vibration was plotted and the results for bearing 11 are shown in Figure 9.2. There are some changes in the 2/rev but they are very small, and at this stage could not be regarded as diagnostic of a crack. There were, however, significant changes on the 1/rev components, suggesting a change in the state of balance. This could possibly be caused by settlement of the end-bell.

During August, there were changes in both the 1/rev and 2/rev vibration. Whilst the 1/rev could possibly be a consequence of a crack, it was at least equally likely to be just part of continuing settlement of the end-bell. The 2/rev showed an accelerating increase from 2.3 mm/sec on 30th July to 3.3 mm/sec on 28th August. All vibration levels however remained well within acceptable levels.

After a brief outage, the unit returned to service on 8th September, and Figure 9.3 shows the 1/rev and 2/rev at bearing 11 on this run-up, together with those in May and July. The 3/rev shown in Figure 9.4 has a very prominent peak at 660 rpm. This is close to one-third of the second critical speed of the generator rotor. Furthermore, the 2/rev shows a clear change at 1,050 rpm, around half the critical speed. Taken together, these two factors were the first clear evidence of a crack in the rotor. Further evidence was observed in the 1/rev traces where a slight downward frequency shift was apparent in all the peaks indicating a reduction (albeit small) in the critical speeds.

The unit returned to high load and on 10th September changes were observed on 1/rev and 2/rev, particularly on bearing 11. The 1/rev did not cause immediate concern as it was considered to be part of an on-going trend, but the 2/rev increases appeared to confirm the provisional diagnosis of a crack. The levels continued to increase and the unit was removed from service on 12th September when the 2/rev reached an agreed limit of 6 mm/sec. The rundown is shown in Figure 9.5 where a number of features are apparent. All vibration components on the alternator bearings had approximately doubled in the four days of service since the outage. The changes in the 2 and 3/rev components that had, by this time been taken to indicate a crack, were more pronounced. Both generator critical speeds showed a reduction of

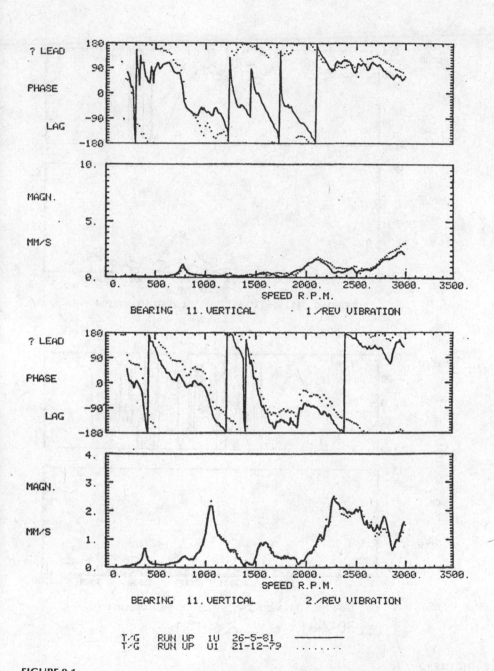

FIGURE 9.1
Comparison of Bearing 11 Vibration Signatures, Dec. 1979 and May 1981.

(Data supplied by and reproduced with permission of EDF Energy, Gloucester, UK)

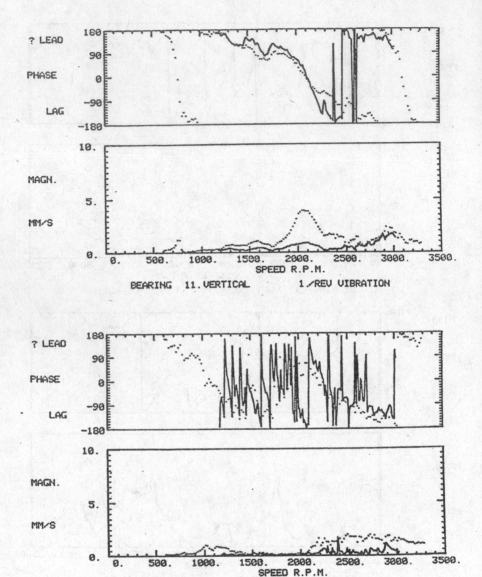

FIGURE 9.2
Plots Showing Changes as Compared to the Signature of May 1981.

(Data supplied by and reproduced with permission of EDF Energy, Gloucester, UK)

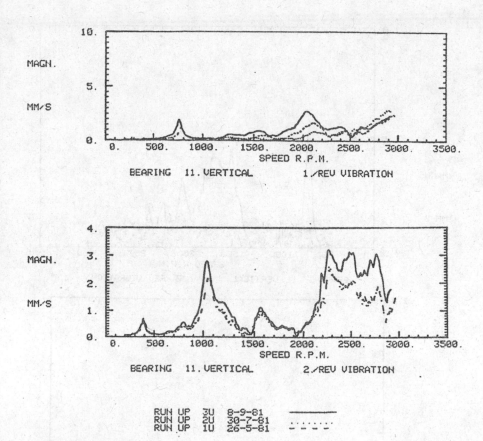

FIGURE 9.3
A Comparison of Run-Up Signatures on Bearing 11.

(Data supplied by and reproduced with permission of EDF Energy, Gloucester, UK)

around 5%, and the horizontal second generator critical speed recorded on bearing 12 had split into two peaks, at 1,970 and 2,060 rpm.

At this stage, other possible causes of the machine's behavior were considered but none were found to be fully compatible with the vibration behavior. The diagnosis of a crack was confirmed, but the remaining issue was to identify its location in order to assist the maintenance staff in their investigation. Consideration of the rundown changes showed that the fault was on the generator or exciter rotors rather than any of the turbine stages. The shift in the critical speeds of the generator rotor suggested that this is where the defect was since the much smaller exciter rotor would not have such a significant influence. On the generator rotor, the most significant changes are associated with the second critical speed; therefore as discussed

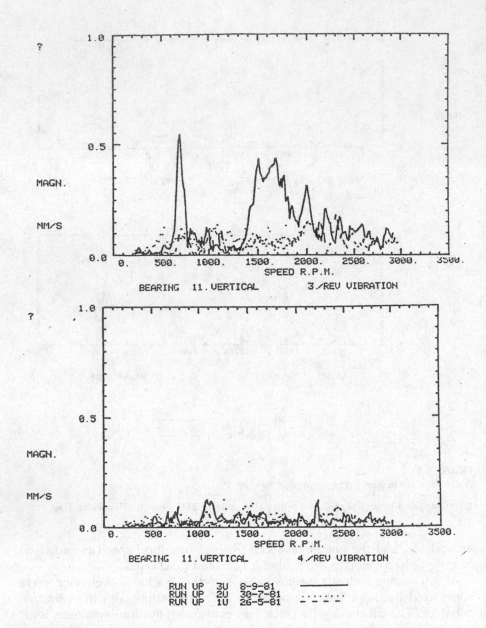

FIGURE 9.4

Higher Harmonics at Bearing 11.

(Data supplied by and reproduced with permission of EDF Energy, Gloucester, UK)

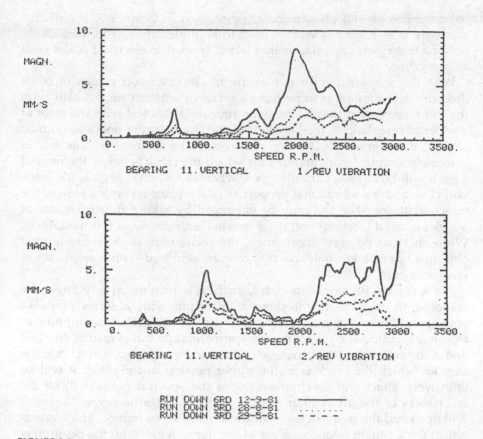

FIGURE 9.5
Bearing 11 Rundown Signatures.

(Data supplied by and reproduced with permission of EDF Energy, Gloucester, UK)

in Chapter 5, the defect is close to a point where the second derivative of the corresponding mode shape is a maximum. This leads to the conclusion that the crack is towards one end of the rotor.

When the rotor was removed from the machine, a large crack was found at the turbine end, close to a change in section. It extended to the bore and had an angle of around 140°.

The timely diagnosis had prevented a possible catastrophic failure.

9.3 Workshop Modal Testing of a Cracked Rotor

Over recent decades, many papers have been written on cracked rotors and this is appropriate, as it is an important practical problem and the basic technology and phenomena are discussed in Chapter 5. The discussion here is of a technique often used to help identify and locate a crack, but one that needs careful attention when applied to electrical rotors such as generators.

With the suspected rotor out of the machine a good means of crack detection and location is to perform a series of hammer impact tests with the rotor suspended from a crane. By repeating this test with the rotor at a series of orientations, the behavior of the crack during operation can be simulated by gradually turning the rotor. The concept of this test is extremely simple: In orientations for which the crack is below the neutral axis, it will be at least partially open, whereas for other angles, the crack will close and the vibrational properties of the rotor revert to those of the intact condition. The test may be enhanced by taking measurements at a series of axial locations and the exercise then becomes a full modal test. While this can be very informative, the focus here is the more limited objective of crack location and that may be achieved with a single set of measurements.

For a rotor with axial symmetry, such as a turbine, at any angle the response to an impact will show a spectrum with a series of peaks corresponding to the free–free natural frequencies. The assumption of free–free conditions is, of course, an approximation but is readily checked and a correction applied if necessary. It is to be expected that at orientation for which the crack is in the upper portion of the rotor, it will be effectively intact and the frequencies of the spectral peaks will be the resonances of the intact rotor. As the angle changes however, the crack will open and the frequencies of all these resonances reduce. The angle at which the resonant frequencies are at a minimum indicates the position at which the crack is fully open (assuming that there is a single crack) and furthermore, the relative changes of the different peak frequencies between intact and crack fully open orientation, can be used to locate the crack axially.

In Chapter 5, there is a brief discussion of the very elegant analysis by Mayes and Davies (1976) which related the change in natural frequency to the crack depth and location. It was shown that

$$\Delta(\omega_N^2) = R y_N''(s_c) \tag{9.1}$$

where $y_N''(s_c) = \left[\frac{d^2 y_N}{dz^2}\right]_{z=s_c}$, that is, the second derivative of the mode shape at the crack location s_c, and R is a function of crack and rotor geometry. y_N is the Nth mode shape and the rotor lies along the z-axis. In Mayes and Davies (1984), the same authors extend their study using dimensional analysis to describe the stress concentration factor at the crack front. Their study leads to the simplified expression (shown in Chapter 5, repeated here for convenience)

$$\Delta(\omega_N^2) = -4\left(\frac{EI^2}{\pi r^3}\right)(1 - \nu^2)F(\mu)y_N''(s_c) \tag{9.2}$$

where μ is the nondimensional crack depth (relative to the local shaft radius), and F is a function independent of all other parameters and is a universal function for a given shape of crack. The other parameters have the same meanings as in Chapter 5.

In principle, the function $F(\mu)$ can be derived from the appropriate stress concentration factor, if this is known. A different approach was taken in Mayes and Davies (1984) and Davies and Mayes (1984) where the authors inferred the values of F from a series of experiments with chordal cracks. It was shown that to a very good approximation, this function is just equal to the fractional change in the second moment of area as measured at the crack face.

In performing this calculation, however, it is important to note that the second moment of the remaining portion of the face is referenced to the new geometric center rather than the original one. Hence, F is a very nonlinear function of μ. Using the studies described in Chapter 5, a representative finite element (FE) model can be produced using Equation 5.30, reproduced here for convenience. If the second moment of area of the element concerned is I_0, then

$$\frac{\Delta I/I_0}{1 - \Delta I/I_0} = \frac{r}{l_{el}}(1 - \nu^2)F(\mu) \tag{9.3}$$

where l_{el} is the length of the section with reduced properties. Note that this parameter is at the discretion of the modeler within a reasonable range but the choice determines the value of second moment of area. For any given chordal crack, there are two values of F, and two orthogonal directions in the plane of the crack. Hence, the parameters of a representative model are now fully specified.

All this means that provided there is a reasonably accurate model of the rotor, and hence knowledge of the (uncracked) mode shapes, the relative effect of the crack on each of the modes can be estimated. The significance of this is that it provides a crucial means of determining the location of a crack without completely dismantling the rotor. It is worth noting here that many industrial rotors are very complicated assemblies and it is often very difficult to locate even a major crack.

Test procedures vary but often eight equally spaced orientations were used and the response to a hammer impact recorded. The frequencies of the first two or three flexural resonances were noted in each position. It is sufficient to measure in a single direction, normally the horizontal, but as discussed next, there are advantages in monitoring both orthogonal directions. With eight orientations there are four pairs of diametrically opposite pairs for which differences in natural frequencies are measured and it is these differences that lead to the crack identification. In its simplest terms, the following statements can be made:

a) The orientation at which the frequencies are minimum corresponds to the angular position at which the crack is at its lowest position and is consequently fully open.

b) The ratio of the frequency differences of the resonances locates the axial position of the crack.

c) Once the position of the crack has been located, the magnitude of the crack can be determined by evaluating the function F in Equation 9.2 and hence the nondimensional crack depth μ.

Of course, this basic procedure may lead to partially contradictory conclusions and it may be appropriate to use some form of least square procedure but this is not discussed here.

It is, however, worth considering some causes of confusion in executing this procedure. One important issue here is that while the crack is modeled as planar and chordal, this will at best be an approximation to reality. It must be emphasized that all cracks are different and the purpose of models is to extract key general features rather than accurately reproduce the behavior of any specific case. Perhaps more importantly, the nature of the rotor itself must be considered: In particular, electrical rotors, such as generators and motors, are not axisymmetric and this leads to considerable confusion.

An electrical rotor has different stiffness in two orthogonal directions and this has some important consequences. Firstly, it is important to appreciate that a crack may be recognized on an operational (horizontal) rotor by a change in the twice per revolution component of vibration. In an axisymmetric rotor, the change of this harmonic component will always be an increase but this is not the case with a rotor which has an inherent degree of asymmetry. This is because the orientation of a crack may be such as to initially decrease the overall asymmetry which essentially drives the twice per revolution vibration: Of course, as a crack grows, the asymmetry will at some stage show an overall increase. The salient consideration is the vector change in the twice per revolution vibration. This will almost always be very much smaller than the synchronous vibration, but it is very "stable" and in most machines shows very little variation making it a very useful diagnostic tool. Secondly, the introduction of a crack introduces cross coupling between the two principal planes and this must be taken into account in interpreting the hammer tests described before. The cross coupling means, for instance, that when excited, predominantly in the "weak" direction, may well produce a displacement in the "strong" axis and due allowance must be made for this, complicating the diagnosis considerably.

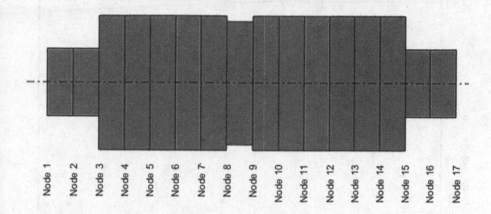

FIGURE 9.6
Rotor Model to Check Cross Coupling.

Consider the example shown in Figure 9.6. A uniform rotor has a 5% difference in stiffness in its two principal directions and consequently a 2.5% difference in the free–free natural frequencies. A crack has been introduced at the midpoint which has a nondimensional depth of $\mu = 1$, that is, the crack effectively removes half of the total cross section and the crack face tip is at an angle of 45° to the principal axis of the rotor. The situation was modeled using the reduction of second moments of area given by Equation 9.3, and the location is identified in Figure 9.6 between nodes 8 and 9. The rotor is considered to be suspended from a crane and hence there are no bearings in the model. The details of the calculation are given by Lees and Friswell (2001).

With the rotor intact, there is no cross coupling of motion in the two directions of the principal axes, but this changes markedly with the crack. Figure 9.7 shows the degree of coupling between the orthogonal direction and the excited direction. In a number of modes, the degree of coupling is pronounced and this makes the interpretation of simple resonance tests somewhat more complicated.

Once this phenomenon is recognized, it is readily taken account of. Indeed, the sensitivity of this cross coupling is an interesting feature, which could possibly provide further diagnostic tools for crack detection.

9.4 Gearbox Problems on a Boiler Feed Pump

The machine in question was the main boiler feed pump at a large coal-fired power station. The main turbo-generators were rated at 500 MW and

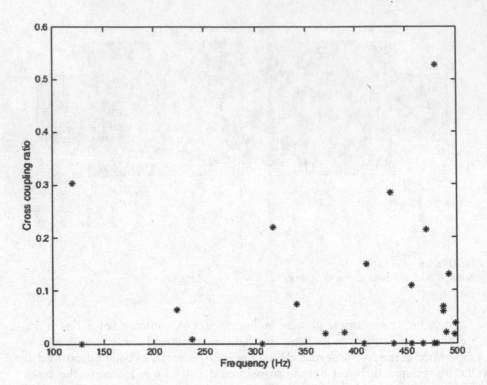

FIGURE 9.7
Cross Coupling of Directions in a Cracked Rotor.

each of the four units has a main boiler feed pump and two start and standby units. At the time of the investigation, the plant was relatively new but the main boiler feed pumps had suffered a number of failures and poor reliability. This was costly because although the turbo-generators could operate at full load using only the start and standby pumps, this mode of operation by-passed part of the feed heating system and consequently efficiency was significantly reduced. The study was reported by Lees and Haines (1978).

The main boiler feed pump supplies 1.6 million kg of water per hour at a pressure of 220 bar and a temperature of 160 °C. Clearly under these conditions, some preliminary pressurization is needed to suppress cavitation at the impeller inlet and for this reason a small so-called booster pump is incorporated into the system. This booster pump is driven from the main pump via nominally 2:1 reduction gearbox (actually 111:56). This gearbox was the prime source of the problems encountered, with several gear failures, broken couplings, and a cracked shaft. The whole arrangement is driven by a single-stage turbine rated at 17 MW with a maximum rotational speed of 5,000 rev/min. Of this power, about 10% is transmitted to the booster pump. The turbine, HP pump, gearbox, and booster pump are all connected by flexible couplings in order to allow for the substantial alignment changes with temperature during operation. A schematic of the complete arrangement is shown in Figure 9.8.

The overall problem was clearly to resolve the many and varied problems on the unit. At the time there were four units and this power station, but another four at two other sites: All exhibited a similar pattern of faults. The challenge was to determine if there was a single underlying cause and thereby rectify the issue. A feature of this design of pumps was that used a balance drum together with an external bearing to balance the thrust, rather than the (at the time) more conventional balance piston arrangement. The only relevance of this was that the design involved a long overhang from the bearing to the coupling, influencing the torsional properties. As the gearbox appeared to be the focus of problems, it was logical to attempt to monitor the transmitted torque, despite the inherent difficulties in making this measurement.

The torque measurement was achieved by bonding strain gauges in the spacer shaft of the double membrane coupling between the HP pump and the gearbox. The signals were transmitted to a stationary receiver by means of a telemetry system – and the observations were both surprising and illuminating. At the time this was a challenging instrumentation exercise and further details are given by Lees and Haines (1978). It proved, however, to

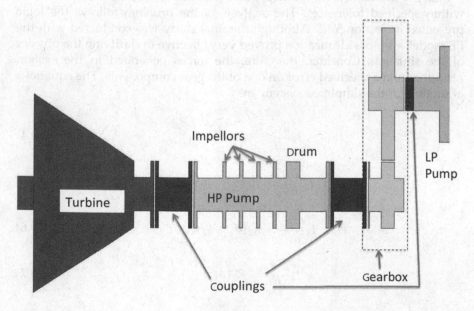

FIGURE 9.8
Idealized Main Boiler Feed Pump and Turbine Layout.

be very worthwhile and led to a solution of this and some related problems, although initial measurement results were puzzling, requiring a mathematical model to interpret the observations.

The measurements showed significant fluctuations in torque over a wide of pump speed and duties. A most important observation was that at under some operational conditions (during run-up to normal speed) the fluctuations were larger than the mean torque – the consequences of which are discussed next. The main objective was to ascertain the origin of these fluctuations. Another important observation was that over a range of speeds torque measurements revealed three distinct frequency components (albeit varying in magnitude and relative importance). The three components were the rotational frequencies of the two shafts together with a fixed frequency of about 39 Hz. It was later established that this fixed frequency term corresponds to the first torsional resonance of the system which is excited by nonlinear terms in the equations of motion.

A FE model of the system was initially developed based on the plant drawings, comprising 52 beam elements. As the investigation developed, however, it was observed that because the coupling between the turbine and the pump was relatively stiff, it was possible to work with the simplified model shown in Figure 5.20 comprising four inertias and two (torsional) springs. This is a purely torsional model: the neglect of coupling between torsion and lateral vibration was judged to be justified in the light of the plant measurements, although this effect can be included in the FE model.

Like any mechanical component, gears are not exact, but are constructed to within specified tolerances. The analysis of the problem follows the logic presented in Section 5.4.1. Although the final study was conducted with the FE model, a simple idealization proved very effective in clarifying the physics of the situation. Consider, therefore, the forces generated in the system resulting from a specified error on one of the gear components. The equations of motion for the simplified system are

$$I_1 \frac{d^2\theta_1}{dt^2} = K_1(\theta_2 - \theta_1) \tag{9.4}$$

$$I_2 \frac{d^2\theta_2}{dt^2} = K_1(\theta_1 - \theta_2) + T \tag{9.5}$$

$$I_3 \frac{d^2\theta_3}{dt^2} = K_2(\theta_4 - \theta_3) + \gamma T \tag{9.6}$$

$$I_4 \frac{d^2\theta_4}{dt^2} = K_2(\theta_3 - \theta_4) \tag{9.7}$$

where γ is the gear ratio. Before any analysis can be carried out however, a constraint must be applied to model the way in which the gears transmit motion. P is the transmitted torque which is unknown a priori, but it must be sufficient to ensure that the gear components can move in such a way that the error profiles can be "absorbed." If the gearbox were perfect, an appropriate equation of constraint would be

$$\dot{\theta}_2 + \gamma \dot{\theta}_3 = 0 \qquad (9.8)$$

However, since some errors are inevitable, the constraint is a little more complicated and may be expressed as

$$\frac{N_1}{2\pi}\dot{\theta}_2 + \frac{N_2}{2\pi}\dot{\theta}_3 = \varepsilon_1(\omega_2 + \dot{\theta}_2)\sin(\omega_2 t + \theta_2) + \varepsilon_2(\omega_3 + \dot{\theta}_3)\sin(\omega_3 t + \theta_3) \qquad (9.9)$$

This is clearly a nonlinear equation. Considerable simplification can be obtained by taking a case in which one of the two error terms dominate, say $\varepsilon_2 > > \varepsilon_1$ and the torsional displacement is "small" in some sense. Then multiplying by $2\pi/N_1$ gives

$$\dot{\theta}_2 + \gamma \dot{\theta}_3 = \frac{2\pi \varepsilon_2}{N_1}\sin \omega_3 t \qquad (9.10)$$

The system of equations may now be solved at each rotational speed. Chapter 5 shows how this can be achieved using general matrix constraints, but in this case, simple manipulation of the equations shows that

$$T = \dot{\theta}_2 \left[K_1 - I_2 \omega^2 - \frac{K_1^2}{K_1 - I_1 \omega^2} \right] = \alpha \dot{\theta}_2 \qquad (9.11)$$

Similarly,

$$T = \frac{\dot{\theta}_3}{\gamma} \left[K_2 - I_3 \omega^2 - \frac{K_2^2}{K_2 - I_4 \omega^2} \right] = \frac{\beta \dot{\theta}_3}{\gamma} \qquad (9.12)$$

These expressions are now substituted into the equation of constraint (Equation 9.10) giving (after cancelling sinusoidal terms)

$$T \left[\frac{1}{\alpha} + \frac{\gamma^2}{\beta} \right] = \frac{2\pi \varepsilon_2}{N_2} \qquad (9.13)$$

Hence, the transmitted torque at frequency ω is

$$T(\omega) = \frac{2\pi \varepsilon_2}{N_2} \frac{\alpha \beta}{\alpha \gamma^2 + \beta} \qquad (9.14)$$

Clearly, this can be expressed explicitly but the added complexity does not contribute to understanding. The important issue to appreciate in this is that the magnitude of the torque oscillations arising, and consequently the stresses in the gear teeth, is dependent on the torsional stiffness of the shaft line – in particular the most flexible part, namely, the couplings. This is because in essence, the process is displacement controlled. In contrast to many situations, the magnitude of the relative motions of the gear components is determined: The difference in the motion of pinion and wheel in the gearbox must be (at least) equal to the tooth error and forces will be generated to ensure that this is the case. Assuming that the gears remain in contact, the relative motion will indeed be equal to the error profile prevailing at the time and point of contact.

There is an assumption here; the gear teeth have been assumed to be absolutely rigid and this is not quite true. In reality, all gear teeth have some flexibility (see Section 5.4.1.1) and this can be easily incorporated into the model although it inevitably makes it a little more complicated. Values for tooth stiffness can be calculated for standard tooth profiles and for this case, it was found that the effect only becomes important for frequencies above 1 kHz and was therefore neglected.

Returning to Equation 9.14, consider the influence of the coupling stiffness values on the torque oscillations. The situation is already complicated, so to illustrate the effect of the displacement-controlled mechanism, consider now the much simpler situation shown in Figure 9.9 in which the driven component has simply inertia and the driving inertia is very large. For this simple system, the equations may be written as

$$I_1 \ddot{\theta}_1 = k(\theta_2 - \theta_1) \qquad (9.15)$$

$$I_2 \ddot{\theta}_2 = k(\theta_1 - \theta_2) + R_2 P \qquad (9.16)$$

$$I_3 \ddot{\theta}_3 = R_3 P \qquad (9.17)$$

P is the force of the gear teeth and other symbols retain their meanings as before.

Again, damping terms have been neglected for the sake of simplicity but these will be included in the calculation of responses. If it is now assumed that there is a single sinusoidal error on the pinion, then the equation of constraint becomes

$$R_2 \theta_2 + R_3 \theta_3 = e \sin(\omega t + \theta_2) \qquad (9.18)$$

Note that e is a dimension on the gear rather an angle. For this illustration, it is assumed that I_1 is very large and hence $\theta = 0$. Then it follows that (neglecting second-order terms)

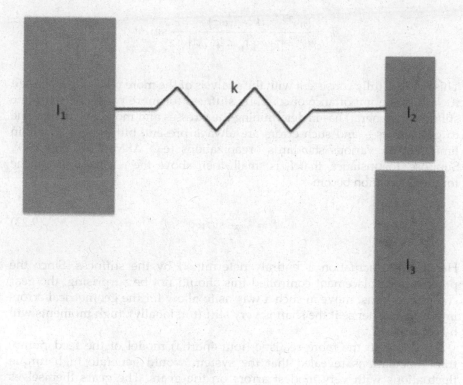

FIGURE 9.9
Simplified System for Illustration of Displacement Control.

$$(k - I_2\,\omega^2)\theta_2 = R_2 P \tag{9.19}$$

Provided θ_2 is sufficiently small, Equation 9.18 can be linearized to give

$$\theta_2 + \gamma\theta_3 = \frac{e}{R_2}\sin\omega t \tag{9.20}$$

The consequences of this linearization are discussed later, but it does make the solution to the problem much more tractable. Using the equation of constraint 9.20 in 9.17 yields

$$T\left[\frac{1}{k - I_1\omega^2} - \gamma^2\frac{1}{I_2\omega^2}\right] = \frac{e}{R_2}\sin\omega t \tag{9.21}$$

Simple manipulation now gives the transmitted torque oscillation due to a sinusoidal error of magnitude e as

$$T = \frac{I_2\omega^2}{\gamma^2} \frac{(k - I_1\omega^2)}{\left[k - \left(I_1 + \frac{I_2}{\gamma^2}\right)\omega^2\right]} \frac{e}{R_2} \sin \omega t \qquad (9.22)$$

This result is fully consistent with the analysis of the more realistic system but it clarifies the importance of coupling stiffness (or more precisely, shaft line stiffness in general) has in determining the stresses and moments arising due to gear errors – and such errors are always present, but hopefully within limits set by various standards organizations (e.g. ASME, BS, and ISO). Suppose, for instance, that I_1 is small, then above the resonant speed, the torque fluctuation becomes

$$T = -\frac{ke}{R_2} \sin \omega t \qquad (9.23)$$

Hence, the fluctuation is entirely determined by the stiffness. Since the process is displacement controlled this should not be surprising; the gear component must move in such a way as to allow for the geometrical errors imposed and hence if the shaft is very stiff (torsionally), high moments will be generated.

Returning to the more realistic (four inertia) model of the feed pump, train calculations revealed that the system would generate high torque fluctuations with very modest errors on the gears. The gears themselves were well within standards but the machine was just too sensitive to these errors. During machine run-up, the torque fluctuations exceed the mean torque on the machine and hence the gears will lose contact, then suffer a significant impact as they re-engage. Although not conclusively proven, it was strongly suspected that this impacting caused the major damage to the gears.

The model was used to compare the plant with proposed solutions given a gear error of 25 μm on the slow slide gear wheel and the results are shown in Figure 9.10. Also shown in this figure is the mean torque of the transmitted through to the gearbox as a function of speed. The solution proposed was a reduced stiffness of the high speed coupling, by installing a spacer with reduced wall thickness, together with the introduction of some torsional damping by the use of a rubber block coupling on the low speed shaft. Note that in these calculations, some damping has been introduced; this is straightforward but was omitted from the derivation above for the sake of clarity. The modifications were carried out some years ago and the result has been highly satisfactory with greatly improved plant availability and reliability. Two features in Figure 9.10 are worthy of note: On the modified system, the peak level of induced oscillation is reduced, but it is also moved in frequency, away from the normal operational conditions. In this particular case, the four inertias are 15, 0.37, 3.38, and 5 kg m^2, respectively, while

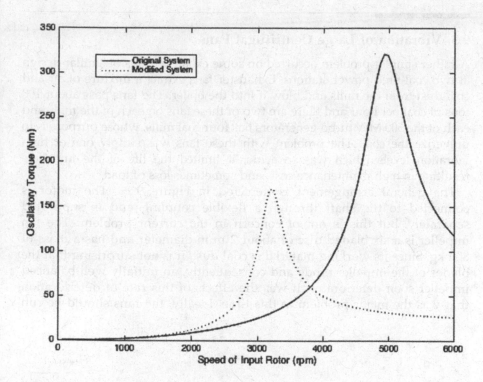

FIGURE 9.10
Torque Oscillations due to Wheel Error of 25 μm.

initially the two stiffnesses are 8×10^5 and 5.8×10^5 Nm/rad. In the modified system, the first stiffness was reduced to 3×10^5 Nm/rad.

One factor remains to be explained, namely, the presence of a component at a fixed frequency (39 Hz) over a wide speed range. This corresponds to the first torsional natural frequency and arises because of the nonlinear term on the right-hand side of Equation 9.9. This is difficult to deal with mathematically but an analysis has been given recently by Lees *et al.* (2011) using modern computational methods. As reported by Lees and Haines (1978) insight may be gained by expressing the displacement as a series of harmonics of the shaft speeds (i.e. of both shafts). The nonlinear term gives rise to products of the two series, which in turn leads to excitation over a wide range of frequencies. Given this broadband spectrum, high excitation will occur at the (torsional) natural frequencies, a finding which is entirely consistent with the observed behavior and the more modern analysis.

The modification, together with the introduction of damping in the second coupling, has proved very successful. The problem highlights the importance of designing a complete system, rather than just an assembly of components.

9.5 Vibration of Large Centrifugal Fans

A rather simpler problem occurred on some exhauster fans of similar design at two coal-fired power stations. Exhauster fans extract a mixture of air and coal dust from the mills and blow it into the boiler. The fans pass about 100 tons of coal per hour and there are two of these fans on each of the mills and each of the 500 MW turbo-generators has four coal mills, whose purpose is to pulverize the coal. The problem with these fans was simply one of high vibration levels which was so acute, it limited the life of the impellers resulting in high maintenance costs and sometimes loss of load.

The general arrangement is depicted in Figure 9.11. The motor is connected to the shaft through a flexible coupling and is supported separately, but this is not of concern in the current problem. The fan impeller is a six bladed disc of about 2 m in diameter and has a mass of 800 kg. Since its working material is coal dust, it is not surprising that the blades of the impeller erode and consequently, an initially well-balanced impeller soon deteriorates. It was the effects of this rate of deterioration that was the major problem at this time. Ideally, the fans should be run

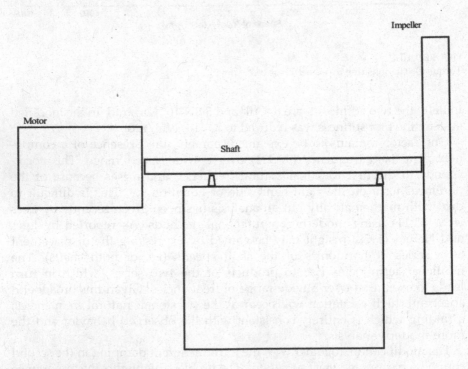

FIGURE 9.11
Exhauster Fan.

for about six weeks, by which time efficiency would be lost owing to the severe wear of the fan blades. This had always been a problem on these fans and so it was decided to review their design. The supplier had assured us that the first critical speed was 1,900 rpm and consequently running at 1,500 rpm should be perfectly acceptable, but plainly this was not the case.

The supporting block was a fabricated steel structure which clearly has some flexibility, but initially it was considered to be rigid for the purpose of calculating the natural frequencies and critical speeds. The overhang of the impeller to the nearer bearing was 600 mm and the shaft diameter was 170 mm. Calculations showed that there is a critical speed well within the fan's running speed range. These results caused some consternation with the suppliers. Their earlier assurances had neglected the effects of gyroscopic moments on the shaft. Whilst these effects are often relatively small for many rotors in which the major masses and inertias are within the bearing span, this is not the case with overhung rotors such as the case on these fans.

It is clear that the design of the fan required some revision. The parameters which could be changed relatively easily were the pedestal stiffness and the length of the overhang between the outer bearing and the impeller. Table 9.1 shows the critical speeds for a few combinations of these variables and it is clear that shortening the overhang has an important part to play. However, while critical speeds are an important consideration, attention must be paid to the excitation. Noting that the predominant concern here is backward whirl modes, it is important to appreciate that reverse modes can only be excited by imbalance if there is asymmetry in the supporting structure as discussed in Chapter 4.

To illustrate the point, consider the response to imbalance in the cases of the table. To calculate the response, however, some assumptions must be made concerning the damping and this, as always, presents some difficulty. Clearly, there will be some damping in the bearings, but logically some damping would be expected from the fan impeller. Any precise evaluation of this influence is difficult, but fortunately it is of little importance to the current discussion. Since the argument here is a

TABLE 9.1

Variation in Critical Speeds

Case	K_x (N/m)	K_y (N/m)	Overhang (mm)	Critical Speeds (rev/min)
A	5×10^8	5×10^8	600	−1,245, +2,250
B	1×10^8	5×10^8	600	−1,178, +1,945
C	5×10^8	5×10^8	300	−1,680, +5,200
D	1×10^8	5×10^8	300	−1,600 +3,000

comparison of different options, provided a consistent damping model is used, appropriate conclusions can be made. Figure 9.12 gives a comparison of the imbalance response of the four cases from Table 9.1. The peak around 1,200 rpm in case B arises from a backward whirl, which disappears for the symmetric cases. This is because the troublesome reverse mode cannot be excited by imbalance forces if the supporting structure (including the bearings) is symmetric in the two directions orthogonal to the rotor. This point is discussed extensively by Friswell *et al.* (2010), but it may be viewed simply as a consequence of the phase relationships between the displacements in the two directions. It should be emphasized that the critical speed still exists in case C as would become obvious if, for instance, an external exciter was applied to the machine in a single direction.

Following a detailed study of the fans, two changes were implemented; the overhang was reduced to 300 mm (the minimum which could be accommodated without major design of the impeller and casing) and the fabricated support was replaced with a reinforced concrete block. Note that the objective of the latter was not an increase in stiffness but rather the achievement of approximately equal stiffness in the two orthogonal directions. Since the modifications, impellers have operated until decline in performance determines a change. Vibration is no longer an issue.

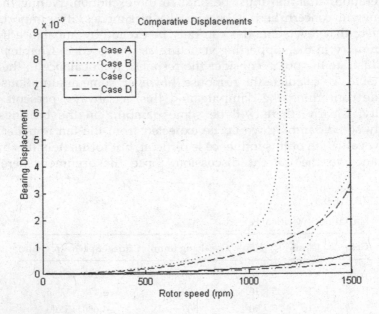

FIGURE 9.12

Response of Options to Nominal Imbalance.

9.6 Low-Pressure Turbine Instabilities

Another interesting problem has arisen in the United Kingdom. A large power station comprising four 500 MW turbo-generators has been subject to blade cracking on the last stage of the low-pressure turbines out of the four units. Measurements revealed an intermittent band of excitation of 7–12 Hz on the LP turbines which was not evident elsewhere on the machine. Furthermore, this low-frequency band modulates the synchronous vibration and its harmonics. A study of this particular problem is given by Hahn and Sinha (2014).

The interaction of the low-frequency band with the 1/rev vibration gives an excitation around 43 Hz, very close to a critical speed calculated as 46.88 Hz. Modulation with the 2/rev gives an excitation at 93 Hz, whilst the natural frequency of the final row blades is 90 Hz. There is an on-going investigation into this problem but early measurements, coupled with model studies, offer some insight into the sources of the difficulties.

The overall problem is complicated and believed to be due to a stall mechanism within the turbine. It occurs at low load, not surprising as under these circumstances, the turbine is operating well away from its design conditions. The flow of steam is, under these circumstances, not collinear with the rotating and static blades and consequently, secondary flow patterns are initiated. This is entirely consistent with the fact that the two units displaying this problem are those which have had significant two-shifting and operation at reduced load.

This is just one manifestation of a problem which is arising across the world. Reiger (2012) gives an overview of blading problems and discusses this stall phenomenon. Owczarek (2011) discusses in detail, the physical basis of the problem. Although problems of stall have been known for a considerable time, concern has risen in recent years owing to the following two contributory and partially conflicting factors:

a) In the drive to improve efficiency, blades tend to be freestanding without the cover bands or lacing wires of earlier designs.

b) Fossil plants tend to two-shift or operate at reduced load as renewable energy take up part of the demand.

Given that nuclear units must generate at high load for both technical and financial reasons, it is inevitable that greater flexibility will be required of fossil plants and this presents a significant design challenge.

Hahn and Sinha (2014) have explored their data using conventional Fourier-based approach and these have proved effective. Wu and Qu (2009) report related stall phenomena on a blower unit using empirical mode decomposition (see 8.6.7). Their approach visually expresses the two-state behavior.

9.7 Concluding Remarks

Several case studies are described in this chapter, some go back a number of years, while others show some current issues. All however, relate to plant recently in service and hopefully give some insight into issues arising on real machines. Over recent decades, both instrumentation and data handling facilities have seen dramatic developments. Analysis tools have also provided much greater prospect for the understanding of processes within machines. It is this understanding of the basic machine operation which remains at the heart of all condition monitoring and related disciplines.

Problems

9.1   An alternator rotor may be idealized as a uniform beam with mass 50 tons and supported on bearings with stiffness 5×10^8 N/m spaced 6 m apart. Determine the first two natural frequencies.

9.2   The rotor of Problem 9.1 has a chordal crack 2 m from the first bearing. Calculate the change in the first two natural frequencies if the crack has nondimensional depth of 0.25, 0.5, and 1.

9.3   A rotor with a mass of 10000kg has a length of 3m. is mounted on bearings with stiffness 108 N/m and stiffness 5x104 Ns/m. There are balance planes 0.5m from each bearing.

9.4   There is an imbalance of 0.002 kg m at the first balance plane and the crack has nondimensional crack of 1. Calculate the synchronous and twice per revolution components of vibration as functions of rotor speed. (Note: Chapter 5 has details of this type of analysis.)

9.5   A single-stage gearbox has a gear ratio of 2.5:1. The pinion has moment of inertia of 100 kg m^2 with a radius of 100 mm, and the wheel has inertia 800 kg m^2. The pinion is connected to a motor of inertia 1,000 kg m^2 through a coupling of stiffness 10^5 Nm/rad and the wheel connects to a pump unit of inertia 500 kg m^2 through a coupling of stiffness 0.58×10^6 Nm/rad. If the wheel has a cyclic error of 10 μm, determine the magnitude of torque fluctuations as a function of input rotor speed. Assume each coupling has damping of stiffness 2,000 Nsec/m.

9.6   The rotor of a fan has an overall length of 2 m and a diameter of 0.2 m. It is supported on two bearings with a constant stiffness of 4×10^5 N/m. The first bearing is 0.2 m from one end of the shaft which is driven via a flexible coupling. The second bearing is 0.5 m from the other end of the rotor. Overhung is an impeller with mass 800 kg and diameter 1.8 m. Draw the Campbell diagram up to a rotor speed of 2,000 rpm.

References

Davies, W.G.R. and Mayes, I.W., 1984, The vibrational behaviour of a multi-shaft, multi-bearing system in the presence of a propagating transverse crack, *Journal of Vibration, Acoustics, Stress and Reliability in Design*, 106, pp. 146–153.

Friswell, M.I., Penny, J.E.T., Garvey, S.D. and Lees, A.W., 2010, *Dynamics of Rotating Machines*, Cambridge University Press, New York.

Hahn, W.J. and Sinha, J.K., 2014, Vibration behaviour of a turbo-generator set, *Proceedings of VETOMAC-X*, Manchester, September 9–11, pp. 155–161.

Lees, A.W. and Friswell, M.I., 2001, The vibration signature of chordal cracks in asymmetric rotors, *19th International Modal Analysis Conference*, Kissimmee, FL.

Lees, A.W., Friswell, M.I. and Litak, G., 2011, Torsional vibration of machines with gear errors, *9th International Conference on Damage Assessment of Structures (DAMAS)*, Oxford, UK.

Lees, A.W. and Haines, K.A., 1978, Torsional vibrations of a boiler feed pump, *Transactions of the American Society of Mechanical Engineers, Journal of Mechanical Design*, 100(4), pp. 637–643.

Mayes, I.W. and Davies, W.G.R., 1984, Analysis of the response of a multirotor-bearing system containing a transverse crack, *Journal of Vibration, Acoustics, Stress and Reliability in Design*, 106, pp. 139–145.

Owczarek, J.A., 2011, On the phenomenon of pressure pulses reflecting between blades of adjacent rows in turbomachines, *American Society of Mechanical Engineers, Journal of Turbomachinery*, 133(2), pp. 1–11.

Reiger, N.F., 2012, Progress with the solution of vibration problems in steam turbine blades, www.sti-tech.com/dl/vibnfr.pdf, Accessed 5 April 2015.

Wu, F. and Qu, L., 2009, Diagnosis of subharmonic faults of large rotating machinery based on EMD, *Mechanical Systems and Signal Processing*, 23, pp. 467–475.

10

Overview and Outlook

10.1 Progress in Instrumentation

Recent years have seen significant advances in the instrumentation available for the monitoring and diagnosis of problems in rotating machinery. Over 60 years ago, the development of the noncontacting eddy current transducer was indeed a major step for the industry and these are now a crucial component of many systems. More recent developments have been the refinements of wide-band AE sensors which are enabling practitioners to progress some way beyond the basic process of counting events, the basis of early AE investigations. Other important developments have the enhancement of telemetry systems with significantly enhanced bandwidth capabilities.

Perhaps more important is the development of wireless technology for instruments. Systems can now be controlled directly from laptop computers and this offers significant potential for industry in general. Cabling has in the past been a very significant part of the cost overheads of monitoring systems and hence the introduction of wireless communications has the potential to transform the economics of supervisory systems. It may, for instance, become a viable option to monitor the behavior of machines which were, hitherto deemed uneconomic. In this context, it is worth emphasizing that all effective progress in this field is ultimately governed by economics. For instance, it is pointless analyzing in detail the cause of a problem and means of alleviation if this cost is more than the replacement of the machine or component but these choices are, for the most part, fairly obvious.

10.2 Progress in Data Analysis and Handing

Of course, the really dramatic changes in recent years have been the availability of vast storage for copious volumes of data at minimal cost. This development has changed perspectives on the way in which plant data is stored, but it also presents challenges as to how best to sort, present, and interrogate the databases. Hopefully, recent trends will continue for some time.

It will be recognized however, that there is not a direct link between volume of data and the information content and this is a theme that has been stressed a number of times throughout the text. This is particularly apparent with regard to on-load plant data. The dilemma here is that there is a vast array of data available, but because the machine is running under quasisteady conditions, the information content is limited. However, because of fluctuations introduced by the control system and external influences, there is some information to be gleaned, but only if some form of model, either statistically or physically based, can achieve this.

10.3 Progress in Modeling

The first numerical models of rotating machinery were based on the transfer matrix approach, a method which made very efficient use of the very limited computer storage available at the time. With the development of finite element techniques around 1960, this soon became the predominant method. The approach being linear is rather simpler than the transfer matrix method, albeit rather more demanding in terms of computer memory. The initial attraction of the transfer matrix approach was its modest storage demands but in recent years computer memory has become extremely cheap and so high storage requirements have ceased to be a limiting factor.

Early studies were undertaken with linear beam models and, as discussed in earlier chapters, these methods remain in use today. More sophisticated approaches using three-dimensional elements are needed for the analysis of some rotors, but the appropriate methods are now well developed and documented but not discussed here. A discussion is given by Friswell *et al.* (2010). Some practitioners advocate the use of 3D modeling for all rotors, but the author disagrees with this view. The argument is that 3D models can be applied to all rotors regardless of profile. While this is true, a very substantial proportion of rotors, probably a majority, are amenable to the simpler approach. The main benefit of this is dramatically faster execution times coupled with simpler input and output. Underlying this is, perhaps, a difference in philosophy; some people aim to feed as much information as possible into a calculation and on occasions this has some merit. The other, perhaps more analytic approach, is to focus on a more basic calculation encapsulating the key physical components.

Finite element modeling (FEM) must now be regarded as a fully mature technology and this represents a major achievement over the last five decades. Methods are available for the representation of many practical machines and limitation is, in most cases, a lack of knowledge of internal details. An example of this is the treatment of rotor rub, as discussed in Chapter 6. There is no difficulty in deriving a theoretical model describing the process, but it is extremely difficult to determine the basic meaningful parameters needed to provide quantitative predictions. Two questions arise at this point, the first being what is possible? Secondly, and perhaps more importantly, what is required?

As discussed elsewhere, the fundamental goal in all condition monitoring activities must be economic. While it may be academically satisfying to uncover details of the internal operation of a machine, there is little point if this does not lead to enhanced, faster or cheaper repairs. For minor plant items, a replacement policy is clearly an appropriate procedure.

Obviously, plant data are fundamental to progressing the understanding of machinery, yet of itself, the value is somewhat limited. It becomes invaluable, however, if it is used to gain insight into the internal processes of the machine, but there are a number of ways in which this can be achieved. This insight is achieved in three main routes, or best of all, a combination of all three. The three contributory routes are as follows:

a) Detailed historical records of trends for each individual machine. At present, these records take the form of vibration amplitudes and phases during rundown. In addition, on-load data in the form of scatter plots form a valuable background. These should include records of any balancing procedures which may be utilized to save time in future operations.

b) Statistical models indicating the levels and variations in both on-load vibration levels and transient data. This will facilitate comparison with any changes observed in behavior. These statistical models may extend to trained artificial neural networks (ANNs) to register the sensitivity of the system to significant operational parameters.

c) A mathematical model based on the physics of the machine's operation and the engineering drawings provide additional insight relating external measurement to external operation.

Of the three components described here, few would argue with the first two, yet there may be some practitioners who doubt the value of numerical models. It is true that with current understanding (and more often, limited data on clearances and other detailed parameters) global models often lack accuracy, but to some extent, this is of secondary importance. Although it is useful to have models that are a faithful reflection of plant under operational conditions, the primary requisite is that the model can yield accurate sensitivities with respect to parameters.

ANNs are discussed in Chapter 8 and offer considerable scope in the understanding of machinery. These can be trained using a wide range of operational parameters and are easily adapted to determine the acceptability of a given machine state. Furthermore, an ANN can be used to categorize fault conditions, but this application, while useful, has some limitations. This is basically because the ANN is a statistical tool and has no physics built into it, unlike a formal mathematical model as outlined before. A consequence of this is that the ANN can only interpolate; it cannot extrapolate. In other words, the ANN can only classify the machine into a fault condition that is known to it (through previous data), whereas a mathematical model may be used to investigate new effects and new fault conditions.

This is not to decry the significant influence of neural computing in recent years. As classifiers of fault conditions, their application may be enhanced by using mathematical models to generate a wider range of pseudo-data for a variety of possible faults. This overcomes the perennial problem that operators have a vast array of data under normal operating conditions but relatively little in the presence of faults. Of course, the use of models always raises the issue of validation but recognizing the importance of sensitivities rather than absolute accuracy, satisfactory qualification can be reached.

A further valuable application of ANNs is in determining the contribution of each of a series of contributory factors. This application was alluded to in Chapter 8 but not described in any detail: where the dependence on parameters is nonlinear, the ANNs provide a much better insight than the linear methods of singular value decomposition.

When a network has been trained for a range of inputs, it is a straightforward and instructive exercise to perform a series of calculations on the network varying one input parameter at a time. In this way, the dependence of the system to each individual parameter may be investigated. Of course, it may be that cross dependences between two or more input variables have an influence, but this too may be clarified by reference to the network. To date, this approach has found limited application yet it appears to have significant potential as discussed by Mayes (1994).

This leads to a larger, and very important, question of how to merge the power of neural computing with the predictive capabilities of a physical model, and here, as yet, no clear solution is available. Perhaps the crucial distinction is that the physical model is constrained by the form of the equations: in Chapter 7, the fitting of a model to too many variables improves the fit to measured data but produces a model with limited predictive capability. In the same way, perhaps current ANNs have too much freedom and although producing excellent data fitting, this is at the expense of predictive capability. This reasoning leads to the notion of some form of guided neural computing, but precisely what form this might take, is unclear at present.

10.4 Expert Systems

A number of researchers have developed various "expert system" concepts to the general problem of machine condition monitoring. The techniques used in these studies range from ANNs (Ben Ali *et al.*, 2015), decision table analysis (Ebersbach and Peng, 2008) and Bayesian methods (Xu, 2012). The aim of these investigators is the encapsulation into software of the expertise of experienced assessors. Some convincing diagnosis has been shown for studies on rolling element bearing faults. For some categories of machine, these systems offer a promising way forward and evidence to date shows them capable of giving correct diagnoses. This will, at the very least, filter the more straightforward problems relieving pressure on the expert engineers.

Although the papers cited indicate promising capabilities, these logic-based systems give only diagnosis rather than any quantitative information as regards location or severity of a fault. This being so, the role of expert systems for large complex machines may be rather limited.

10.5 Future Prospects

There is currently a substantial array of tools available for the diagnosis of machinery ranging from instrumentation through data capture and storage to analysis and evaluation. It is worth considering the likely direction of trends in the foreseeable future, while recognizing the dangers of prediction. In the previous section, the distinctive powers of physically based models and neural method have been discussed and a long-term objective must be to at least aim for the automatic diagnosis of machine faults. It is, of course, questionable whether this will ever be achieved, but any substantially reduction in the number of problems referred to specialist engineers would be a worthy objective. A range of possible developments now emerge and these are discussed in the following subsections.

10.5.1 Machine Diagnostics

The first of these possible developments relates to the optimization of information of the internal condition of a machine. As discussed elsewhere, this inevitably relies on adequate instrumentation and recording of data but equally important is the interpretation of the measurements. This is where some form of model is needed, wherever possible a physics-based idealization is preferred, but on occasions, statistical approaches can help. ANNs are certainly of help in checking various parameters against previous behavior but are limited in extrapolating. However, there is a possible way around this difficulty.

A problem in the use of statistical approaches (including ANNs) to automatically detect and categorize fault condition is that there is insufficient data on faults; despite the continual occurrence of fault conditions, the proportion of available machine data under fault conditions is actually minimal. Using plant models to artificially generate appropriate data can readily rectify this shortfall. To do this, the model used would need to be validated appropriately but this is an example of complementarity between physical and statistical approaches to the area of machine diagnostics.

This begs the question of how far the technology can or should go. Any answer to this will be highly machine dependent. Clearly, a small machine that can easily (and cheaply) be replaced cannot justify major expenditure of fault detection hardware and software. On the other hand, a large turbo-generator represents a major capital investment and loss of output is in itself very costly and so the argument for effort toward fault detection and rectification is very significant. Several instances are discussed in other chapters which have illustrated the value of such methods. Models tuned to individual machines offer the prospect of much more sensitive tools coupled with the ability to infer more meaningful information from on-load data. While the general trend of this is clear enough, any prediction of how far progress will reach is a matter of speculation. For high-value machines, it is not unreasonable to anticipate that a tuned model will run in parallel with the machine in real time being constantly updated. This would then provide a constant monitor with the capability to infer internal machine conditions and hence a rapid dissemination of faults. In effect, the model becomes "virtual instrumentation" and yields operational information.

Whether or not we will ever reach the ultimate point of truly automatic detection of faults remains a valid point. It seems to the author that this is perhaps a little doubtful, but really this is not a great concern. Provided diagnostics can reach a point to significantly reduce the number of cases needing referral to a human expert, then a significant objective will have been achieved. Furthermore, in the referred cases, the tuned model would provide the assessor with internal information not available at the current time without extensive investigative studies. Coupled to this would be the facility discussed in Chapter 7 whereby rather than merely saving vibration records, the refined system would record physically meaningful parameters of the machine, such as imbalance profile, bearing loads, and the dynamic properties of the foundation structure. This in itself would significantly accelerate the resolution of problems.

To go beyond this takes us to the second strand of the vision, the concept of a "smart machine."

10.5.2 Self-Correction ("Smart Machines")

Having developed and used models to determine the condition of an operating machine, the obvious next step is the possible application of forces to rectify a fault condition. This concept has been applied to structures with some success, but the application to machinery is in its infancy. Nevertheless, it is an area of considerable potential and it is likely to gain attention from researchers in the foreseeable future. While it is likely that early progress will focus on smaller machines, the effect will inevitably spread in due course.

Consider the ways in which an operating machine can be influenced by external factors; the possibilities are summarized in the following categories.

10.5.2.1 Magnetic Bearings

Magnetic bearings are, perhaps, the most developed of the technologies which may form part of future "smart machines." Here, the primary interest is in autonomous bearings in which the magnetic is controlled directly from the monitored shaft motion. Extensive work has been carried out on various aspects of their performance (see e.g. Schweitzer, 2009). Many are in operation on a very wide range of installations, ranging from small precision machines to large gas line compressors. They have many attractions, but high cost. A limitation to their usage is the limited load carrying capacity which is substantially lower than that of a comparable oil journal bearing. It is worth noting that this limitation is really one of size: in principle, an active magnetic bearing (AMB) could be designed to take any load, but because of magnetic field saturation effects, capacity can only be increased by using larger conductors and this reaches the point where further increases become impractical.

10.5.2.2 Electrorheological Bearings

At the end of the twentieth century, there was considerable interest in journal bearings that had particles of dielectric material immersed in the oil. Application of an electric field transforms the fluid from Newtonian to pseudo-plastic with very much higher viscosity. This is another route by which the application of an external influence modifies the machine's behavior. A number of papers on this subject were published (see e.g. Nikolakopoulos and Papadopoulos, 1998) with promising results. Some further studies have been reported more recently but progress appears to be slow.

10.5.2.3 Externally Pressurized Bearings

This represents another established technology with some potential for smart machinery insofar as external forces can modify the bearing properties. As the name suggest, in these bearings the pressure within the oil film is generated by an external pump rather than by the viscous flow within the journal. For this reason, these bearings are suitable for low-speed application but the added complexity does add to cost. The interest in the context of smart machine is that the external pump in effect modifies the bearing properties and any components having this feature are of potential interest for the control of machines.

10.5.2.4 Other Approaches

Lees *et al.* (2007) designed a prototype bearing housing, whose stiffness could be varied by the application of a voltage. This was achieved by mounting the bearing in a collar of an elastomer material, the compression on which can be varied. Because the elastomer has highly nonlinear properties, it becomes stiffer when it is compressed. The compression was imposed via two coils of shape memory alloy (SMA) which has the remarkable (and useful) property that it shrinks when it is heated and affects phase changes within the material. The heating is achieved simply by passing a current through the coil. The discussion given in Lees *et al.* (2007) is at an exploratory stage but it does raise the possibility of smart machinery, albeit for a fairly limited range of machines.

There are, however, a number of difficulties with this approach – an important one being the rate of response. While there is some scope for applying this technology, difficulties arise in the cooling phase which demand some form of forced cooling for any usable response characteristic and although this can be achieved, the complexity of the system increases. Although the response time is much slower than the rotational speed of the shaft, control is possible provided that the rate of speed change is slow.

Figure 10.1 shows the manner in which the maximum amplitude may be very significantly reduced during a machine run-up by varying the stiffness of the bearing support during the transient. While this technology has some potential, SMA material performs most effectively in noncyclic operations and can be sensitive to a range of parameters. The other difficulty is that modification of bearing stiffness will be a less effective means of control for machines with flexible rotors. Figure 10.2 shows the way in which a rolling element bearing has been mounted surrounded by an elastomer whose nonlinear properties are modified by forces exerted using the shape memory coils.

10.5.3 Shaft Modification

This is rather different to the three possible routes mentioned in the preceding subsections. Whereas the previous three actions exert influence via the bearings, here consideration is given to exerting a force, or combination of forces, directly on the shaft. This is important as it opens up the possibility of application to machines with flexible rotors. While changes to bearing properties will always have some influence, if a rotor is flexible, the effect will be significantly reduced. Over recent decades, there has been a general trend toward the use of flexible rotors as this facilitates fast, smaller, and more efficient installations. It is therefore highly desirable to develop technologies to promote smart machines so here consideration is given into the ways in which this might be achieved.

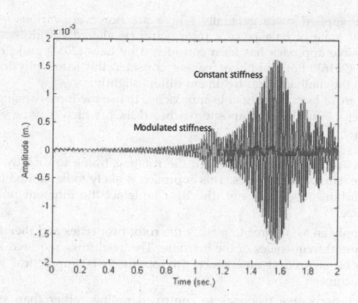

FIGURE 10.1
A Controlled Run-Up.

(Reproduced with permission from Society for Experimental Mechanics, IMAC-XXV, Lees *et al.*, 2007)

FIGURE 10.2
A "Smart" Support Structure.

(Reproduced with permission from Society for Experimental Mechanics, IMAC-XXV, Lees *et al.*, 2007)

It makes sense to focus attention on the leading fault in all rotating machines, namely, rotor imbalance, discussed extensively in Chapter 4. Automatic balancing devices are currently in use on, for example, washing machines (Rodrigues *et al.*, 2011), but it is not clear that devices of this type could be applied more generally. There are however, various ways in which the state of balance of a rotor could be altered during operation. One possible approach has been considered by Lees (2011) and Lees and Friswell (2016). Both of these papers consider the same physical idea although the mathematical treatment differs slightly.

There are at least five possible approaches to use the forces arising in the piezoelectric multifiber composite patches (MFC), which may be outlined as follows:

- Use proportional feedback in the rotating frame to directly change the natural frequencies. This approach is likely to be limited to small machines and does not directly counteract the inherent unbalance force.

- Apply an axial force to modify the rotor properties and therefore the natural frequencies of the machine. The axial force required is likely to be significant and hence this approach is impractical in most machines.

- Use derivative feedback to control damping rather than stiffness. This will improve the transient response but will not significantly affect the unbalance response, unless the machine is operated close to one of its critical speed.

- Use the patches to induce rotor bends, which can counteract the unbalance. This may be viewed as a standard balancing approach where the moments from the piezoelectric patches replace the forces from the balancing masses.

- Use integral feedback in the rotating frame. Since the unbalance response in the rotating frame can be viewed as a steady-state error, applying integral feedback can reduce this error and hence balance the machine.

The first three approaches are unlikely to be practical and hence this discussion will concentrate on the last two approaches.

Essentially, the piezoelectric actuators apply moments to the shaft that may be easily controlled. The determination of the best moments to apply is identical to a balancing procedure, with the essential difference that the moments that may be applied take a particular form dictated by the position of the actuators on the rotor. The great advantage is that the moments applied may be continuously adjusted due to changes in the machine balance or changes in the rotor spin speed. Hence, existing methods of balancing and automatic balancing may be used. Methods of automatic balancing based on changing the mass distribution of the rotor do have the advantage that a balanced machine will be balanced at all speeds, although in practice the range of spin speeds where the balance is satisfactory is limited.

Integral feedback is well established to reduce the steady-state errors in a wide range of control systems. The error of interest in this case would be the response at the location of the sensors in the rotating frame. Note that this error would have components in both transverse directions and this is counteracted by two orthogonal bimorph actuators. Since the measurements will be taken in the stationary frame and consist of proximity probe or accelerometer measurements, they must be transformed into the rotating frame, in the directions of the actuators. Most rotating machines will have a phase reference that gives both the rotor spin speed and also the phase correction; this reference enables the transformation matrix to be calculated. If the response in the rotating frame is aligned to the directions of the piezoelectric actuators, then each actuator voltage is simply proportional to the integral of the corresponding response.

Piezo-electric patches are small devices normally comprising lead zirconate titanate (PZT) and some electrode connections. These are capacitive devices which exert a force when polarized by the application of a voltage. Originally used in the fine movement of components, these devices are now used extensively in "smart structures." They have several advantages over SMAs, the most important being a very much more rapid response and enhanced ability to work in a cycle as they are not dependent on thermal properties. The disadvantage is the high voltages required but this should be judged against the negligibly small currents demanded. They also have a rather smaller strain range than SMAs – no more than 2%, about one-third of that which a SMA can tolerate.

The concept is to apply a bend to the rotor to compensate the imbalance. This is in direct contrast to the much more usual procedure of "balancing out" a rotor bend. In the case studied by Lees and Friswell (2016), a shaft bend was formed by forces exerted by two pairs of piezo-electric patches mounted centrally on the rotor. Similar concepts have been studied by Horst and Wolfel (2004) and by Sloetjes and de Boer (2008). The work in this area shows promise but clearly significant development work is required in this area. The piezo-patches are activated by signals transmitted to the rotor by means of either slip rings or telemetry. After extensive discussion in Chapter 4, illustrating the important distinctions between a bend and imbalance it is, perhaps a little surprising that such an apparent equivalence is a possible route to smart machinery, but the important point is that the force imposing the bend is controllable and readily changed to suit rotor speed and conditions. As set out in Chapter 4, the big distinction of the two faults is the frequency variation. The studies carried out to date focus on the correction of a single mode and while this is entirely understandable as a research investigation, it is likely that a real machine would need at least four patches per mode within the running speed range. More relevant, however is a brief discussion on the connectivity of such a system, or how to get signals into the rotor.

Early studies have used piezo-electric MFC patches which are now widely used in smart structures where they have excellent high-frequency characteristics. Because of this they are, perhaps, the natural choice for development. The high voltages (but minimal currents) can easily be input to the rotor by means of either slip rings or telemetry. These technologies are readily available but there would inevitably be some cost implication. An alternative strategy would be to transmit the signal at lower voltage with some amplification on the rotor. Again, this would appear feasible since the power involved is limited. There are various options that await a comprehensive investigation.

However, other possibilities offer significant potential. Rather than MFCs, the bending moments on the rotor could be applied using SMAs. The characteristics of SMAs are rather different than those of MFCs; the important differences are that they operate over a substantially greater range of strain but they have a much slower response, particularly in the cooling phase. A significant advantage however, is a much lower voltage requirement (typically a few volts and against the 1.5 kV needed for some MFC activation), and this may be important in some applications.

The disadvantage of SMAs is the slower response time but in many cases this need not be a problem: although the vibration is relatively high frequency, the induced bend to bring about control is quasistatic. Therefore, the requirement for frequency response is governed by the rate of change of running speed, rather than the speed itself and this means that SMAs would be acceptable for some applications. Again, more research studies are needed here.

An intriguing possibility would be the use of magnetic shape memory alloys (MSMAs), a relatively new technology in which there have (at the time of writing) been few, if any, studies in their application to rotor systems. These have similar properties to SMAs in that phase changes give rise to geometric distortions but the changes are activated by magnetic fields rather than by temperature. The key benefit of this is that it raises the possibility of a noncontacting transmission to the rotor that may be rather simpler than telemetry. Of course, the potential benefits may be limited to certain types of machine.

The potential for transmission of signal to and from the rotor does raise other possibilities; for instance, readings from strain gauges could yield direct alignment data and information on bearing performance. Of course, the extent to which these possibilities reach fruition is a matter of assessing the added value against cost of implementation. Some basic machines will not justify the investment but developments in the directions suggested are likely for some major plant areas.

This raises many and varied questions which hopefully will be addressed in the coming years.

It should be appreciated that the ability to automatically apply correcting forces would represent a major step for rotating machinery and would completely change the constraints placed on designers. For instance, the ability to compensate for imbalance would negate the need to avoid critical speeds in the running range. Often it is the need to avoid resonances (or critical speeds) which dictate the dimensions of rotors. Freed from the need to introduce rigidity, some rotors may become much lighter, needing only suitable dimensions to transmit the required torque. This could have great influence, as in many machines the overall dimensions are chosen to avoid critical speeds rather than to convey the declared torque loading and of course this will have important repercussions on machine mass. Reduced mass will lead to reduced losses and higher efficiency, in addition to better usage of natural resources.

10.6 Summary

"Smart machines" are at the research stage and in the very early stages of development. However, they represent a culmination of work over recent decades in the interrelated disciplines of instrumentation, modeling, and control.

The related fields of condition monitoring, rotor dynamics and control have seen significant strides in the last 50 years and these have produced great economic benefits. The coming years should be equally productive.

References

Ben Ali, J., Fnaiech, N., Saidi, L., Chebel-Morello, B. and Fnaiech, F., 2015, Application of empirical mode decomposition and artificial neural network for automatic bearing fault diagnosis based on vibration signals, *Applied Acoustics*, 89, pp. 16–27.

Ebersbach, S. and Peng, Z., 2008, Expert system developments for vibration analysis in machine condition monitoring, *Expert Systemms with Applications*, 34, pp. 291–299.

Friswell, M.I., Penny, J.E.T., Garvey, S.D. and Lees, A.W., 2010, *Dynamics of Rotating Machines*, Cambridge University Press, New York.

Horst, H.-G. and Wolfel, H.P., 2004, Active vibration control of a high speed rotor using PZT patches on the shaft surface, *Journal of Intelligent Material Systems and Structures*, 15, pp. 721–728.

Lees, A.W., 2011, Smart machines with flexible rotors, *Mechanical Systems and Signal Processing*, 25, pp. 373–382.

Lees, A.W. and Friswell, M.I., 2016, Active balancing of flexible rotors using strain actuators, *Journal of Vibration and Engineering Technologies*, 4, pp. 483–489.

Lees, A.W., Jana, S., Inman., D.J. and Cartmell, M.P., 2007, The control of bearing stiffness using shape memory, *Society of Experimental Mechanics, Proceedings of IMAC-XXV*, Orlando, FL.

Mayes, I.W., 1994, Use of neural networks for on-line vibration monitoring, *Proceedings of the Institution of Mechanical Engineers*, 208, pp. 267–274.

Nikolakopoulos, P.G. and Papadopoulos, C.A., 1998, Controllable high speed journal bearings, lubricated with electro-rheological fluids. An analytical and experimental approach, *Tribology International*, 31(5), pp. 225–234.

Rodrigues, D.J., Champneys, A.R., Friswell, M.I. and Wilson, R.E., 2011, Two-plane automatic balancing: A symmetry breaking analysis, *International Journal of Non-Linear Mechanics*, 46(9), pp. 1139–1154.

Schweitzer, G., 2009, Applications and research topics for active magnetic bearings, *Proc. IUTAM-Symp. on Emerging Trends in Rotor Dynamics*, March 23–26, 2009 Indian Institute of Technology, Delhi, India: Springer-Verlag.

Sloetjes, P.J. and de Boer, A., 2008, Vibration reduction and power generation with piezo- ceramic sheets mounted to a flexible shaft, *Journal of Intelligent Material Systems and Structures*, 19(1), pp. 25–34.

Solutions to Problems

Chapter 2

2.1 Maximum resolution sample over 71 revolutions = (initially 0.71 sec) and the maximum resolution is

$$\Delta f = \frac{1}{0.71} = 1.4\ \text{Hz}$$

Overall sampling rate must be at least 800 Hz, block size 800 × 0.71= 570 samples.

2.2 If the acquisition is fixed by rotor position, then the resolution follows. Equation 2.14a and the resolution improved as the rotor decelerates. If it is at fixed time intervals, then the resolution stays fixed at its initial values assuming that the block size is constant.

2.3 Clearly, there are two important components.

Case (a), time span 0.5 sec, shows both components but with low resolution and the 25 Hz component has suffered leakage.

Case (b), time span 1 sec, no leakage, both components accurately reflected.

Case (c), time span 2 sec, sampling interval now 0.01 sec, and so cut-off frequency is 50 Hz and so the higher frequency component does not appear.

Case (d), time span 4 sec, both components are aliased to zero owing to an inadequate sample rate.

2.4 The resolution must be at least 2.5 Hz, preferably 1.25 Hz which means a sampling period of at least 0.8 sec. The two frequencies are 25 Hz and 27.5 Hz and to avoid leakage there must be an exact number of cycles of both signals, this suggests a 10 sec sample, sampling at 55 Hz. But this gives a cut-off frequency of 27.5 and an aliased term at 30 which is confusing. So acquire at a higher frequency, say 100 Hz.

2.5 This question just examines understanding and recall of material in the text.

2.6 The standard deviations (semiaxes of the ellipse) are given by

$$\sigma_1 = \sqrt{38.14} = 6.18 \qquad \sigma_2 = \sqrt{89.29} = 9.45$$

The angle of axis orientation is given by

$$\theta = \frac{180}{\pi} \times \cos^{-1}(0.5544) = -56°.$$

Probability is 0.2%.

2.7 A slow rundown shows clear peaks at 850 and 2450 rev/min. The rapid rundown still gives peaks but they are more diffuse – see solution manual for details.

Chapter 3

3.1 Natural frequencies are 17.51, 15.76, 13.72 Hz.

3.2 First with four elements, F = 32.3246 105.1589 Hz and with eight elements, $F2$ = 32.3213 105.0695 Hz.

3.3 Using Timoshenko elements, F = 32.3114 105.0193 Hz and with eight elements, $F2$ = 32.3079 104.9252 Hz.

That is, very little difference as the shaft is slender.

3.4 For simply supported rotor $k = \frac{48EI}{L^3}$.

Mass of discs: 0.1286, 0.5881, and 1.3539

Mass of the rotor: 3.6757 kg

Natural frequencies: 17.38, 15.65, and 13.64 Hz

3.5 The Campbell diagram shows that the critical speeds are (in rpm) 070, 4,842, 5,157, and 6,581.

Chapter 4

4.1 17.66 g at 200 mm and an angle of 148°.

4.2 Balancing with a bend

In this case, the parameters are W = 19 g at 20 cm radius and 170°.

Note, however, that at the balance speed, the vibration will now be higher, but will never exceed the magnitude of the bend.

4.3 The result for the two imbalances is

80 g at 10 cm on disc 1, at 178°;

168 g at 10 cm on disc 2, angle −96°.

Hence, the correction masses will be applied 180° opposite in phase.

4.4 Modal balancing is explained, with an example in Section 4.3.2.

The required correction masses are:

On disc 1 0.0103 kg m at 167°. and on disc 2 0.0047 kg m at 53°.

4.5 Required correction masses 0.0103 kg.m. at 347°on disc 1 and 0.0047 kg.m at 233° on disc 2. The similarity to Problem 4.4 is not significant.

Chapter 5

5.1 Large peak at 550 rpm of 120 μm, second lower peak at 2350 rpm of 60 μm.

5.2 Displacement angle is 0.17 min = 5×10^{-4} rad.

So first bearing must be raised by $2 \times 0.5 \times 5 \times 10^{-4}$ m = 0.5 mm.

Second bearing must be raised by $2 \times 4.5 \times 5 \times 10^{-4}$ m = 4.5 mm.

To ensure the outer bearings are at the same height, rotate the whole arrangement by 5×10^{-4} rad.

Brg 1 0, Brg 2 $-4 \times 5 \times 10^{-4}$ = −2 mm, Brg 3 −2 mm, Brg 4 0.

5.3 Moment is 1,170 Nm.

5.4 This is just a straightforward two (torsional) mass problem. The analysis follows that in the text and the rigid body mode is not relevant to the problem. Torsional natural frequencies are 13.6 and 21.4 Hz. Peak response to the errors 700 Nm.

5.5 The analysis of this problem requires the use of asymmetric rotor elements and the changes to the cross section as given by Equation 5.29.

Frequencies of intact rotor 77.95, 31.69, and 70.92 Hz

Crack Mu = 0.25, frequencies are 7.92, 7.93, 31.59, 31.6, 70.877, 70.879 Hz.

Crack Mu = 0.5, frequencies are 7.88, 7.9, 31.39, 31.49, 70.79, 70.82 Hz.

Crack Mu = 0.75, frequencies are 7.79, 7.88, 30.98, 31.36, 70.61, 70.77 Hz.

Crack Mu = 1, frequencies are 7.51, 7.87, 29.84, 31.34, 70.05, 70.77 Hz.

5.6 Given the large signal at 300 Hz, the crack in this case is probably close to the centre of the rotor, but not at the centre, since that would give no response in the second mode. Given values of the response, an estimate can be formed of crack axial location. No calculations are required for this question

Chapter 6

6.1 This is readily solved from Equation 6.2 of the text. However, a numerical solution is required as it is difficult to deal with analytically. With viscosity 0.04, eccentricity = 0.41. For $e = 75\%$, viscosity of about 0.006 is needed, a very low value. This is best solved by repeated solutions of Equation 6.2.

6.2 This is a straightforward application of the modelling software. Of course, a variety of calculation routines may be used following the logic in the main text. Peak response 1 μm at 1,490 rpm. With increase clearance, the resonance drops to around 800 rpm and displacements go up. With both clearances doubled, the response becomes.

6.3 At 5,000 rpm, eigenvalues are 1.0e+02 × −0.0065 − 2.8165i, −0.0065 + 2.8165i, −0.1800 − 2.8838i, −0.1800 + 2.8838i, −0.0605 − 7.8751i, −0.0605 + 7.8751i, −0.1239 − 8.1285i, −0.1239 + 8.1285i.

6.4 With reduced pressure this becomes (rad/sec) 0.53 − 241.92i, 0.53 + 241.92i. Note these two are unstable! −13.72 − 248.64i, −13.72 + 248.64i, −4.24 − 774.42i, −4.24 + 774.42i, −8.80 − 799.73i, −880 + 799.73i.

6.5 To answer this problem, we use Equations 6.24–6.28.

At 50% cut-off, $f = 38$ Hz, @75% $f = 71.3$ Hz, and @90% $f = 81.9$ Hz.

Chapter 7

7.1 The results are

$$k = 10^6 \begin{bmatrix} 1.85 & -0.72 \\ -0.56 & 0.79 \end{bmatrix} \qquad M = \begin{bmatrix} 8.35 & -0.29 \\ -0.29 & 6.43 \end{bmatrix}$$

$$C = \begin{bmatrix} 408 & 118 \\ 118 & 440 \end{bmatrix}$$

These are not precise answers as the truncation of the input (to two decimal places and phase to the nearest degree) has introduced significant noise.

7.2 Revised estimates are

$\mathbf{K}$ = [3.3 −1.33; −1.33 0.9] × 10^6, $\mathbf{M}$ = [16.3 −1.7; −1.7 49], $\mathbf{C}$ = [989 49; 49 323]

These are in substantial error, because not enough of the frequency range has been considered. In particular, only the first resonance is included.

7.3 The estimates with a diagonal damping matrix are

$\mathbf{K}$ = [1.85 −0.72; −072 0.79] × 10^6, $\mathbf{M}$ = [8.35 −0.29; −0.29 6.43], $\mathbf{C}$ = [488 0; 0 439]

7.4 The minimum eigenvalue for the regression matrix is as follows:

−0.0000 −0.0000 0.0000 0.0004 0.0019 0.0068 0.0307 0.1950 0.2808 0.3188

Note the significant increase at the second resonance.

This is reasonable even though it slightly distorts the original system.

7.5 Loads are 1,608, 1,114, 1,670, and 1,161 N, respectively.

Chapter 8

8.1 10 μm limit is taken up at 1850 rpm.

8.2 Maximum velocity is 113.4 m/sec, occurring at 22.2 sec and the instantaneous frequency is 45.36 Hz.

8.3 Form the autocorrelation function, then Fourier transform to get the power spectral density. Components at 30 and 50 Hz are dominant.

8.4 Three intrinsic mode functions emerge – see Solutions Manual for details.

8.5 $S = uLv'$ where u = [−0.3863 −0.9224; −0.92224 0.3863]; L = [9.5080 0 0; 0 0.7729 0],

V = [−0.4287 0.8060 0.4082; −0.5663 0.1124 −0.8165; −0.7039 −0.5812 0.4082]

Also note $S = vL'u'$.

Chapter 9

9.1 The two natural frequencies are 42.5 and 101 Hz.

9.2 Cracks give the following frequencies for nondimensional crack depths of

0	0.25	0.5	0.75	1	
42.4624	41.9213	40.8130	38.7215	33.5258	mode 1 x
42.4624	41.9369	41.1645	40.6617	40.5705	mode 1 y
101.0152	100.4137	99.2497	97.2694	93.2518	mode 2 x
101.0152	100.4307	99.6095	99.0974	99.0063	mode 2 y
214.9278	213.9426	211.9073	207.9549	196.8865	mode 3 x
214.9278	213.9710	212.5561	211.6270	211.4578	mode 3 y

All frequencies are in Hz.

9.3 The once per revolution excitation is determined in the usual manner. Twice per revolution has a peak at half the critical speed, where amplitude rivals that of the synchronous signal.

9.4 Torsional resonance 1.16 and 6.89 Hz. Peak fluctuation is 6 Nm at 200 rpm of input shaft.

9.5 First critical speed is 1,299 rpm which is a backward mode.

Index

A

Accelerometer, 6–8, 10, 343
Acoustic emission (AE), 4
    acoustic activity, 233
    frequency content, 233
    Hertzian contact, 233
    high-frequency oscillations, 232
    molecular scale, 233
    processes, 235
    rotor rub, 235
    signal, 233
Active magnetic bearings, 211
AE, see Acoustic emission (AE)
Alford's force, 19, 222
Alternator rotor
    amplitude and phase, 308
    comparison, 308–309
    diagnosis, 307, 311
    exciter rotor, 311
    harmonics, 312
    high-pressure turbine, 307
    low-pressure turbine, 307
    plots, 308, 310
    signatures, 308, 311, 313
    vibration components, 308
Analysis methods
    autocorrelation function, 278
    categories, 276
    empirical mode decomposition,
        299–300
    FFT, 279–280
    Gaussian random noise, 277
    Kurtosis function, 276
    low-value plant, 275
    periodic structure, 278
    physical interpretation, 279
    physics-based approaches, 275
    power spectral density, 279–280
    root-mean-square vibration, 275–276
    safe operation, 276–277
    sinusoidal components, 277
    statistical approaches, 277
    substantial fluctuations, 278
ARMA approach, see Autoregressive
    moving-average (ARMA)
    approach
Artificial neural networks (ANNs), 19
    Bayesian methods, 337
    fault conditions, 335
    graphs, 287
    input and output layers, 281
    linear methods, 336
    machine behaviour, 283
    MATLAB, 286
    Moore–Penrose inverse, 286
    multilayer perceptron, 281
    non-linear curve fit, 11
    parameters, 280
    with physics-based models, 284
    positive and negative values, 282
    re-scaling, 283
    scatter diagram, 280, 283
    strength and weakness, 283
    trained network, 281
    turbine/generator, 281
    vibration levels, 281
Autoregressive moving-average
    (ARMA) approach, 243

B

Beam models
    bending beam, 69
    elements, 68–69
    Euler–Bernoulli beam equation, 70
    neutral axis and uniform beam, 69–70
    radius of curvature, 70
    Rayleigh–Ritz approach, 68
    shear force, 70
    strains, 70
    stress, 70
Bearings
    active magnetic, 211
    externally pressurised, 212
    foil, 212
    oil journal, 206, 207–208

rolling-element, 209–211
types, 13–14
Bernoulli's equation, 215
Boiler feed pump
  balance-piston arrangement, 319
  displacement-controlled mechanism, 322
  error profile, 321
  finite element model, 320–321
  gearbox, 317
  gear components, 320
  matrix constraints, 321
  non-linear terms, 320
  sinusoidal error, 322
  stiffness, 322
  torque measurement, 320
  torsional displacement, 321
  transmitted torque, 321
  and turbine, 320
  turbo-generators, 317

C

Campbell diagram, 81, 82–83
Carpet plots, *see* Scatter plots
Centrifugal fans
  coal-fired power stations, 326
  natural frequencies, 327
  nominal imbalance, 328
  parameters, 327
  reinforced concrete block, 328
  six-bladed disc, 326
  variation, speed, 327
Cepstrum technique, 295
Computational fluid dynamics (CFD), 1
Constraints
  dissipative force, 261
  iteration, 259
  magnitude and phase, 261
  oil-film journal bearings, 261
  ordering convention, 260
  parameters, 259
  stiffness matrix, 259–260
  turbo-generator, 261
Cracked rotors
  behaviour of, 177
  dynamics of, 175
  forced response

ax-symmetric rotor, 185
equation of motion, 185
gravity deflection at shaft speed, 185–186
non-linear analysis, 183
solution at rotor speed, 185
stiffness matrix, 184
time-independent displacement, 184
vibration at twice rotation speed, 186
natural frequencies
  chordal crack, 180
  compliance function, 179–180
  crack depth and change, 177–178
  crack effects of, 181–182
  energy, 177
  equivalent crack, 181
  non-dimensional crack depth, 179
  Rayleigh type of approach, 179
  stress intensity factor, 178
  un-cracked portion, 178
rotating machines, 176
self-weight bending, 176
transverse crack, 176–177
vibrational behaviour, 176
Crack growth rates, 6
Cyclostationary methods, 296

D

Damped systems
  classical/proportional model, 88–89
  coefficients, 86–87
  complex mode, 89
  damped natural frequency, 87
  degree-of-freedom system, 88
  disturbance, 87
  eigenvalue equation, 86
  first-order equations, 84
  matrices, 84
  modal contribution, 89
  multi-degree-of-freedom system, 88
  natural frequencies and mode shapes, 81
  oil-journal bearings, 88
  phase, 89
  rotor speed, 83

state space, 84
stiffness and mass matrices, 88
3D plot, 88
three matrices, 84
time dependence, 89
un-damped second-order
    calculation, 86
Data presentation
calculations, 61–62
detection and diagnosis process, 62
dynamic behaviour, 24
gyroscopic effects, 24
rotating machinery, 23
scatter plots (*see* Scatter plots)
time and frequency, 24–26
transient operation
    acquisition hardware, 38
    amplitude and phase, 36
    exponential speed decay curve, 42
    FFT of sinusoidal signal, 37–38
    finite element models, 36
    Fourier transform, 37
    frequency resolution, 39
    Kalman filters, 40
    machine's dynamic properties,
        36–37
    pair of plots, 36
    polar plots, 58–59
    rundown frequency resolution, 42
    rundown plot, 36
    shaft orbits, 57–58
    spectrograms, 60–61
    speed decay curve, 40
    speed of machine, 38
    system response, 52–56
    time interval, *N* cycles, 43
    vibration, 36
    Vold–Kalman method, 51
types of, 24
waterfall plots (*see* Waterfall plots)
Down time, 16

E

Electro-rheological bearings, 339
Error/uncertainty
corner joints, 246
modelling, 245

portal frame, 245–247
shear characteristics, 247
three-dimensional elements, 246
vibrational behaviour, 245
Euler–Bernoulli beam equation, 70
Expert systems, 336–337

F

Failure modes and effects analysis
        (FMEA), 15
Finite element analysis (FEA), 1
Finite element method (FEM)
element formulation
    Euler–Bernoulli formulation, 74
    kinetic energy, 73
    mass matrix, 75
    nodal variables, 71–72
    orthogonal motions, 74
    potential energy, 72–73
    shear and rotary inertia effects,
        74–75
    stiffness matrix, 74
    Timoshenko beam theory, 74
    transverse displacement, 71
    vector, 72
Euler beam, 71
geometry, 71
node condensation, 71
Rayleigh–Ritz variational
    method, 71
uniform beam, 71
Flexible couplings, 149, 159
angular accelerations, 153
Hooke's joint, 153
mass and stiffness matrices, 152
natural frequencies, 152
principle, 150–151
sample system, 152
speed variations, 154
transformation matrix, 151
types of, 153
Fluid forces predominate, 227
FMEA, *see* Failure modes and effects
        analysis (FMEA)
Foil bearings, 212
Fourier analysis, 16
Fourier transform, 37

G

Gears, 188–193, 195, 295, 320, 322, 324
Gear error, 188, 195, 324
Gyroscopic effects, 24

H

Hand-held monitors, 15
Harmonic effects, 236
Hertzian contact principles, 209, 233
Higher-order spectra (HOS)
    bi-spectrum, 297, 298
    diagnosis, 296
    linear system, 298
    non-linear stiffness, 298
    power spectral density, 297
    tri-spectrum, 297, 298
Hilbert transform (HT)
    Cauchy integral, 289–290
    contour integral, 290
    machine monitoring, 291
    physical system, 290
    re-expression, 291
    signal processing, 288–289
    signum function, 291
    sine wave, 289
    specific requirements, 288
HOS, *see* Higher-order spectra (HOS)
HT, *see* Hilbert transform (HT)
Huang–Hilbert transforms, 19

K

Kalman filters, 40
    bearing properties, 262
    categorical statements, 263
    convergence and solution efficiency,
        263
    essentials, 265–266
    modal analysis, 264
    multi-rotor systems, 229
    oil pressure, 262
    parameters, 264
    quasi-linear approach, 264
Kernel density estimation, 19
    analysis techniques, 300
    bandwidth, 302

    distribution curves, 302–303
    Gaussian curve, 301
    gearbox wear, 303
    life prediction, 303
    naïve estimate, 302
    smoothing parameters, 303

L

Least squares
    axial locations, 256
    calculation, 255
    dynamic stiffness, 256
    identification process, 258
    non-linear estimation, 255
    pedestal vibration, 256–257
    physical co-ordinates, 255
    rotor vibration, 258–259
    shaft location data, 258
    sinusoidal excitation, 256
Linear voltage differential transformer
    (LVDT), 8–9
Low-pressure turbine instabilities, 329

M

Machine faults
    asynchronous vibration,
        200–201
    cracked rotors (*see* Cracked rotors)
    mass imbalance (*see* Mass imbalance)
    misalignment
        adjacent bearings, 149
        bearings, 148
        coupling bolts, 172–175
        description, 147
        flange, 160
        flexible couplings (*see* Flexible
           couplings)
        imbalance excitation, 161
        moments due to imbalance,
           161–162
        numerous consequences, 150
        oil whirl, 150
        rigid couplings, 160
        rotating machinery, 147–148
        rotors, 148
        simplified rigid coupling, 160

solid couplings (*see* Solid
    couplings)
turbo-generators, 158
types, 148–149
multi-bearing machine, 201
non-linear effects, 135–137
non-linearity, 175, 193–195
practical machine, 175
rigid and flexible rotors
    amplitude, vibration, 112–113
    centre of mass, 107–108
    definition, 105–106
    degree of flexibility, 108–109
    degrees of freedom, 106
    equations of motion, 107
    form of balancing, 113
    free-free frequency, 110
    free-free mode, 110
    frequencies, 110
    gyroscopic terms, 108
    idealised bearings, 110
    inertia components, 107
    machine imbalance, 105
    modal displacement ratios,
        112–113
    movement, 106
    orthogonal directions, 106
    parameters, 108
    rocking mode, 106–107
    rotor stiffness, 109
    spring supports, 106
    suspension system, 110
    unbalance and torsional motion,
        106
    uniform rotor, 110–111
rigid short bearings, 202
synchronous excitation, 199
torque transmission, 175
torsion (*see* Torsion)
twice-per-revolution excitation,
    199–200
Machine identification
    beneficial applications, 242
    Bugey Power Station, 254
    error criteria, 250–251
    FE approach, 242
    laboratory and operational
        plants, 241
    least-square process, 243

magnitude, 242
physics-based mathematical
    description, 242
plant behaviour, 242
regularisation
    constraint, 253
    diagonal matrix, 252
    flexible low-tuned supports, 254
    pseudo-inverse matrix, 253
    re-scaling, 251
    SVD, 251
system identification
    frequency component, 249
    inverse problems, 248
    measurements, 249
    Moore–Penrose solution, 250
    sinusoidal force, 248
    square matrix, 249–250
Machine's operational behaviour, 11
Magnetic bearings, 339
Mass imbalance
    damping, 115
    direct solution, 114
    eccentricity, 113
    eigenvalues and eigenfunctions, 115
    equation of motion, 113
    features, 114
    Kronecker delta, 115
    mode shape estimates, 116–117
    orthonormality conditions, 115
    physical interpretation, 114
    proportional model, 115
    rotating machinery, 113
    steady-state solution, 114
    synchronous vibration signal, 115
    system identification techniques, 137
    transient response curve, 117
    turbo-machines, 113
MATLAB®, 6, 21
Matrix assembly
    damping, 76
    degrees of freedom, 76–77
    Euler–Bernoulli beam theory, 77
    gyroscopic term, 76
    mathematical approach, 76
    natural frequencies, 77
    plant data, 78
    rotor modelling, 77–78
    system matrices, 76

uniform beam, 78
Modal balancing
   equal correction masses, 133
   flexible rotor, 132, 134
   influence coefficient method, 132
   lack of orthogonality, 134
   magnitude and positions, 132–133
   non-linearity, 134
   oil journal bearings, 134
   range of speeds, 133
   rigid rotor, 134
   rotor speeds, 133–134
   vibration levels, 141
Mode shapes
   classical/proportional damping, 95
   convenient approach, 95
   equation of motion, 95
   Fourier transform, 94
   imbalance response, 96–97
   linear equations, 93–94
   mass matrix, 94
   mass stiffness and gyroscopic
      terms, 93
   mathematical treatment, 96
   natural frequencies, 94
   non-zero solution, 94
   orthogonal properties, 98
   rotor, imbalance, 96
   sinusoidal excitation, 95
   system's symmetry, 97
Monitored parameters, 3–4
Moore–Penrose inverse, 286
Morton effect, 235

N

Newkirk effect
   complicated chain of events, 226
   cyclic pattern of behaviour, 223
   effective unbalance, 224
   mathematical analyses, 225
   non-linear functions, 226
   non-uniform temperature profile, 225
   rotor and stator, 224–225
   steady-state motion, 225
   stresses and strains, 226
   theoretical treatments, 224
Normalised force variation, 206

O

Oil journal bearings
   closed-form solution, 206
   modified Sommerfeld number,
      207–208
   modulus of f, 206
   normalised force variation, 206
   Ocvirk number, 207
   radial and tangential forces, 206
   running position of shaft, 207–208
   stiffness and damping, 209
   system parameters, 206
On-load data, 17

P

Perturbation techniques
   equation of motion, 98
   vs. exact solutions, 101
   first-order frequency changes, 100
   gain, 98
   lambda, 99
   natural frequency, uniform beam,
      100–101
   power, 99
   quarter and mid-span, 102
   single numerical calculation, 103
   stiffness/damping term, 98
   turning on effect, 98
   un-damped natural frequencies and
      mode shapes, 97–98
Piezo-electric, 6, 10, 343
Primary components
   global model, 244
   journal bearings, 244
   monitoring/fault diagnosis, 245
   representation, 243
   shrink-fit stresses, 243
   supporting structure, 244–245
Proportional model, 115
Proximity sensor, 8

R

Rayleigh–Ritz approach, 68
Rolling-element bearings
   Hertzian contact principles, 209
   hydrodynamic journal bearings, 209

Root cause, 5
Rotating machinery
  analysis methods
    Campbell diagram, 81–83
    dynamic stiffness, 90
    harmonic excitation, 90
    imbalance response, 80–81
    non-dimensional form, 91
    plant measurements, 80
    quantity of data, 80
    root locus and stability, 89–90
    second-order differential
      equations, 91–92
    steady-state response, 92
    3D plots, 91
  dynamic properties, 67
  equipment, 67
  Euler beam theory, 79
  FEM, 67–68, 80
  machine's behaviour, 67
  normalised frequency calculations, 79
  plant measurements, 68
  shaft rotational speed, 67
  Timoshenko's beam theory, 79
Rotor balancing
  influence, unbalance, 117
  linearity, 117–118
  single plane
    angular orientation, 119
    balance mass, 122
    equation of motion, 120
    magnitude and phase, 119, 121
    magnitude, imbalance, 121
    original imbalance, 121
    phase measurement, 119
    resonance influences, 119, 123
    sensitivity, system, 120
    single-plane balancing, 119
    sinusoidal form, 120
    vibration, 119–120
  two plane
    'as-found' condition, 123–124, 128
    Campbell diagram, 126–127
    and degrees of freedom, 123
    dynamic behaviour, 126
    flexible rotor, 126, 130
    imbalance, 125
    influence coefficient method,
      131–132

    linearly independent, 123
    measurements, 130
    modulus and phase, 123
    optimum balance conditions, 129
    parameters, 126
    rigid rotor, 125–126, 130
    rocking motion, 128
    single speed, 129–130
    system's modal behaviour, 129
    vectors, 123
    vibration level, 126, 127–128
    and vibration measurements, 124
Rotor bends
  axisymmetric rotor, 138
  balanced out, 140
  basic layout, test model, 140
  bearings, 142
  bend responses, 141
  damping, 140
  equation of motion, 139
  Fourier components, 139
  frequency range, 140
  and imbalance, 141
  kinetic energy, 138
  Lagrange's equation, 138
  machine fault, 137
  mass eccentricity and vibration level,
    137–138
  physical validity, 138
  steady angular speed, 139
  unbalanced rotor characteristics, 138
Rotor rubbing, 235
Rotor–stator interaction bearings (*see*
    Bearings)
  categories, 205
  collision and recoil
    acoustic emission, 232–233, 235
    physical effects, 226–227
    simulation, 227, 230–232
  extended contact
    Newkirk effect, 223–226
    physical effects, 222–223
  fluid regions, 205
  harmonics of contact, 236–237
  internal operation, 237
  'light' rub, 222
  Morton effect, 235
  working fluid (*see* Working fluid)
Rotor–stator rub, 223

S

Scatter plots
  advantages, 31
  analysed scatter plot, 35
  description, 29
  diagonal matrix, 34
  Gaussian random numbers, 32
  orthogonal directions, 34
  orthogonalised data, 34
  psuedo-static parameters, 30
  re-plot these data, 31
  steady load conditions, 36
  transformation, 34
  2D Gaussian noise., 32
  typical carpet plot, 31
  variance matrix, 33
  vibration velocity, 29, 31
SCILAB package, 21
Shaft modification
  automatic balancing devices, 341–342
  critical speeds and imbalance
      response, 345
  flexible rotors, 340
  frequency variation, 343
  integral feedback, 342
  magnetic shape memory, 344
  measurements, 343
  phase reference signal, 343
  piezoelectric MFC patches, 342
  potential benefits, 344
  rotating machinery, 344
  single mode, 343
  slip rings/telemetry, 343
  SMAs, 344
  thermal properties, 343
  torque, 345
Shannon's sampling theorem, 16
Shape memory alloy (SMA), 340,
      343–344
Shorttime Fourier transform (STFT), 48,
      291–293
Singular value decomposition (SVD)
  matrix, 284
  orthogonal matrices, 253
  parameters, 286
  rectangular matrix, 253, 284
  re-expression, 285

SMA, *see* Shape memory alloy
      (SMA)
Smart machines, 20
  controlled run-up, 341
  cooling phase, 340
  elastomer material, 340
  electro-rheological bearings, 339
  externally pressurised bearings, 339
  magnetic bearings, 339
  prototype bearing housing, 340
  rolling-element bearing, 340
  SMA, 340
Solid couplings
  bearings, 154–155
  catenary
    alignment, 156
    axial positions of bearings, 157
    correction, 159
    high pressure turbine and
        alternator, 158
    rotors, 155
    uncoupled rotors, 156
    vertical position, 158
  high-power machines, 154
Steam whirl, 222
STFT, *see* Shorttime Fourier transform
      (STFT)
SVD, *see* Singular value decomposition
      (SVD)

T

Torsion
  constraint equation, 190
  electrical machines, 187
  excitation, 187
  FE methods, 191
  flexible couplings, 191
  fluid-handling machines, 187
  Hook's joint principles, 191
  idealised torsional model, 188
  lateral motion, 189
  non-linear components, 192
  occasions, 191
  rotating shaft, 187
  single-toothed wheel, 188
  torque fluctuations, 189
  torsional vibration, 187

transmitted torque on rotor, 188
variations, 187–188
Turbo-set foundations, 247

V

Velocity transducer, 7
Vibration problems
ANNs, 336
components, 335
data analysis and handling,
333–334
expert systems, 336–337
FEM, 334
instrumentation, 333
linear beam model, 334
machine diagnostics, 337–338
neural computing, 336
physical model, 336
pseudo-data, 336
range of inputs, 336
rotor rub, 334–335
3D modelling, 334
transfer matrix approach, 334
Vold–Kalman method, 43
amplitude, 45
angular rotation, 44
error vector norm, 45
Fourier methods, 43
frequency response, 47
higher-order filter, 45
magnitude, 46
Nyquist frequency, 48
Prony analysis, 46
rapid rundown, 50
100 s rundown, 47–48
second-order filter, 50
structural equation, 44–45
synchronous excitation, 49
trigonometric identities, 44
variation, 50
vibration signal, 44
weighting effects, 48

W

Waterfall plots
description, 26–27
FFT, 26
non-linearity, 27, 28
shaft speed, 28
single channel, 29
Wavelet analysis, 293–294
Wigner–Ville distribution, 291–293
Working fluid
forms of excitation, 220–221
idealised system, 218–220
pump bushes and seals
Bernoulli's equation, 215
cyclic forces, 214
efficiency of, 213
high-pressure fluid, 212
mini-bearings, 213
modern boiler feed pump, 213–214
in power plant, 212
pressure drop with leakage flow,
215
single pump stage, 214
steam whirl, 222
system pressure distribution, 216–218
Workshop modal testing
axial symmetry, 314
concentration factor, 314
crack
depth and location, 314
face tip, 317
identification, 316–317
cross coupling, 316–317
diagnostic tools, 317
FE model, 179
as generators, 314
hammer impact tests, 314
orthogonal directions, 315, 317
planar and chordal, 316
resonant frequencies, 314
test procedures, 315
twice-per-revolution component, 316

Printed in the United States
by James & Taylor, Publisher Services

Printed in the United States
by Baker & Taylor Publisher Services